P9-BYM-311

ALSO BY KATHY D. SCHICK

STONE AGE SITES IN THE MAKING

KATHY D. SCHICK

AND NICHOLAS TOTH

MAKING SILENT STONES SPEAK

HUMAN EVOLUTION AND THE DAWN OF TECHNOLOGY

A TOUCHSTONE BOOK
PUBLISHED BY SIMON & SCHUSTER

NEW YORK LONDON TORONTO
SYDNEY TOKYO SINGAPORE

TOUCHSTONE
ROCKEFELLER CENTER
1230 AVENUE OF THE AMERICAS
NEW YORK, NEW YORK 10020

FIRST TOUCHSTONE EDITION 1994

TOUCHSTONE AND COLOPHON ARE REGISTERED TRADEMARKS
OF SIMON & SCHUSTER INC.

DESIGNED BY KAROLINA HARRIS
LAYOUT BY LINEY LI
MANUFACTURED IN THE UNITED STATES OF AMERICA

1 3 5 7 9 10 8 6 4 2

1 3 5 7 9 10 8 6 4 2 (PBK)

LIBRARY OF CONGRESS CATALOGING-IN-PUBLICATION DATA
SCHICK, KATHY DIANE.
MAKING SILENT STONES SPEAK : HUMAN EVOLUTION AND THE DAWN OF
TECHNOLOGY / KATHY D. SCHICK AND NICHOLAS TOTH.
P. CM.
INCLUDES BIBLIOGRAPHICAL REFERENCES AND INDEX.
1. TOOLS, PREHISTORIC. 2. FLINTKNAPPING. 3. STONE AGE. 4. TOOL USE
IN ANIMALS. I. TOTH, NICHOLAS PATRICK. II. TITLE.
GN799.T6S35 1993
930.1'2—dc20 92-35337
CIP
ISBN: 0-671-69371-9
87538-8 (PBK)

ACKNOWLEDGMENTS

The results of our research represent some pieces in the very large jigsaw puzzle of the major patterns and processes of human evolution. Paleoanthropology (the study of human origins) has developed into a truly multidisciplinary field, incorporating archaeologists, paleontologists, geologists, geophysicists, animal behaviorists, primatologists, ecologists, anatomists, neurologists, cognitive psychologists, biochemists, molecular biologists, geneticists, and social anthropologists. It has been our privilege and pleasure to work with people in many of these fields and to be part of a large network of scientists interested, primarily or peripherally, in the quest for our biological and technological origins. Our explorations have been shaped and nurtured by training, assistance, and inspiration we have received from a number of them.

Our three major professors at the University of California, Berkeley, all of whom have been remarkable people in their professional as well as private lives, have served as role models for a generation of students. They have shown us what the field of paleoanthropology is: an integrated, multidisciplinary endeavor fundamentally open-ended in its approach. We would like to thank

1. The late Glynn Isaac, who was one of the leading figures in the archaeology of human origins. Glynn first invited us to join the Koobi Fora Research Project in Kenya to study the stone technologies there from an experimental perspective and to try to decipher how those early Stone Age sites formed. We subsequently worked together at Lake Natron in Tanzania, as well. His suggestions, advice, and encouragement throughout the course of our early research were critical to the development of our methodology and our ideas.
2. J. Desmond Clark, who has been a teacher and then professional colleague and close friend. It has been an honor to work with him on research projects in Ethiopia, Zambia, and China. As the acknowledged world's expert in African prehistory, he has been for many years a constant source of information and advice.

3. F. Clark Howell, whose encyclopedic knowledge of fossil hominids and human evolution is second to none. As codirector of the Omo Research Expedition in Ethiopia, he was the first researcher to organize an integrated, multidisciplinary paleoanthropological expedition. We were very fortunate to have been able to work with him at the Spanish site of Ambrona.

Also critical to formulating and carrying out these investigations have been Tim White, Department of Anthropology, University of California, Berkeley; Donald Johanson, director of the Institute of Human Origins, Berkeley, with which we have previously been associated as researchers; Richard Leakey, former director of the National Museums of Kenya and codirector of the Koobi Fora Research Project; and many of our colleagues and students at Indiana University.

Indiana University has provided us with a research facility—the Center for Research into the Anthropological Foundations of Technology (CRAFT), specifically to continue our investigations of the origins and development of human technology through time—as well as laboratory facilities to develop specific aspects of research and training.

Various facets of this research were funded by a variety of different organizations, including the National Science Foundation, the L.S.B. Leakey Foundation, the Wenner-Gren Foundation for Anthropological Research, the Wilkie Brothers Foundation, the Harry Frank Guggenheim Foundation, the Henry Luce Foundation, the Council for the International Exchange of Scholars (Fulbright Fellowship), the Social Science Research Council, and the Ligabue Center for Study and Research, Venice.

We would like to give our very special thanks to Margaret Barrier and Gordon and Ann Getty for helping us get our research center, CRAFT, off the ground, and to Leighton A. Wilkie—who has been a seminal influence in recognizing and studying the technological foundations of our species—for funding research and helping support this book project. We are also greatly appreciative to Henry and Alice Corning, Del and Marty Fuller, Dorothy Schick, and Virginia Kubis for their interest in and support of studies of the human past.

Finally, we would also like to extend our thanks to John Brockman and Katinka Matson for helping make this book a reality. The cogent comments of our editor, Robert Asahina, and the help of Sarah Pinckney at Simon & Schuster have been invaluable.

At this moment we are writing on a portable computer, traveling hundreds of miles per hour through a dark and starry night in the last row of the upper deck of a 747 airliner, on our way to yet another archaeological research project to explore some of humankind's oldest tool-making traditions. There is something philosophically pleasing about exploring the world's most ancient and rudimentary technologies with the assistance of some of the most sophisticated contemporary ones . . . and, as we hope the reader will come to appreciate, there is no accidental relationship between the two.

Kathy Schick and Nicholas Toth

Somewhere over the Atlantic Ocean

TO OUR EARLY STONE AGE ANCESTORS,
FOR MANAGING TO SURVIVE
AND FOR LEAVING BEHIND
SUCH A FASCINATING TRAIL.

CONTENTS

	ACKNOWLEDGMENTS	7
	INTRODUCTION	15
1	BEFORE THE DAWN	25
2	THE STONE AGE CONSIDERED	48
3	DAWN BREAKS: THE FIRST STONE TOOL MAKERS	77
4	FASHIONING OUR FUTURE: THE MAKING OF EARLY STONE TOOLS	108
5	THE ROLE OF ROCK: USES OF EARLY STONE TOOLS	147
6	THE NATURE AND SIGNIFICANCE OF EARLY STONE AGE SITES	187
7	THE HANDAXE MAKERS AND THEIR CONTEMPORARIES	225
8	THE HUMAN THRESHOLD	285
9	BRAVE NEW WORLD?	312
	AFTERWORD	319
	FURTHER READING	325
	INDEX	335

INTRODUCTION

The sun has just emerged from beyond the gray-black clouds that delivered the short but intense morning shower, enhancing the smells and sounds of the savanna. Through the retreating sounds of thunder, the distant, resonant grunts of a lion can be heard. The sun's rays highlight the diagonal pattern of rain falling from the thunderheads onto the distant highlands in the east, still in the retreating shade. A broad rainbow spreads across the volcanic cones to the west on the other side of the lake.

The grassy plains are teeming with wildlife: enormous herds of wildebeests stretching toward the horizon in a dotted blur of brown; topis standing on old ant mounds to survey their feeding places; hyenas loping through the bush in search of new carcasses; warthogs trotting with their tails up; baboons and colobus monkeys foraging along the stretch of riverine forest that dissects the plains and winds toward the shimmering lake.

A new group emerges from under the flat-topped acacia trees. Even from a great distance there is no mistaking these animals: vertical figures in a predominantly horizontal, quadrupedal world. These slowly approaching figures betray their lineage by their distinctive bipedal gait and graceful, swinging arms. Their dark skin and hair are still glistening from the rain. There are about twenty of them, and the adult males are somewhat larger than the females, several of whom are carrying infants. Two of the juveniles are chasing each other and occasionally wrestling as the group moves through the grasslands.

The territory of these protohumans is some thirty square miles, stretching mainly along the major river, which appears on the volcanic plateau in the east and flows westward toward the slightly saline lake. The landscape is dominated by grassland and shrubbery, with a taller gallery forest along the river and its tributaries. They may travel up to six miles in a day, foraging for food, eating some and carrying any leftovers. One of them climbs a tall termite mound and sees miles of grazing wildebeests, gazelles, and topis, with giraffes browsing from the acacia treetops. A herd of elephants feeds under the trees along the river.

The hominids or protohumans are searching for foods they can find on their way to one of their favorite midday resting places along the river. Along a small tributary they come across numerous bushes of a purple berry with a spicy-sweet taste, as well as shrubs laden with small orange berries with a sweet, pumpkinlike flavor, which becomes more astringent as the hominids suck on the kernels inside. They feed here for about an hour, until the bushes have been denuded of their bounty.

As they leave the bushes, they sense movement in the sky ahead, over the plain about a mile away. A few large vultures are making lazy swoops through the air, circling rhythmically over a dark spot on the ground. The hominids' pace quickens as they head in the direction of the birds, their eyes straining to see the focus of this patient, expectant vigil. As the hominids approach, the scene comes into view: a large buffalo dead on the ground, with three female lions finishing their meal, starting to move off in search of shade and rest. The hominids can see some red meat the lions have left behind, especially on the head, back, and parts of the legs of the downed animal. The older males and females move toward the front of the hominid group, while the young and the pregnant or nursing females drift toward the rear. In the hominids' clenched fists can be seen dark objects—cobbles and flaked stones, which catch the light and glisten in the morning sun as the hominids approach the carcass.

> "I think Dr. Franklin's definition of Man a good one—'a tool-making animal.' "
>
> **James Boswell, citing Benjamin Franklin, in *Boswell's Life of Johnson*, 1791.**

Old Ben Franklin got it right, just about. In the animal world we are the consummate, though—as we now know—not totally unique, tool makers. We envelop ourselves in technology, using tools in incredible, almost unlimited, and ever-complex ways. During the past four million years we have evolved from a relatively modest, small-brained bipedal (two-legged) animal into a relatively pretentious, large-brained bipedal animal. In the process we have acquired new technological skills, novel means of procuring our foods, new types of social organizations, and new levels of intelligence and communicative skills. This human odyssey (an appropriate term, since there has been a considerable amount of wandering during the course of human evolution) is a saga in which tools and technology have played a crucial role.

Humans have proved to be the great manipulators of the planet. With our technology we have shaped our own lives and those of many of the animals and plants on the earth. We have reached far into our atmosphere, deep into our oceans, and even beyond our planet into and beyond our solar system. But despite the dramatic impact we have today, it was only a short time ago in our evolutionary history that our ancestors made the critical first step down the path toward our modern world. They did this with a novel idea—picking up pieces of stone, making them into tools, and using them in ways that became crucial to our adaptation and survival.

Who were the earliest tool makers? What did they look like? What types of implements did they have? How were these tools made, and how were they used? What was life like for our early tool-making ancestors? Most important, from our perspective, what was the role played by technology during the course of human evolution? These are fascinating questions with great relevance for understanding where we have come from, how we got where we are today, and where we might be going in the future with our use of technology.

This is not just a biological story. Human evolution concerns not just physical change, but also behavioral transformations. Changes in our biology are perhaps easier to see—in our bones, our teeth, our brain, and even our genes themselves. Changes in our behavior, however, even if many are strongly linked to our biology, are less easily visible but equally important. These include transformations in our sexuality, our diet, our ecology, our social organization, and our reliance upon culture. While the biological side of the story is fairly well-known—fossils have a compelling presence—the overarching importance of these behavioral shifts is less widely appreciated. This is unfortunate, since one of our species' most drastic evolutionary events, which has profoundly affected what we are today, was a major behavioral change: a shift to using tools and technology to adapt to our environment.

Culture and technology, the learned or acquired ways humans deal with the world and shape it to their own ends, are crucial parts of human evolution. In fact, technology is probably *the* most significant element in determining what we are today, not just in forming modern "civilization," but in directing the course of our evolution from a distant apelike ancestor.

Genetically, anatomically, behaviorally, and socially, we have been shaped through natural selection into tool makers and tool users. This is the net result of more than 2.5 million years of evolutionary forces working upon our biology and behavior. Our main focus here is the evolution of human beings as profoundly technological creatures. A great deal of this book is a personal account of our exploration of this subject, which has consumed much of our time during the past two decades.

We will show the archaeological origins of tool making and tool use and the adaptive significance of technology throughout human evolution. We will include diverse lines of evidence, both the hard facts of archaeology—the data, or materials retrieved through excavation—and the insights gained through experimentation and through explorations of modern human and primate behavior and adaptation. We will incorporate our personal experiences in making and using stone tools in East Africa and elsewhere exploring how and why ancient protohumans diverged from the rest of the animal world, most recently from the rest of the apes, on an evolutionary trajectory that finally led to *Homo sapiens* (Latin for "man who thinks")—the species in the order Primates to which all members of the human race belong. It is a unique and unparalleled chapter in the history of life, and from a human perspective, it is of the utmost interest. From the planetary perspective as well, it may be of supreme importance.

Why and how did our ancestors begin to crack rocks together, creating the first recognizable archaeological record? How did this most basic technology expand in complexity through time, sometimes along with profound biological changes in our bodies? Ultimately, what was the role of technology in shaping the course of human evolution?

The first glimmerings of the dawn of human technology can now be traced back to approximately 2.5 million years ago in Africa. These signs are not, at first look, all that impressive: a few lumps and fragments of broken stone, often found with fossil bones and teeth of extinct forms of animals that lived during this time. However unimpressive looking to the casual observer, these shattered rocks bear witness to a new behavior pattern among early hominids, the percussive flaking of rock, a behavior that was to have profound repercussions. Fortunately for archaeologists, stone tends to be fairly indestructible, (as Shakespeare noted in *Romeo and Juliet*: "will ne'er wear out the everlasting flint"), so that unlike most organic materials, stone artifacts have an excellent chance of surviving in the prehistoric record.

It was at this time that some of the early hominids equipped themselves with tools and moved into new evolutionary niches that proved to be enormously successful. This was the start of a new adaptation, seemingly insignificant at first, that continued and evolved over the next few million years and finally led to what and where we are today.

This first technology coincides roughly with the appearance of a new and different biological form of hominid: *Homo habilis*. The most outstanding distinguishing feature of this creature—so exceptional that we place it in our very own genus, *Homo*—is its larger brain. This is the earliest species of *Homo* that we can see in the fossil record so far. Its brain size is only about half the size of modern humans, but it is significantly larger than that of the earlier hominids who preceded it and those who coexisted with *Homo* for a million years or so.

Over 99 percent of the story (measured in years) of human technology took place in the prehistoric Old Stone Age, or Paleolithic, before the advent of written ("historical") records. The bulk of our evidence of these events comes from underground—from materials that have lain buried for thousands or millions of years, and have subsequently been discovered and excavated, analyzed, and interpreted by scientists from diverse disciplines: specialists in archaeology (the study of ancient human behavior), paleontology (the study of ancient, usually fossilized, life forms), and geology (the study of the earth and its ancient record). These disciplines are essentially *observational* sciences, whose practitioners look through various keyholes into our prehistoric past. Our success as scientists of this prehistoric past depends upon our skill in discovering, identifying, and explaining the nature of ancestral creatures, lifeways, environments, and technologies that existed long ago.

EXPERIMENTING WITH THE PAST

Before we look at the earliest archaeological sites and their stone tools and explore what role these implements played in early hominids' lives, it is important to understand what constitutes a stone artifact. What is so distinctive about stone tools? How is it that archaeologists can pick out stone "artifacts," made by the human or protohuman hand, from the profusion of other rocks out there in the natural world? And what is this nebulous thing we call "technology"? To answer these questions, we have to look at how humans (and less often other animals) go outside of their own bodies, using or modifying materials and objects outside of themselves, to achieve their goals. Archaeology is not a laboratory science such as physics or chemistry. We archaeologists are, in a sense, the voyeurs of the prehistoric past: we look, we see, we observe patterns, but we cannot control any of the variables or events that have already occurred.

In its essence, archaeology simply requires that we try to uncover evidence of our past and then interpret it to the best of our abilities. This almost always means going into the field, walking, digging, and usually getting very hot and dirty. In the past, making reasoned, informed interpretations about the evidence we dug up relied largely on the experience and intuition—and luck—of the archaeologist. For the most part, this phase was conducted in the library or study, as part of what has sometimes been called armchair archaeology. But interpreting the patterns we see in prehistory is the most challenging task for an archaeologist, and lately it has been taking us out of our armchairs and plunging us into the midst of various problems faced by our prehistoric ancestors.

In order to fully appreciate these patterns, archaeologists can use the present as a key to the past to understand how the archaeological record was formed. We study the physical remnants of past activities, but what we really want to understand is what *produced* those physical traces, what prehistoric activities and behaviors left those cryptic records behind. That is, we try to see the relationships between real-life *processes* and the *products* they leave behind in the prehistoric record.

The term *actualistic studies* is sometimes used for this approach. Examples of actualistic studies include examining a river bank after a flood to fathom its sedimentary structure; studying a group of modern hunter-gatherers to understand how they use their landscape and where they leave evidence of what they do; studying modern African habitats to explore the range of foodstuffs that our protohuman ancestors could have exploited in the past; and, most important for this book, engaging in an approach that has been dubbed "experimental archaeology."

Experimental archaeology is an attempt to re-create aspects of ancient life-styles by using the same materials, techniques, and strategies believed to have been employed by those ancient peoples. In this way we can explore in detail how early peoples made their tools, how difficult this task was and how much time it took, what by-products were generated, how they used their tools, which types were useful for which tasks, how well they worked, how they became damaged in use, and, overall, how their technologies fit into their lives.

Examples of such experiments have included using ground-stone axes to clear tracts of forest; dragging gigantic stones across the Salisbury Plain to study how Stonehenge was built; using primitive furnaces to smelt iron ores and produce forged tools and weapons; constructing a Bronze Age–style palisaded-timber fort; or sailing a balsa raft across the Pacific.

These experiments may be designed on several levels of scientific inquiry. They may be intended simply to test the feasibility of an idea —for example, whether it is possible to produce a particular Stone Age tool found in antiquity, such as a flint handaxe, by a certain technique, such as hard hammer percussion. Experiments can also be designed to try to understand the relationships between living processes and their archaeological products—for example, between the manufacture of this handaxe and the waste products produced at different stages of its manufacture. The results of such explorations can then be used to help identify the materials we find in the archaeological record. Or the experiments may be designed to test a carefully stated proposition, or hypothesis, predicting the outcome of the experiment and then seeing if the results match up with the prediction.

This experimental approach actually had its origins in the nineteenth century, as archaeology emerged out of other antiquarian pursuits. (In fact, the nineteenth century also witnessed an early martyr to the experimental archaeological approach: a Dr. Ball of Ireland was exploring the range of sounds he could produce on a Bronze Age horn called a *lur* when a blood vessel in his neck burst.) A thorough exploration of this approach can be seen in John Coles's book *Archaeology by Experiment*. Experimental archaeology has now become a major tool that archaeologists can use to interpret and understand prehistoric evidence.

EXPERIMENTING WITH STONE: ARTIFACT MANUFACTURE AND USE

One of the earliest experimental stone tool makers was an Englishman named Edward Simpson, better known as "Flint Jack," a man who was clearly fascinated with stone tools. He became remarkably proficient at making replicas of the flint tools then being found in ancient river terraces of the Thames River and in other parts of England. These he sold in the late nineteenth century to museums and an unsuspecting Victorian public fascinated by these signs of bygone prehistoric times. He normally used nothing more than a steel hammer to fashion these artifacts, and by employing chemicals and a lapidary tumbler he was

An early experimental archaeologist. The notable English forger Edward Simpson (AKA "Flint Jack"), who replicated stone artifacts and sold his wares to a Victorian market. Such unscientific activities tended to give experimental replicative studies a dubious name for many years.

even able to simulate chemical patination and physical abrasion, often used as obvious signs of a Paleolithic artifact's antiquity.

It has only been in the last three decades that archaeology by experiment has become a widespread approach to archaeological problems. In Europe, François Bordes gained a reputation as a superb flint knapper as well as a world authority on the Paleolithic of France. In the United States, Don Crabtree astounded archaeologists with his ability to replicate some of the finest stone spear and arrow points ever found at prehistoric sites.

In Africa, famed Kenyan anthropologist Louis Leakey and University of California archaeologist J. Desmond Clark also made and used stone tools in order to understand their manufacture and use in the past. From these early but very informative beginnings, experimental study of the Stone Age has blossomed into a lively field with people in many parts of the world involving themselves in very systematic experiments in Paleolithic technology (as we shall see in chapter 4). These entail not only making the stone tools as our ancestors did, but also using them in real tasks to explore their functions. We have carried out a wide range of experiments using stone tools, even going so far as to butcher the carcass of a dead elephant, showing what was possible with the simplest technology from two million years ago. At the same time, we have been examining the ecology and behavior of nonhuman primates such as chimpanzees to see how they use their environment and what technologies they employ.

Two pioneers of prehistory and experimental archaeology in Africa: J. Desmond Clark, emeritus professor of anthropology at the University of California, Berkeley, and the late Louis Leakey, who was curator of the National Museum of Kenya.

Our investigations into the early Stone Age have led us through many experiences that have helped us understand the origins of human technology. These have included

- archaeological fieldwork in Africa, Europe, and Asia, excavating early Stone Age sites;
- studies of experimental manufacture of early Stone Age tools;
- experiments in using early stone tools for such activities as animal butchery (from small mammals to elephants), woodworking, and hide working;
- simulating early Stone Age sites in a range of environments and seeing how they slowly get buried and incorporated into the geological record, in order to see how artifacts made and used in the past "become" archaeological sites;
- examining prehistoric materials with high-powered technology, including the scanning electron microscope;
- visiting some of the last stone tool makers in the almost inaccessible mountains of central New Guinea to study some of the last vestiges of stone tool traditions on the face of the earth;
- testing the tool-making capabilities of a bonobo (pygmy chimpanzee), a member of the genus *Pan*—the closest relatives we have still existing on the planet to compare these abilities to those of the earliest protohuman tool makers.

Traditionally we humans have been rather focused on our species, or *anthropocentric*. Culture, learning, rational thought processes—and tools—were thought to be the exclusive domain of human beings, setting us off very clearly from the rest of the animal world. They helped define quite neatly what we were and how we were different from, perhaps even beyond or outside of, the rest of the natural world. Now we know that these boundaries are not so clearly marked. Many of these features, including tools, are shared to some degree with a great many other animals. In a sense, we have lost our profound "specialness" and have found that many of our differences from other animals are in degree rather than in kind. This is especially true when viewed from an evolutionary perspective. We must now deal with a more complicated but more accurate—and more interesting—view of our place in the world.

CHAPTER 1

BEFORE THE DAWN

I can trace my ancestry back to a protoplasmal primordial atomic globule.

William Gilbert (1836–1911),
The Mikado

Here lies a man, who was an ape.
Nature, grown weary of his shape,
conceived and carried out the plan
by which the ape is now the man.

Humbert Wolfe,
"Epitaph for *Australopithecus,*"
***The Spectator,* 1925**

ORIGINS: A SERIES OF BEGINNINGS

These two quotes help illustrate the complicated, nested sets of origins for all living creatures. We humans could trace our beginnings back to the origin of life itself on the planet or to a much more recent landmark, such as our evolutionary split from our closest extant relatives, the apes. There are, of course, myriad other milestones between these two events and many additional ones since we diverged from the African apes. We cannot, however, see and identify all of them in the prehistoric record, as it is not complete enough. But we can identify many, with increasing resolution as we proceed through time, and recognize some that have been critical in making us what we are today.

Since most of us will not even reach the century mark, the earth's history is unimaginably long. The concept of thousands, millions, and even billions of years can be very difficult to comprehend. Yet the story of the events that have led to the modern human condition can be traced back almost to the time when organic molecules happened to recombine in elegant new ways to form novel types that could replicate themselves and start "living."

Based upon our present state of scientific knowledge, the major events and appearances in this story would include

15,000,000,000 years ago: the big bang and origin of the universe
4,500,000,000 years ago: the origin of the solar system and earth
4,000,000,000 years ago: the first complex organic molecules
3,500,000,000 years ago: the first unicellular life (prokaryotes)
1,400,000,000 years ago: the first cells with a nucleus (eukaryotes)
800,000,000 years ago: the first multicellular sponges and algae
550,000,000 years ago: the first chordates
500,000,000 years ago: the first vertebrates: fishes
370,000,000 years ago: the first amphibians
300,000,000 years ago: the first reptiles
200,000,000 years ago: the first mammals
65,000,000 years ago: the first primates
35,000,000 years ago: the first apelike forms
4,000,000 years ago: the first (known) bipedal hominids
2,400,000 years ago: the first stone tools
2,000,000 years ago: the first evidence of brain expansion
1,200,000 years ago: the first evidence of humans in Eurasia
400,000 years ago: early archaic *Homo sapiens*
100,000 years ago: anatomically modern humans
30,000 years ago: the first representational art
15,000 years ago: the first humans in the Americas
9,000 years ago: the first farmers
8,000 years ago: the first metallurgy and wheeled vehicles
6,000 years ago: the first cities, writing, and civilization
200 years ago: the Industrial Revolution
30 years ago: the first human space travel
5 years ago: the first human-made artifact (spacecraft) to leave our solar system

This list of evolutionary landmarks presents an overtly biased viewpoint, one that becomes increasingly human-oriented over time and successively excludes hosts of other organisms that have emerged from the primordial soup. The story begins to concern us most directly only a short sixty-five million years ago. All the previous 3.5 billion years of the history of life had laid the essential foundation for, in an anthropocentric view, this ultimate event: the evolution of our lineage, Primates (which include ourselves, apes, monkeys, and prosimians, or premonkeys), among the rest of the mammals. Each of the earlier events formed a base and a link, a foundation and a connection, for the next one in line, so in a very real sense each one has been essential to our evolution.

THE FIRST PRIMATES: PRECURSORS OF BIPEDAL HOMINIDS

MAKING SENSE OF THE FOSSIL RECORD

For the paleontologist (literally, student of ancient life) to make sense out of the fossil record—to understand evolutionary relationships among ancient forms of life—requires recognizing similarities and differences among extinct and modern forms of organisms. From our understanding of the processes of evolution, we know that populations of plants and animals change over time as they adapt to shifting environmental conditions and opportunities, at both the genetic and the phenotypic (outwardly visible) level.

Today we can examine in great detail the anatomical and genetic similarities and differences of modern organisms, which give us important clues to their evolutionary histories and relationships. For most of the prehistoric record, however, we are confined to bones and teeth, which are sometimes preserved as mineralized fossils. For paleontologists of vertebrates, this is essentially comparative anatomy, a matter of comparing pieces of animals' skeletons—the hard, bony, durable parts of animals' bodies, which are most likely to fossilize and endure over long periods of time. Thus, they need to know how best to read and interpret the rare information these fossils hold.

A key step in constructing evolutionary histories, or "phylogenies," is establishing which anatomical features of fossil skeletons are *primitive*, or stemming far back in the ancestry of the animal, and which are *derived*, or evolving more recently in that animal's evolutionary group. Human beings, for example, have five digits (fingers and toes) on our hands and feet, as did our mammalian ancestors and their reptilian ancestors before them some 250 million years ago. Horses, on the other hand, similarly had five-toed, protomammalian ancestors but have developed into essentially one-toed creatures (a derived condition) during the past sixty million years. Thus the five-digit extremities of our limbs are essentially primitive—they retain the basic, ancient, ancestral hand and foot pattern—while the limbs of horses (and many other animals that have become specialists at running) are a new type derived and quite different from the ancestral one. But we cannot say from this that humans are primitive and horses derived in an overall sense. It is not the animal but the feature that is primitive or derived.

We humans are both primitive and derived: some traits show what we inherited from very ancient ancestors, others show what we developed later on as we took advantage of the unique evolutionary opportunities encountered at various times in our past. Some of our derived characteristics have been important in defining and directing our evolutionary path. For example, our ability to walk upright is a very recently derived trait: it shows up only about four million years ago in the fossil record as changes in the hip and leg bones, and it is unique in the mammal world. Another derived feature, a significantly expanded brain, began to appear on the scene even more recently, only about two million years ago.

Each organism is its own blend of primitive and derived traits, which can tell an intricate story of its evolutionary journey. Some features indicate where it came from, a long distant past shared perhaps with many other animals, while others reveal the particular twists and turns taken more recently. But to read this story takes some doing: we have to develop a good sense of before and after, of the order of appearance of our cast of fossils, and we must recognize ancestral similarities and evolved differences among animals living and extinct.

One of the main jobs of the paleontologist is to put the fossil discoveries into chronological order, using available dating techniques, and to decipher the pattern of primitive and derived traits that reveals the evolutionary history for a given lineage of animals. Because the fossil record has significant gaps, there is often a great deal of disagreement about the precise interpretation of the evolutionary significance of the fossil record. Nevertheless, we can still pick out many of the major, key steps in the emergence of our ancestors from very primitive, ancestral mammals to early primates, apes, and protohumans.

DATING THE PAST

To understand evolutionary history it is essential to establish a chronological framework in which fossils, Stone Age sites, and geological events can be ordered. Today a number of methods can potentially be used in dating early fossil hominid or archaeological sites. These tend to be divided into absolute dating methods, which are usually based upon the radioactive decay of an unstable isotope and give an estimated age in terms of numbers of years before present, and relative dating methods, which determine whether something is older, younger, or the same age as something else. Following are some of the most useful methods for earlier hominid evolution in Africa.

RELATIVE DATING

Stratigraphy

The fact that geological layers or strata are sequentially deposited one on top of the other through time means that strata higher in a stratigraphic sequence will be younger than the strata below (called the law of superposition.) This method gives a relative sense of how old different deposits in a sequence are relative to one another but does not give an absolute age in terms of numbers of years.

Faunal Dating

Animals evolve through time; some animals, such as pigs, elephants, and horses, evolved fairly rapidly in certain periods and can be used as indicators of approximate time by studying changes in their morphology, especially tooth structure, throughout a sequence of sediments. Once an evolutionary sequence has been established for an animal lineage, newly discovered fossils can then be placed within this sequence. Absolute dates can be derived for such a faunal sequence by one or more of the absolute dating methods.

Paleomagnetic Studies

For reasons not yet fully understood by scientists, the North and South poles have reversed positions at various times in the earth's past. The last major paleomagnetic reversal was about 780,000 years ago. These paleomagnetic reversals can be ordered through time into a sequence and dated in absolute terms by other methods (such as potassium-argon dating). Analysis of a geological layer produces a reading of either normal (North Pole, where it is at present) or reversed (the prehistoric North Pole being where the South Pole is today). This method is a useful means of cross-checking other means of dating and correlating sites; for instance, if two sites are believed to be contemporaneous due to stratigraphic or faunal correlation, they should share the same alignment of magnetic minerals in their sediments.

Volcanic Ash (Tuff) Correlation

Geological studies have shown that most volcanic eruptions produce a diagnostic chemistry or chemical signature particular to that eruption, and that some eruptions distribute air-borne ashes over large areas, sometimes covering hundreds of miles. By matching the chemistry of one volcanic ash to another at a different locality, geologists often have been able to establish that widely separated deposits were roughly contemporaneous and have thus been able to make correlations over large areas.

Whenever possible, such relative dating methods are used in conjunction with one or more of the absolute dating methods that fix an event more precisely in time.

ABSOLUTE DATING METHODS

Potassium-Argon Dating

One of the most useful approaches to dating African sites has been the potassium-argon dating method. Radioactive potassium (K-40) decays into argon (Ar-40) at a known rate, and the radioactive clock begins ticking when a molten rock cools and solidifies. Volcanic rocks such as lava flows or ash deposits can be dated by comparing the ratios of radioactive potassium to argon isotopes, and by a mathematical formula the geochronologist can arrive at an approximate age in years before the present. Fortunately for prehistorians, there was a lot of volcanic activity in the East African Rift during much of the period of human evolution, so many sites contain volcanic deposits datable by this method.

Other Absolute Dating Methods

Radiocarbon (C-14) dating, which measures how long radioactive carbon in a plant or animal has been decaying since that organism's death and hence gauges how long ago it died, is a well-proven, reliable method but is applicable only to the last fifty thousand years or so of the human evolutionary story. Other methods, such as thermoluminescence, electron spin resonance, uranium series dating, and amino acid racemization, are still to a large extent in their developmental stages and have not yet proved as reliable as the potassium-argon or radiocarbon dating methods. Each is widely used, though, when other methods are not applicable and to confirm dates obtained through other methods.

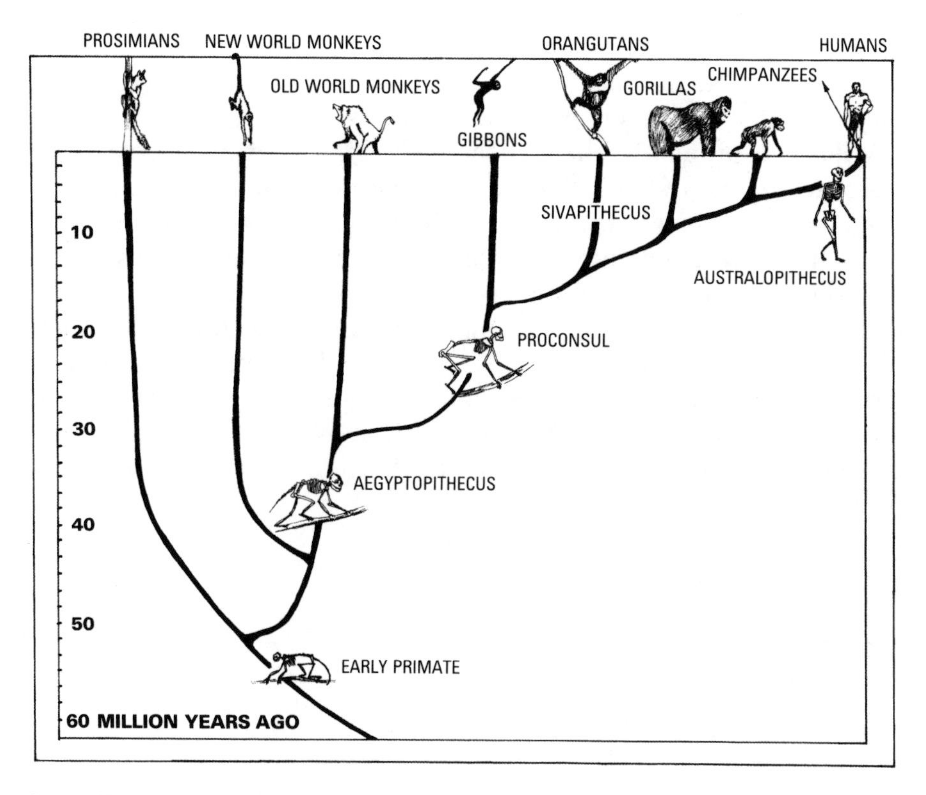

A simplified chart of primate evolution.

ENTER THE PRIMATES: MAMMALS OF DISTINCTION

Our destiny to develop into tool-making humans may well have begun as an accidental by-product of a major, catastrophic event in the earth's history, at a time when the dinosaurs still roamed the lush forests and plains: a dramatic turnover in animal and plant forms toward the end of the Mesozoic (popularly known as the Age of Reptiles), about sixty-five million years ago, when at least three-quarters of the species died out.

The precise reasons for these massive extinctions are still debated. Prominent ideas are that a massive extraterrestrial object such as an asteroid crashed into the earth, throwing up pulverized debris that caused a worldwide brownout of the sun lasting for years, or that a global warming killed off vast numbers of animals, especially the larger ones. A largely vacated planet was left, with many open niches for mammals to radiate into. One group of mammals to emerge at that time consisted of the very early primates.

What is a primate? When in the eighteenth century Carolus Linnaeus wrote his *Systema Naturae*, the first major attempt to organize and classify life hierarchically, he fully recognized that humans share many structural similarities with some nonhuman animal forms. So he placed us and many other animals in a single group, an "order" he first named Anthropomorpha but changed to Primates by the 1758 edition of his work. (In earlier editions Linnaeus had, in fact, correctly grouped together monkeys, apes, and humans in this group but had erroneously also included the sloth . . . a noble first attempt.) Members of the order Primates are separated from the rest of the mammals by a suite of traits. These normally include

1. an enlarged brain
2. an erect trunk
3. stereoscopic vision
4. color vision
5. grasping hands (and often feet)
6. nails, replacing claws

This is quite a package: enhanced vision for seeing small or mobile foods, dexterous hands for grasping them, and a larger brain to coordinate the new visual system with the motor capabilities. Most of these abilities appear to have evolved originally as adaptations for an arboreal (tree-dwelling) life-style—scrambling along tree trunks and branches in search of suitable prey such as quick-moving insects and small lizards or, for some, searching out plant resources such as seeds and fruits in the forest canopies. All primates show this specialization to some extent, an emphasis upon visual acuity combined with manual dexterity, which sets them off from most other mammal orders.

We might say, then, that the primate credo is first to see an opportunity and then to seize it. Overall, primates emphasize visual perception of the world in order to survive—to see food, predators, various possibilities for survival—and also a hands-on approach for probing the environment. When the Age of Reptiles drew to a close, our mammal ancestors were small, insignificant-looking, insectivorelike animals scurrying about in the forests looking for insects and other good things to grab and eat.

The geophysical evolution of the continents played a major role in the origins and dispersal of animals over the last two hundred million years. The gradual breakup of the supercontinental landmass of Pangaea, with drifting continental plates being divided by great waterways and seas, has presented ever-increasing barriers to animal migrations, separating many animal groups and setting them forth on their own evolutionary pathways. On the other hand, some connections and land bridges between the continental masses lingered for millions of years, or new ones formed, allowing some groups to continue to migrate and radiate over large areas.

North America was still connected to the Eurasian mass by a land bridge for much of the first twenty million years or so of primate evolution. But by the time the primates set off to take part in the great mammalian radiation, these masses were separated from South America and, for much of that time, largely also from Africa (Africa was finally relinked with Eurasia only about fifteen to twenty million years ago). Nevertheless, some animals may have managed to cross some of the apparent barriers, perhaps across the growing ocean between Africa and South America (the "raft" theory) or along tenuous land connections between North and South America.

The first recognizable primates to emerge in the fossil record show characteristics of both modern prosimians (such as lemurs) and insectivores (such as modern tree shrews). Like many mammalian orders, these primitive primates began to radiate into a range of new niches, evolving different types of characters as they adapted to these new environments and providing an all-important foundation for the course we humans have charted in more recent evolutionary time. By about fifty million years ago, some of these primates had evolved into prosimian forms similar to modern lemurs and lorises.

BRANCHING OUT WITH THE MONKEYS AND APES

By approximately thirty-five to forty million years ago, in the Oligocene geological epoch, some of the primates in Africa were beginning to exhibit features linking them to modern monkeys and apes but separating them from ancestral prosimians. These features involved certain changes in the skull and teeth, such as a deeper, lower jaw, fused in the middle and strengthened by two shelves of bone, and lower-cusped molars. These may seem esoteric but are in fact very important in defining this major, emergent evolutionary group: they show a step made by the *group*—ourselves, the apes, and all of the monkeys—through an ancestral species we all share in common. This is often called the "anthropoid grade" of primates, and we had finally made this grade after twenty to thirty million years of primate evolution.

Then, by thirty million years ago, we see another important split, or fork, in the evolutionary tree: the development of small, monkey-like, probably fruit-eating creatures, some of which seem on their way to evolving into apes. These primates are clearly related to ourselves, to apes, and to Old World monkeys in Asia and Africa (such as the macaques and baboons), but *not* to New World monkeys (such as spider monkeys and capuchins). Their limbs suggest that they were arboreal quadrupeds at this stage. This group shows some crucial changes, particularly in their teeth, that clearly set it apart from the New World monkeys. If you peer into the mouth of a modern human, you will find symmetrical rows of teeth, the top row mirroring the bottom row and the left side mirroring the right. From front to back on either side, upper or lower, you will see two incisors, then one canine, then two premolars (bicuspids), and finally three molars, or cheek teeth (as long as your subject retains the wisdom tooth, or third molar). This is called our "tooth formula," and it had evolved by this time in our evolutionary past, thirty million years ago, in the ancestors of the Old World monkeys, apes, and ourselves. By contrast, the New World monkeys show the earlier, more primitive condition of three premolars on each side of the jaw.

An important collection of fossils showing this new, two-premolar trait is called *Aegyptopithecus*, found in the Fayum Depression of the Egyptian desert. These thirty-odd-pound creatures may be a common ancestor of all modern apes: the gibbon and siamang ("lesser apes"), the orangutan, gorilla, and chimpanzee ("great apes"), and humans (the bipedal ape), either before or after the split with the Old World monkeys.

ALMOST FAMILY: THE APES, OR HOMINOIDS

Finally the apes appeared on the scene and at first seemed to have flourished. Early in the next geological epoch, called the Miocene, which lasted from about twenty-three to five million years ago, the first distinctive apes and monkeys evolved, but they were apparently originally restricted to Africa. The first ten million years of this epoch was the era of the apes or hominoids. The tropical forests were ripe for the picking, and the apes went forth and multiplied, becoming abundant and quite varied, while monkeys, on the other hand, were relatively rare. Over time, however, climates became cooler and drier, and forests began to dwindle in size. The apes started declining in number, and monkeys began taking over the new habitats. Before the apes went into their decline, though, they managed to proliferate into a variety of forms, and some managed to continue their evolutionary trek up to their present forms, modern apes and humans.

In fact, there was a major radiation of apes, or hominoids, first within Africa and then, beginning about seventeen million years ago, into Europe and Asia, producing a bewildering number of fossil forms, about which paleontologists are struggling to make evolutionary sense. During this period the sea separating Africa from Europe and Asia started shrinking to form the Mediterranean, and Africa linked up with southwest Asia through its Arabian shoulder. This opened up routes from northeast Africa to the Near East and from northwest Africa to Spain, allowing the migration of many animals, including African hominoids, into Europe and Asia.

Just what were the "promising apes" of the Miocene? And can we identify among them more of the important splits in the primate evolutionary tree, perhaps even see individual branches that led to specific modern apes: gibbons, orangutans, chimps, gorillas, or humans?

Some of the major figures in the fossil cast during this phase were

Proconsul. This group emerged in Africa early in this period, about twenty million years ago, and had already developed a skull that was very apelike in many ways. This includes a larger brain, a shorter snout, and eye sockets positioned fully forward in the skull. The different species ranged in size from that of a small chimp to a small gorilla. It now appears that all of the living apes could have descended from a *Proconsul*-like creature.

Dryopithecus. A later, European group of apes the size of chimpanzees, which probably descended from *Proconsul* and still walked on all fours and lived in the trees.

Ramapithecus. During the 1960s and early 1970s, *Ramapithecus* was highly touted as the earliest known hominid, that is, as a distinct ancestor to humans, already separated from the rest of the apes. The rationale for this argument, based on the small number of jaw fragments and teeth known at that time, went something like this:

1. *Ramapithecus* fossils had small canine teeth and thick enameled teeth, as did later hominids;

2. Since they had small canines, they must have defended themselves with the use of tools;
3. Such a use of tools would suggest that their arms were free for manipulation of their environment;
4. Freed arms would indicate bipedal locomotion;
5. By definition, bipedalism would qualify this fossil form as a hominid and a likely candidate for human ancestry.

In hindsight, this scenario seems somewhat simplistic, and now, after two more decades of research yielding a wealth of splendid fossil finds from this period, *Ramapithecus* and related species are now thought by many to be apes, variants of an important, diverse group found primarily in Europe and Asia but rarely in Africa. Although they were certainly not bipedal, their large, thick-enameled chewing teeth (molars), along with small canines were perhaps an adaptation for grinding up hard plant foods in the expanding seasonal woodlands, ten to fifteen million years ago. *Ramapithecus* is now generally considered to be one member of a larger group of fossil forms assigned to *Sivapithecus*.

Sivapithecus. This group of apes, which lived from fifteen to eight million years ago at rich sites in Turkey, Pakistan, and India and also in Europe, China, and Africa, has generated a great deal of interest and controversy in recent years. At least one species has been championed by some as a distinct ancestor of the orangutan, sharing many of the same facial features, which would give us at least one early branch out of this "bush" of Miocene apes leading to a modern ape form.

Where and when did the lines toward each of the modern apes emerge, then, and when did the pathway toward humans split off from our ape relatives? Studies of DNA in modern primates indicate that the Asian apes must have split off earlier (first the gibbon and then the orangutan), and that for some time after that split the African apes (chimps and gorillas) and humans shared a common ancestry. At present, however, there are no clear candidates for a late Miocene ancestor —a combined human-chimp-gorilla line—before it split into these different evolutionary pathways.

Most DNA studies indicate that the gorilla was the first to break away from the ancestral group, leaving a human-chimp line that split again sometime later, probably between five and eight million years ago. But again, fossils for this period are annoyingly sparse, and we just don't yet have in hand the precise ancestors for these individual lines. Undoubtedly, new discoveries during the next several decades will help clarify the evolution of the great apes and humans.

The paucity of specific ape ancestors may also highlight the hard times apes were facing by the late Miocene. Forests were shrinking, grasslands were expanding, and apes were having a major decline in numbers and diversity while monkeys were proliferating. There may not have been many apes around, and they were trying desperately to adapt to a changing environment, scrambling (in an evolutionary sense) to make it. The profusion of ape species in the Miocene largely died out, the monkeys radiated, the existing apes retreated to the dwindling forests (which are today continuing to decline), and our ancestors were in the process of adapting to these changing conditions. At any rate, we see that by the Miocene the apes had finally emerged, and they actually had a very good time of it for several million years, only to find themselves in serious evolutionary trouble toward the end of this epoch, by five to ten million years ago.

FAMILY CONNECTIONS: HOMINIDS AT LAST

THE FAMILY CREST: WHAT MAKES A HOMINID?

The term *hominid* is usually used to refer to humans, our immediate ancestors, and those ancestors' cousins—creatures either directly on our line or at least on our evolutionary branch, as opposed to the branch of the rest of the modern apes. (Technically, hominid designates animals included in our biological family in the Linnaean classification of animals. How one arrives at a definition of the word *hominid* is a matter of some contention. Historically, it has been defined in a number of ways:

1. From an evolutionary perspective, if we could trace the line leading to modern humans *ever since* our last common ancestor with the African apes, all animals directly on that line, or on side branches that lead to extinction, could be considered hominids.
2. From an anatomic perspective, bones (or other types of fossils such as petrified footprints) that indicate a two-legged walking mode among these evolving apes would represent that animal as an undisputed hominid. This view assumes that the common ancestor of apes and humans was some sort of quadrupedal ani-

mal, and that the derived, bipedal form represents the divergence of the human line.

3. From a behavioral perspective, the presence of flaked stone tools represents a technological pattern possessed only by the human line, as it is not found in the wild among modern primates, including the African apes, nor is it seen in the prehistoric record prior to the appearance of bipedal primates.

Interestingly, prehistoric evidence now suggests that these three different perspectives—the evolutionary (our split from the apes), the anatomic (evolution of bipedal walking), and the behavioral (development of flaked stone technology)—represent events that did not happen all at once but were spread out over some time in our evolutionary development. It is in a sense a moot question which of these better defines a hominid: any consideration of human evolution must somehow recognize all three of these criteria. Most anthropologists have tended to use bipedalism as a convenient hallmark of the hominid family.

HIDDEN BRANCHES IN THE FAMILY TREE

The crucial split from the last ancestor we had in common with the African apes (most recently the chimpanzees) probably occurred more than five million but no more than eight million years ago. This branching of our two lineages, the chimpanzees going off to pursue their evolutionary destinies and we our own, is really very recent in evolutionary time.

So far, this crucial missing link in a long series has yet to be found, and there may be several good reasons for this. Besides the rarity of apes during this evolutionary period, there is the problem of preservation. Our joint ancestors may have been adapted primarily to lush, tropical forest environments in Africa, where bones do not always seem to survive well. Tropical forests tend to be wet and humid with acidic soils that soon destroy organic remains, including the bones of dead animals. Thus far, absolutely no fossil record has been uncovered in Africa that would indicate the immediate ancestors of our nearest living relatives, the chimpanzees and the gorillas. In fact, the oldest known examples are only a few hundred years old.

We also face the problem of limited windows into this period of time: known fossil-bearing deposits of this age are relatively scarce in Africa. Many of the sediments laid down during this time period were eroded away long ago, while others still lie deeply buried in the Great Rift valley in Africa under younger sediments and have yet to be uncovered by erosion. The next several million years of water and wind etching away at these sediments should uncover more fossils, which might solve some of the prevailing mysteries surrounding the human-ape split. In the meantime, though, we have to deal with what we can find now and to accept that our understanding should improve as more fossil finds are made in the future.

EARLY CANDIDATES FOR FAMILY STATUS

What does the fossil record yield between four and eight million years ago—the crucial years during and immediately after the human-ape split? Only a few fossil bits and pieces are possible candidates for hominid ancestors.

From near Lake Baringo in central Kenya:

- a lower jaw fragment from a locality called Tabarin within the Chemeron Formation of sediments, dated between four and somewhat more than five million years ago. This is similar to jaws of hominids found elsewhere in East Africa in a slightly later time range, between three and four million years ago.
- part of the upper end of a humerus (the upper arm bone) also found within the Chemeron Formation. This seems to represent the arm of a hominoid, or ape, if not an early hominid, and may be as old as the jaw fragment.

From farther north in Kenya, off the southwestern end of Lake Turkana:

- another fragment of a lower jaw from the Lothogam locality. Radiometric dates of this fragment, and other animals found here, indicate that the age for this jaw is possibly five to six million years, and it could belong to either an ape or a hominid.
- the lower end of a humerus, from the Kanapoi locality, thought to be four to five million years old. This appears to be a fragment of a hominid arm.

From further to the north in the Afar triangle of Ethiopia in a region known as the Middle Awash (along the middle reaches of the Awash River):

• a fragment of the frontal bone of a skull from the Belohdelie locality. This appears to be from a hominid and was found below volcanic ash dated to four million years ago, suggesting a probable age of four to five million years for this fossil.

This is a tantalizing but really rather sparse assortment of fossils for such an important time period in our evolutionary past. The search is on for more fossils within this time range, but until we find more hard evidence we are largely in the dark. Luckily, however, enough fossils are being discovered from the ensuing time period to keep paleontologists occupied—studying the fossils and arguing about their meanings—for some time. Recent research has unearthed a remarkable array of fossils between three and four million years old representing early members of our family, the bipedal hominids.

ENTER THE AUSTRALOPITHECINES: UPSTANDING HOMINIDS

Beginning about four million years ago, the hominid fossil record picks up dramatically. With the end of the Miocene and beginning of the Pliocene geological epochs about five million years ago, worldwide cooling and drying continued with further shrinking of the tropical

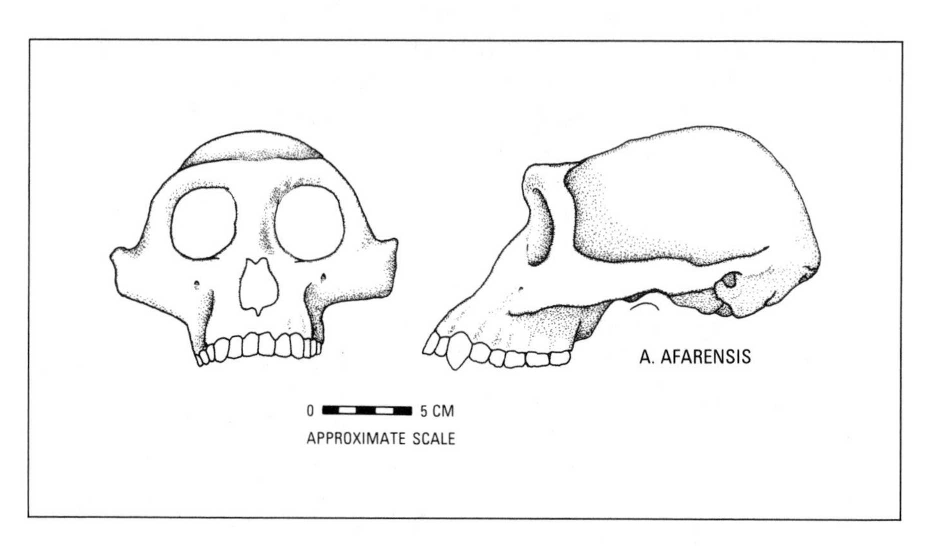

Australopithecus afarensis. The reconstructed cranium of a male "Lucy," done by Tim White and William Kimbel, was based on fragments of different individuals from the Hadar region of Ethiopia. Fossils assigned to this species are between three and four million years old and represent the oldest known bipedal hominids from Africa. No stone tools are known from this time period.

forests and expansion of a drier type of vegetation, creating a combination of woodland and grassland. The hominids appear to have adapted to this drier environment—in fact, bipedalism may have emerged in response to it—and their fossils start appearing in sediments preserving a record of this time.

In the Great Rift valley of Tanzania, Kenya, and Ethiopia, researchers have found exceptional evidence of our first known bipedal ancestors: the earliest known members of the genus called *Australopithecus*. These fossils have come from a number of sites representing a variety of settings: ancient lake fronts, river floodplains and channels, volcanic ash–coated highlands. The most important sites from this period are those in the Hadar region of Ethiopia and the Laetoli region of Tanzania. The fossil hominids from these sites have been grouped by most researchers, at least for the time being, into one species: *Australopithecus afarensis*.

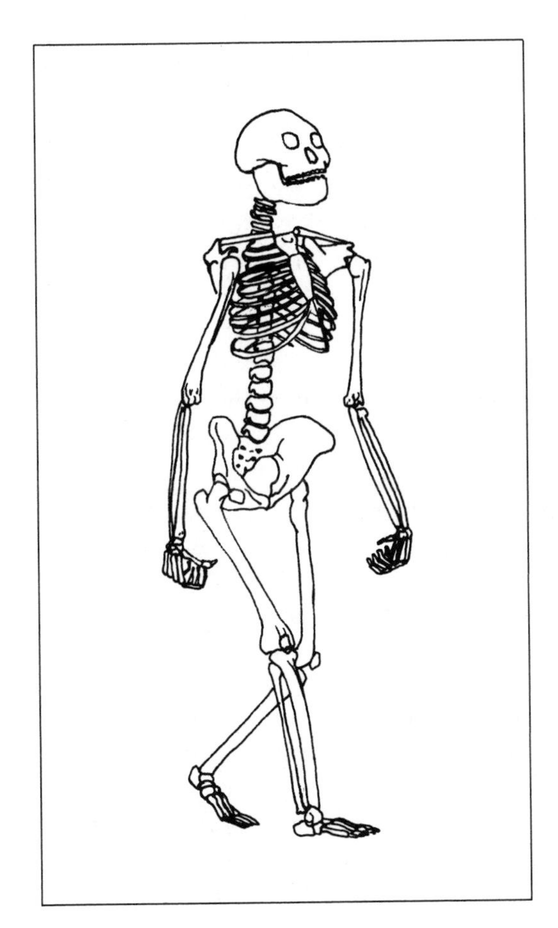

Reconstruction of the skeleton of *Australopithecus afarensis* by Owen Lovejoy. This creature had an ape-size brain, a combination of apelike and humanlike characteristics in the face and teeth, yet an upright posture, with relatively long arms and short legs. There is no evidence of artifacts or tools from this time period (four to three million years ago).

Let's look at four sites that have yielded evidence of early bipedal hominids.

Hadar, Ethiopia

During the 1970s, Donald Johanson, Maurice Taieb, and Yves Coppens led expeditions to the Afar triangle region of Ethiopia to search for early hominid fossils and understand their geological and environmental context. This research has dramatically increased our sample of hominid fossils and contributed to our understanding of the course of human evolution between 3.0 and 3.5 million years ago.

Over 240 fossil specimens were recovered from Hadar localities during the initial burst of research in the 1970s, and new fieldwork, started in 1990, is already adding substantially to these. Among the first lot of fossils found in the 1970s, which represented at least thirty-five individuals, were not only large numbers of jaw fragments and teeth, but also bones from the trunk and limbs, which have been crucial in identifying bipedalism. These include parts of femurs (thighbones), a knee joint, and the extraordinary find of part of the skeleton of one individual, a female classified as *Australopithecus afarensis*, catalog number AL (Afar Locality) 288-1, better known by her nickname, "Lucy."

To the general public this fossil find is perhaps the most famous discovery in the history of paleoanthropology. This skeleton suggested a bipedal creature only about 3.5 feet tall, with longish arms relative to leg length and a combination of both human and apelike characteristics in the skull and teeth. The completeness of this specimen was unprecedented in the early hominid fossil record and allowed anthropologists to reconstruct overall body proportions with relative precision.

Another Hadar locality, AL 333, was, in many ways, an even more sensational discovery. Here, at a single locality (numbered 333), excavations in 1975–1976 yielded fragmentary remains of at least thirteen individual hominids, with no other species of animal bones nearby. This so-called First Family, a phenomenal find showing such an unexpected degree of diversity, is particularly interesting for two reasons.

First, it seems likely that this collection of bits and pieces of at least thirteen hominids represents some sort of social group of closely related, contemporary individuals. In addition, the presence of infants and adults gives clues to patterns of growth and development of these creatures. The marked difference in the estimated body sizes of the adults has suggested to many paleoanthropologists the tremendous degree of sexual dimorphism (differences in size and morphology between the sexes) in *Australopithecus afarensis*, with males approximately twice the body size of females (as with such primates as baboons and gorillas today).

Second, although the causes of death are unknown, it would appear that these hominids died in very close temporal proximity. They might have been victims of some catastrophic epidemic, casualties of a lightning strike as they huddled in or under a tree during a thunderstorm, unfortunate prey of some group of social predators in a surprise attack, or mass casualties in some other scenario. In any case, there were no modified or natural stones in the sediments in which these creatures were found to suggest that they were habitually carrying around or collecting materials for potential tools. Although the absence of evidence is not *necessarily* evidence of absence, there is nothing at this unique and well-preserved site that would clearly indicate any technology.

Archaeologists who joined Hadar research teams have searched for evidence of early Stone Age archaeological sites from the geological strata (layers) that have yielded fossils of *Australopithecus afarensis*, but so far they have come up blank. In geological deposits above these strata (making them more recent in age), however, Paleolithic archaeological sites *were* discovered, which will be discussed in chapter 3.

Laetoli, Tanzania

Fieldwork started in the 1930s on the Serengeti Plain in northern Tanzania, about forty-five kilometers south of Olduvai Gorge, which has produced fossils of many extinct animal forms. It was here, in 1935, that the first discoveries of *Australopithecus afarensis* fossils were made, though they were not recognized as such until the 1970s, when additional hominid specimens were found. Potassium-argon dates and correlations of fauna indicate that these hominids are about 3.5 million years old. There is other compelling evidence of the ancient life-forms that lived here in these times: footprints of a multitude of animals have been found, from the size of millipedes to immense extinct elephants, fossilized in hardened volcanic ash deposits, which would have been muddy ground at the time the impressions were made.

In 1976 members of Mary Leakey's research team discovered multiple sets of what are clearly bipedal hominid tracks, all walking in the same direction: a set of small footprints to the left, with a distinctive splaying out of the feet during strides, and just to the right of this, two sets of larger footprints with a straighter orientation, one set overprinted inside the other as if one individual were walking behind the other. These prints demonstrated graphically that bipedal hominids with a characteristic humanlike foot structure existed 3.5 million years ago, bringing an immediacy and certainty to the paleontological record that captured the imagination of scientists, the news media, and the general public alike.

Mary Leakey, trained in Paleolithic archaeology and responsible for most of the Stone Age excavations and analyses at Olduvai Gorge, also searched with her crew for evidence of early stone artifacts from the footprint levels. But the returns at Laetoli were the same as at Hadar from the time of A. *afarensis:* they found no stone tools.

Middle Awash

In the region just to the south of Hadar in Ethiopia, not far from where the older skull fragment had been found at Belohdelie, a paleoanthropological team headed by J. Desmond Clark and Timothy White discovered another fossil of a bipedal hominid, believed to be *Australopithecus afarensis*. This is the proximal femur (upper thighbone) from Maka, dated to between 3.5 and 4.0 million years ago.

Omo

In the late 1960s an expedition to this important locality in southern Ethiopia, led by F. Clark Howell of U.C. Berkeley and Camille Arambourg and Yves Coppens of the Collège de France, found hominid teeth within deposits about 3.3 million years old. Although these had originally been thought to represent *Australopithecus africanus*, a different hominid form well-known from later deposits in South Africa, some researchers now see similarities in some of the fossils to those of A. *afarensis* hominids discovered elsewhere in East Africa during the following two decades. So future fossil finds in this area, particularly if more complete fossils can be found, could put A. *afarensis* on the map in this locality as well during this time period.

SIGNIFICANT FEATURES OF THE EARLIEST KNOWN HOMINIDS

Several specimens, especially from the Hadar, have enough of the braincase represented to allow anatomists to estimate the brain size, or cranial capacity. These estimates range from approximately 350 cubic centimeters (cc) in the smaller adult skulls to about 500 cc in the larger specimens. The average for modern humans is about 1350 cc, for chimps and gorillas about 450 cc. Thus, the early australopithecines, the earliest known hominids, had a brain size very much like that of an ape, coupled with a lower body and a way of walking much like that of a human.

Put more simply, the evolution of our posture, our way of walking, appreciably preceded the great expansion of the human brain. In fact, as we discuss in more detail below, bipedalism seems to have been basically established before our brain even started its rapid expansion. A major set of questions concerns why (or, from a more evolutionary point of view, how) such a creature evolved. To what was it adapted? Why did this particular lineage of apes develop bipedalism—what advantage did it give them? What could this small-brained ape have been doing that it somehow selected upright walking? Overall, what made these creatures so successful?

Just how much time these bipeds were spending on the ground is a matter of debate at present. Anthropologists such as Owen Lovejoy from Kent State University and Donald Johanson from the Institute of Human Origins, Berkeley, maintain that the structure of the pelvis, leg, and foot bones of *Australopithecus afarensis* show a creature already fully adapted to a terrestrial life-style. Lovejoy sees the evolution of bipedalism as an adaptation related to a unique hominid reproductive strategy: by freeing the hands, it allows individuals, especially males, to carry food to share with a female mate (in this model, a monogamous relationship) and their young offspring. Such assistance in the care and feeding of a hominid baby or child would permit the female to give birth more often, increasing their overall reproductive success, thus giving an adaptive advantage.

Others, such as Randall Susman and Jack Stern of SUNY Stony Brook, point to anatomical features like curved finger and toe bones (present in modern chimpanzees but absent in modern and recent humans) as evidence that early hominids, despite their bipedal features, still spent substantial amounts of time in trees as well as on the ground. Keeping to the trees, at least some of the time, could have been important for feeding on some plant foods as well as finding safe sleeping places away from dangerous predators, which is what modern chimpanzees and baboons do in the wild.

Another interesting feature of these early bipedal hominids is what they *lack:* there is so far absolutely no evidence of tools preserved with their fossils. Since the time of Charles Darwin, many evolutionary scholars have proposed that manipulative use of the hands and tool use were underlying reasons for the origins of upright walking—to free the hands. So far, however, we can see absolutely no sign of tools in the fossil record of *Australopithecus afarensis*. Although it could be argued that perhaps their tools were merely made out of perishable materials that have not survived, we must also contend with the other striking feature of these early hominids: their essentially apelike brain. The most conservative position would seem to be that early australopithecine tool using and tool making was not appreciably advanced beyond what is seen in modern chimpanzees. It seems that tools became particularly important for protohumans only sometime after the development of bipedalism.

In sum, *none* of the geological deposits and sites yielding fossils of the earliest bipeds *(Australopithecus afarensis)* have so far produced conclusive, recognizable stone artifacts. It would appear that at this stage of human evolution a flaked stone technology had not yet become a habitual or critical part of the adaptation of bipedal hominids. It is not until significantly later than the earliest bipeds that we begin to see the earliest stone tools in the archaeological record. This may have been one of the most important and pivotal changes responsible for our nature and success today (discussed in chapter 3).

Between two and three million years ago, the hominid record in East Africa becomes very sketchy. For this time period the great majority of fossils come not from the Great Rift valley of Africa, but instead from limestone cave deposits in South Africa. Close to Pretoria, ancient limestone caves and sinkholes were filled up with sand, silt, and other debris, including the bones of many of the animals living in the region during this time. Over the ensuing millions of years, these deposits hardened into rock with calcium carbonate from the groundwater in the region. Contained within these cave deposits (called breccias) are thousands of fossils, including hundreds of hominids, which have been patiently and meticulously unearthed for more than half a century. These fossils will be discussed in chapter 2.

THE FIRST STEP: UPWARD AND ONWARD

There is thus far no direct sign of flaked stone tools in Africa before 3 million years ago. The origins of upright walking could have been partially related to an increased reliance upon technology, but not a technology that we can recognize in the early hominid prehistoric record. Also important here, we can see no significant increase in the size or complexity of the brain over the first million years or so of bipedal walking by these small-brained australopithecines. Basically there is no evidence that early hominid material culture was more complex than, if even *as* complex as, that observed among modern chimpanzees today.

This would suggest that bipedalism—one of the first giant leaps toward what we are today—probably did not evolve for tool using per se. It may have developed for carrying (foods and so forth) or as part of a feeding posture, or as an efficient adaptation for long-distance travel, giving them some important advantages in the changing world of the late Miocene and early Pliocene. On the other hand, bipedalism would certainly have enhanced the ability to make, use, and carry tools, since the hands did not have to be structured or used all the time for locomotion. In a sense, bipedalism could be seen as a preadaptation for using tools, but tool use was possibly very limited in these early stages and did not engender a great increase in the brain or intelligence in this phase of hominid evolution.

So what were these early australopithecines like? The overall pattern emerging from the fossil evidence shows an evolutionary trajectory involving bipedal posture and locomotion (with some possible retention of arboreal adaptations), a small brain size relative to body size, a jutting face, large jaws and teeth, and appreciable differences between the size of adult males and females. These creatures were probably omnivorous, eating a large variety of foods, but concentrating on vegetable resources in the tropical and subtropical woodlands and grasslands of Africa.

The australopithecines we have looked at so far had truly made some great strides away from their ape ancestry. They had managed to survive and adapt in the growing woodlands and grasslands on the African continent, and they had continued for at least one million years. (Their ape relatives, though invisible in the paleontological record for this time, were presumably also coping with survival and adapting in the dwindling forests of the continent.) But these hominids were apparently still fairly limited in where they could live (both ecologically and geographically).

To continue their evolution toward what humans are today, some of these australopithecines made another critical evolutionary shift. This change seems to begin as we see at least two separate hominid lines emerge and also the origins of flaked stone technology in the prehistoric record.

C H A P T E R 2

THE STONE AGE CONSIDERED

The situation in which these [flint] weapons were found may tempt us to a very remote period indeed; even beyond that of the present world.

John Frere, 1799. Report to the Society of Antiquaries, London, regarding ancient stone implements and fossil animal bones found at Hoxne, England.

REACHING BEYOND THE SELF—FOR A TOOL

Archaeologists dealing with traces of ancient human culture have usually defined "tools" as objects that have been used, either modified or unmodified, by human beings (or ancestral protohumans). In practical terms this is a useful concept, since archaeologists focus on sites created by humans. But to understand tool use as a larger phenomenon, and to understand its evolution in the distant human past, we must also look at tools in the animal world at large.

Science has now documented a number of bona fide cases of tool use by nonhuman species, such as apes, sea otters, birds, and insects. We will look at these in more detail in a moment, but for now we might ask: What constitutes a tool in the animal world? In all of these instances, a tool is some external object, beyond an animal's body, used to achieve some end.

In its simplest form a tool is used for some very basic, essential purpose: to acquire food, for instance, or to fend off competitors or predators. An early human ancestor throwing a rock as a missile to hunt an antelope would be an example of using a simple unmodified tool to acquire food. In most cases nonhuman species use rocks, sticks, or leaves just as they are found in nature. A tool doesn't have to be *made* in the sense of modifying or shaping it—it just has to be *used*.

Objects that have been modified by humans, either intentionally or unintentionally, are called *artifacts*. The emphasis here is upon the *alteration* produced by human action; the object itself may or may not be used. For instance, when humans make a spear out of a wooden branch, a lot of wooden shavings are incidentally created as a by-product. But both the spear and the debris generated in making it are artifactual: they have been modified by humans. (When chimpanzees use stones or wooden branches to break open nuts, their stone or wooden hammers incidentally become pitted from pounding the hard nut shells, and in a sense they become artifacts. Chimpanzees also deliberately modify their tools: for instance, by trimming twigs to fish insects out of termite nests or anthills. Although by archaeological definition an artifact should be modified by a human or protohuman, we should recognize that this concept could be extended to other animals.)

To illustrate the difference between tools and artifacts in the classroom, we sometimes wake up an introductory anthropology class by throwing a rock—or what *seems* to be a chunk of granite—at a student accomplice sitting in the front row (it is actually a sponge prop bought at Universal Studios for just this purpose). The unmodified rock is a tool, since it has been used as a missile, while the hypothetically fractured skull of the student would be an artifact, since his cranium would have been modified by the human action.

Technology is something much larger than the tool itself. It refers to the system of rules and procedures prescribing how tools are made and used. In a broader sense, this term can be used for the systems of tool-related behavior of nonhuman species as well. To have a technology per se, there should be some agreed-upon ways of doing things in a social group—that is, there should be some learned, *cultural* aspect to the tool use or artifact manufacture.

LEARNING AND TOOLS: A CATALYTIC COMBINATION

Just how important are learning and culture to the development of tools and technology in the animal world? Here we use the term *culture* in an ethological sense, meaning behavior learned by one animal from another, rather than in the traditional anthropological sense, as the system of "knowledge, belief, art, morals, law, custom, and other capabilities and habits acquired by man as a member of society" (Edward Tylor, 1871, *Primitive Culture*). Learning and tools: can you have one without the other? We know humans are not unique in either of these. Many other animals learn information and even modes of behavior, not only on their own, but also culturally or from others in the animal's group or community, and several animals besides humans have tools. But do cultural creatures generally tend to develop tools and technology? Put the other way around, does tool making and tool use require culture or learned behavior?

Learning adds an extraordinary dimension to a creature's adaptation to the world. If behavior is flexible enough to adjust to new or changing circumstances and challenges, to calibrate and regulate behavior through trial and error, it gives wonderful advantage beyond genetically programmed behavior. This is particularly so when animals are adapting to complex or changing environments through more elaborate networks of behavior. Thus learning has gained increasing importance throughout the evolution of many animals, particularly among the vertebrates (although some invertebrates, such as certain insects, appear to have at least some learned component in their behaviors).

Learning can be very individual, through trial and error by experimenting until a suitable or superior response is discovered, or it can be social, gained from experience of others in the animal's society, either through pure imitation or with some guidance, especially by a parent. Learning becomes an extremely powerful tool for adaptation when it becomes cultural (in an ethological sense), when it steps outside the experience of the individual—when it is shared among a group of individuals who learn different advantageous behaviors from one another. We know humans today rely upon culturally learned behavior much more than any other animal, but what is the link between technology and culture in the animal world at large—and in our own evolutionary past?

Culture is relatively widespread among animals, but it does not automatically generate technology. There are large numbers of animals with cultural behaviors, adopted and actually shared among a group of animals through learning, but without tools. John Bonner in *The Evolution of Culture in Animals* (1980) gives many fascinating examples of cultural transmission of information in a wide variety of animal societies, both among vertebrates and invertebrates, such as insects. Animal culture may involve learned systems of communication, ways of finding or processing food, or defining territories and locations to live or feed.

Though tool use is still relatively rare in the animal world, a small number of nonhuman species are now known to use simple tools.

Particularly in nonmammalian species, however, technological behaviors are more along the lines of what we call "instinct"—stereotypic of the species or population, relatively inflexible, bound to a set of stimulus-response interactions, and apparently imbedded in the animal's genetic makeup. But among our primate relatives the chimpanzees, learning and culture appear to play a much larger role in technology. When their tool-using behaviors are examined in detail, some aspect of learning, by trial and error and imitation, is evident.

It is only in the human lineage, however, that culture and technology are coupled and fundamental to our existence. After infancy, we rely essentially upon learning as we grow and adapt to our environment, and most of what we learn is acquired from and somehow shared with our society or culture. Eminent among our learned behaviors is our use of tools and technology in everyday activities—to acquire and process food, to defend ourselves against predators or aggressors, to travel through our environment, to shield our bodies from the extremes of the environment, and to communicate. This is the way we are today, but what was the situation in our remote past? How did it happen that we are so tied not only to learning, but to a very special learned behavior, our technology, and why has this been so important in our evolutionary development?

SHAPING THE WORLD: CULTURE AND TECHNOLOGY

A very critical aspect of learned, cultural behavior—with or without technology—is how it improves an animal's ability to adapt to its environment. It allows an animal to explore alternative ways of achieving its ends—getting food, finding mates, transmitting information. Carried to an extreme, it can allow an animal population to adapt more and more by manipulating and transforming its environment, rather than its biology, in order to survive and reproduce. As evolution proceeded, and the anatomical structures and behaviors of certain animals increased in complexity through time, some—our ancestors—began to place more and more emphasis on modifying their environments as a means of adaptation.

A significant development of culture was necessary for the technological burst we can see in our past. This must have involved an intensified ability to teach and learn. There is good evidence that culture was part of our heritage long before we started down the special human evolutionary path. We must have developed some reliance on culturally learned and shared behavior relatively early in our adaptation. We can look around at our primate relatives today and see the important role of learning and culture. If we look closely at modern chimpanzees, for instance, it is obvious that culture plays a significant role in their lives: in their social relationships, their feeding behaviors, and their use of tools. Culture is part of our primate heritage.

TOOLS ARE US

It was on this substrate of culture that humans were able to develop complex technological traditions. But this was an unusual way to go, one not taken by chimps, gorillas, or any other animal on earth. Two paths did diverge—whether in the forests, on the grasslands, or somewhere in between—and our lineage took the less traveled one, and that *has* made all the difference. A short while ago in evolutionary time, just a few million years in our past, our ancestors occupied a small part of equatorial Africa. Now we have expanded to virtually every habitable part of the planet (and "habitable" here is defined by our technology), and we even make brief forays beyond. But back in those dwindling forests of Africa, our closest relatives are still fighting what seems to be a losing battle for survival over the long run. Why?

It is the technological path we humans took that has separated us most profoundly from our primate ancestry and from our extant primate relatives. Without an increased reliance on technology, hominids might still be striving for survival in the grasslands and woodlands of Africa or might have already gone extinct (as did several of our ancestors' bipedal cousins). Our technological adaptation has been shaping our evolutionary trajectory in crucial ways for the past several million years. But why did tools matter that much? How do our earliest tools compare with technologies among nonhuman animals, and how did they give us such an edge—an advantage in terms of survivability and increased reproductive success that was selected by evolution?

TOOLS IN THE ANIMAL WORLD

It is essential that we try to discriminate between those instances of tool use and tool making that are instinctive and those that are cultural. Instinctive behavior is genetically programmed and stereotypic and does not rely upon transmission of information from one individual to another to maintain that behavior over time.

Let us think back for a moment to our definition of a tool: an object used for some purpose. A tool must be *something outside* the individual that is used to *modify or manipulate something else* in the environment. This might seem an arbitrary distinction, but it is actually a very important one. This sets off the tool as something very special—and excludes things such as nests, spiderwebs, insect cocoons, and myriad other ways animals shape the environment in some way. Building activities such as these are admittedly very useful, sometimes even crucial, in an animal's adaptation, even if they are instinctive and inflexible within a particular species. *Object manipulation*, however, usually involves greater behavioral and organizational complexity than these building activities, which helps explain the relative rarity of tool use.

Tool use can unleash a world of potential for manipulation: there are countless objects out there and almost infinite ways of using one to act on or modify another. When animals begin to tap this potential with intelligent, intuitive, exploratory behavior, tool use becomes diverse, interesting, and wonderfully adaptive. Our ancestors became specialists at just this.

Examples in the animal world of innate tool-using abilities include a small but astonishingly varied list of species in nature. In *Sociobiology: The New Synthesis*, Edward O. Wilson compiled a fairly comprehensive list of known instances up to 1975, and no significant increase has been noted in recent years. Prominent cases of nonprimate tool use include:

1. the mud wasp, which holds a tiny, unmodified pebble in its jaw to tamp down mud to construct its nests.
2. some Galápagos Islands finches, found in these Pacific islands off the coast of Ecuador, which tear off and use a spine from a cactus plant to probe and "fish" for burrowing insects inside a tree. The insects sense an invading foreign body and latch on to the cactus spine, and the finch withdraws the spine and eats them.
3. the Egyptian vulture, which will carry unmodified rocks in its beak and drop them on thick-shelled ostrich eggs in order to break them open for food.
4. the California sea otters, which crack open the hard shells of clams and abalone using unmodified stone hammers or anvils. (A visitor to the Monterey coast probably will have seen these mustachioed aquatic mammals diligently rapping away at their molluskan prey.) The sea otter has two ways to accomplish this: floating on its back, the otter either places the shellfish on its chest

and hammers away at it with a rock held in its paws or else it places the rock on its chest and then bashes the shellfish against this anvil.

Some examples of tool use in the animal world. A mud wasp using a pebble to tamp mud; a Galápagos finch using a cactus spine to probe for insects; and a California sea otter cracking a shellfish on a stone.

But despite these examples, tool use among animals other than primates is rare. Only a few other cases could be included here, most of them variations on a theme: a few more birds that also use twigs as extensions of their beaks to get food; another bird that bombards eggs with rocks; some insects that catch prey by throwing sand to knock them down into their pits; and one fish that spits water at flying insect prey. Several birds, some insects, a mammal, and a fish: altogether fewer than twenty nonprimate animals that use tools in the natural world.

But within this small sample of animals, there is nevertheless remarkable diversity of species. These examples crosscut major groups of animals separated by very long periods of evolutionary time, even hundreds of millions of years. And each of these tool-using animals also has many closely related species that do not use tools. Simple tool use seems to have developed independently in a number of completely unrelated species. So we cannot make sweeping generalizations about tool use or lack of it based simply on the evolutionary history and relationships in the animal world at large.

Another interesting feature of all these examples is that with the exception of the mud wasp, all these behaviors involve some improvement in the species' ability to get food, to improve their use of food resources, or even to change or expand their ecological niche. Thus, very strong selection pressures could operate for these behaviors to survive and become fixed in a species.

In addition, there are drastically different levels of intelligence among these tool-using animals, so we cannot use intelligence to predict tool use in the animal world at large. In most of these cases, it would appear that the particular tool-using behavior is part of the species' instinctive behavior, rather than variable according to the individual and its circumstances.

The sea otter may be the major exception to this, as more flexibility and innovation seem evident in its use of tools in bashing mollusks. This is understandable in view of the greater emphasis upon learning and culture among many mammals than among birds or insects and might be particularly expected in such a dexterous, agile creature as the sea otter (although the precise roles of instinct and culture in their tool use are not yet established). The relatively fixed tool-using behaviors observed in insects and birds may have developed originally through alteration or modification of preexisting behavior patterns that were selected for and became fixed in the species. This is much different from the reasoning and insight learning characteristic of tool use among chimpanzees—and developed to a fine art by humans.

THE MANIPULATIVE PRIMATES

The order Primates includes prosimians (such as lemurs and tarsiers), New and Old World monkeys, apes, and humans. As a group, the 200 or so living primate species show outstanding dexterity and manipulative ability: they rely strongly upon their hands in exploring and using their environment, particularly in their feeding activities. They use their hands in plucking, probing, perusing, and processing a variety of foods—nuts, seeds, fruits, and leaves—but most of them do not normally use tools. Their dexterity and manipulative skills are a preadaptation for competent tool use—it has evolved for specific purposes, primarily for feeding, and has great potential for tool use, untapped by most species. Although there have been unconfirmed reports of capuchin monkeys in South America using stone hammers in the wild to crack open nuts (and documented laboratory experiments of nut-cracking behavior in these animals), at present only one nonhuman species, the common chimpanzee *(Pan troglodytes)*, is known habitually to manufacture and use tools in the wild.

A chimpanzee "fishes" for termites with a carefully trimmed blade of grass.

In the 1960s Jane Goodall shocked and fascinated the scientific world with her documentation of simple tool use and manufacture among chimpanzees she was studying in Africa. Some individual instances of tool use by captive and wild chimpanzees had been reported before this, but her work finally showed the richness and diversity of tool use by chimps in their daily lives. In the Gombe Reserve in western Tanzania, chimpanzees modify a stick or piece of grass by trimming off any side stems and breaking it to a desirable length in order to fish for termites and ants: they insert the probe into a hole in the nest and wiggle it around. The insects think it is an invading enemy, and attack the probe. The chimp then withdraws the probe from the nest and carefully licks off any adhering insects. Other forms of tool use documented by Goodall and others include chewing wads of leaves to make sponges to clean themselves or to draw water out of a hollow in a tree trunk; using stones or branches as missiles or clubs during confrontations with other animals or as a form of display to assert dominance; using sticks as prying levers; and even using twigs as toothbrushes. Thus chimps have shown the largest and most varied set of tools yet seen in any animal other than human beings.

Some of the most remarkable recent accounts of chimpanzee tool use have come from West Africa, where chimpanzees in several areas have been observed cracking open hard-shelled nuts with stone or wood hammers. This phenomenon, first observed in the early 1970s, was studied in fascinating detail during the 1980s by researchers such as Christophe Boesch. When the chimpanzees want to open a type of nut whose shell is too hard to crack with their teeth, they rest the nut on a tree root or stone to serve as an anvil and bash it with another stone or a stout wooden branch. Most interesting, these nut-cracking tools are coveted items in the chimp society, often retrieved and carried about from one nut grove to another.

Such tool use among chimpanzees is remarkable in many ways. First, chimpanzees use a fairly large variety of tools in the wild and for a variety of purposes, not only for the immediate goal of acquiring a morsel of food, but also for defense or offense, for drinking, for washing their bodies, and even for cleaning their teeth. Furthermore, they not only *use* tools, but are often known to *make* them, to modify a twig, a branch, or a leaf to do a job better. Most important, perhaps, tool use and tool making by wild chimpanzees, unlike other animals, appear to be based profoundly upon learning. Individuals sometimes invent innovative uses of tools in solving some problem or, more casually, in the process of exploring their environment. Play behavior among younger chimpanzees probably plays a large role in these inventions. And even more crucial, tool use seems to be routinely learned by watching other chimps. Tool use in chimps is a cultural phenomenon, with regional tool "traditions" (such as nut cracking, termiting) observed in some populations but not others.

Recently, comedian Jay Leno pointed out that if chimpanzees were *really* like humans, they would just *borrow* tools and not return them! Upon viewing Christophe Boesch's film of chimpanzees hammering nuts in the Ivory Coast, we found this joke to be remarkably prophetic. Since rock is a fairly rare commodity in the chimpanzees' habitat, a premium is placed on the large stones used for nut-cracking activities. These rocks are sometimes begged (the needy chimp making a supplicating gesture with an outstretched hand), borrowed (particularly within a family group of a mother and her offspring), or even stolen (generally by a dominant male or when the owner's attention is diverted). The earliest inception of technology thus may have had some profound repercussions: it could have required new sets of rules of social behavior and patterns of communication to help share the first real personal possessions.

This level of culture or shared learned behavior in chimpanzee tool use is unique in the nonhuman world and gives us a glimpse of what the primeval stages of culture may have been like in our own evolutionary past. We do not know how far back in time chimpanzee tool making and tool use goes; it could predate the chimp/human evolutionary split, or it could be quite recent. It is interesting to speculate whether, if left alone without human interference for several million more years, chimpanzees or other primates would have taken a similar evolutionary course of tool use and brain expansion to what occurred in human evolution. In our ancestry, rather than tool use and tool making remaining incidental, and perhaps unessential, behavior patterns, as may be the case even among the chimpanzees, they became central to our mode of adaptation.

THE IDEA OF A STONE AGE

It is fascinating from our modern vantage point to look back and observe earlier cultures coming to terms with the evidence of their own prehistory and their relationship to primates. Greek and Roman scholars had a dim awareness of an ancient, Stone Age prehistory, when people lived differently and used strange tools. From Herodotus to Hesiod, from Tacitus to Lucretius, many early philosophers and historians were intrigued with "savages and barbarians" living alongside their own "civilizations" and sometimes imagined successive stages in the early history of human society. These stages sometimes included an age of bronze or copper preceding their times (to which Hesiod dismally referred as the Age of Iron and Dread Sorrow). This is not too surprising, for the demise of local Bronze Age societies was relatively recent, only several hundred years in their past, probably just within reach of oral tradition and folk memory.

But classical speculation on even earlier times was fuzzy or even fanciful. Hesiod invoked an Eden-like age of gold as a sort of original state for humans (whose condition worsened progressively thereafter), and Lucretius advocated a less auspicious beginning, with humans struggling for survival using their teeth, stone, and fire before advancing into the ages of copper and iron. (This was a somewhat more realistic vision, but it still wasn't based on any hard archaeological evidence.)

Interestingly, these philosophers usually conceived human society's "original" condition to be either far better or far worse than their prevailing times. Our course since the very beginnings of human society was thought either to entail some sort of fall from grace or else to testify to the surging progress of humanity over time. Even today in our society we can see a similar opposition between advocates of a back-to-nature philosophy and champions of technological progress.

Knowledge of early Stone Age tools goes back at least to the Middle Ages. One medieval French painting portrays Saint Etienne solemnly holding what appears to be a flint handaxe, probably about three hundred thousand years old. In medieval Europe ancient stone tools were sometimes carried as amulets. Some mystical or religious significance seems to have been attached to these objects.

In European folklore such handaxes were known as "thunderstones," in the belief that these were cosmic projectiles of lightning bolts that hit the ground. The flint arrowheads found in early farming communities were sometimes called "elfshot," referring to the fanciful notion that these were the points of tiny arrows fired by elves from their little bows. This idea also served to explain the occasional, mysterious death of livestock grazing in areas where such artifacts were found.

By the time of the Renaissance, Western society fully embraced as its "past" the classical ages of Greece and Rome, as well as the biblical ages of the Near East, but an even earlier, prehistoric Stone Age was intellectually still beyond reach. During the age of exploration, western voyagers encountered many societies—in the Americas, the Pacific Islands, and southern Africa—that still used flaked or ground stone as a major component of their technology. These living, stone-using peoples served as a valuable clue to a way of life long forgotten in the urban circles of Europe, of a past universal to all human societies but forgotten by most of them. This provided a historical connection for people in areas of the world where metal technologies prevailed, but where puzzling evidence was found scattered about their towns and villages—stone artifacts littering a plowed field or found at the base of a freshly dug foundation or well. But it took some time, a lot more evidence, and even more argument, to confirm the connection to their own ancestry.

In the late seventeenth century, in Grey's Inn Lane, London (near Temple Bar, the seat of British jurisprudence), an apothecary and antique dealer named Conyers found a stone implement lying with the fossil bones of an elephantlike creature, probably an extinct mammoth, uncovered during road construction. Conyers suggested that the tool was produced by an ancient inhabitant in a time before the use of metals, while antiquarian John Bagford interpreted these as evidence of an attack by a stone spear–wielding Briton against a Roman war elephant during the reign of Claudius.

About a century later, in 1797, British country squire John Frere made similar observations regarding some stone objects found more than six feet underground in the process of digging up clay to make bricks at Hoxne in Suffolk, England. These objects were, again, flint handaxes, and they too were found with the bones of extinct mammals. Frere took them to be "weapons of war, fabricated and used by a people who had not the use of metals."

Frere drove his point home respectfully but with compelling earnestness, rationality, and attention to detail. He suggested that these stone implements were more than mere curiosities because they were found deep in the ground and covered with three different layers of sediment showing different environments and because the overlying layer seemed to have been on a seabed or seashore, though no modern sea even approached the site. Frere ended his amazingly short but remarkably canny report with a humility atypical of some modern scientific reports: "If you think the above worthy the notice of the Society you will please to lay it before them. I am, Sir, with great respect, Your faithful humble Servant, John Frere."

Although Frere's observations were precociously astute, they did not make much of an impact upon the British intellectual community at the time. Nevertheless, he merits the distinction of having made one of the first rational, well-substantiated arguments regarding a long, deep record of Stone Age human activity buried in the ground, much more ancient than any recorded in historical or biblical texts.

THE NINETEENTH CENTURY: CONFRONTING HUMAN ANTIQUITY

A French customs officer stationed at Abbeville, on the Somme River, was to play a pivotal role in the acceptance of a Stone Age before recorded history. Jacques Boucher de Perthes was fascinated by the antiquities that could be found in the river gravels of the Somme's terraces (ancient river deposits). These gravels were dug for road work and other forms of construction, and workmen would often uncover the fossil remains of strange animals and curiously shaped pieces of flint. Boucher de Perthes began to frequent these excavations to study and collect the materials that were being exposed. He became convinced that these represented a long-ago period when metals were unknown, stone was used for tools, and humans coexisted with strange animals unknown in the modern or recent world.

When Boucher de Perthes published his study in 1836 and began to display his discoveries at meetings and expositions, many were still skeptical of his conclusions. Nevertheless, people were for the first time directly confronted with a disturbing body of evidence showing human beginnings in a strange, long-distant past. Significantly, at this time people were coping seriously with vast, accumulated evidence for the tremendous length of geological time as well.

Meanwhile, in another part of Europe, further strides were being made in developing a stronger sense of order to our prehistory. A series of archaeological discoveries and collections in Denmark in the early decades of the nineteenth century had uncovered a wealth of artifacts that showed a variety of materials with no known connection to modern societies or to ones recorded in history. Soon thereafter, in the 1830s, the significance of these was to be realized when another customs officer, Christian Thomsen of Denmark, became the director of the Museum of Northern Antiquities in Copenhagen.

In his new museum position, Thomsen was immediately faced with a perplexing problem: displaying the finds of these excavations in a way that made sense of them. He came up with a system that suggested that prehistoric societies in Europe had gone through three major technological stages: a Stone Age, followed by a Bronze Age, and then an Iron Age. This "three age system," as it came to be known, had a profound influence on antiquarian studies in Europe, as it formally acknowledged a period when stone was the principal material for fashioning tools.

THE DARWINIAN AGE: ACKNOWLEDGING OUR PREHISTORY

By the mid–nineteenth century the time was ripe for realizing and accepting the fact that we had had a very long prehistoric development. The groundwork had been laid; what was now needed was critical reexamination and final acceptance of the mass of prehistoric evidence accumulated for over a century. In 1859 Charles Darwin published *Origin of Species*, in which he presented his theory of evolution by natural selection, arguably history's most important contribution to biology as well as philosophy. In retrospect, however, this momentous publication overshadows another important, but less well-known, event in that year. Two prestigious members of the Royal Society of London traveled to France to inquire into Boucher de Perthes's long-standing claims of Stone Age tools.

The visitors were archaeologist John Evans, one of the members of the Society of Antiquaries in Britain (and father of Sir Arthur Evans, who later dug the Minoan civilization site at Knossos in Crete), and geologist John Prestwich. They were following on the heels of a visit the year before by paleontologist Hugh Falconer, who had been quite impressed with the stone tools found at Abbeville. As Evans set off for France on this investigation and anticipated its possible implications, he wrote:

> Think of their finding flint axes and arrowheads at Abbeville in conjunction with bones of Elephants and Rhinoceroses 40 ft. below the surface in a bed of drift . . . I can hardly believe it. It will make my ancient Britons quite modern if man is carried back in England to the days when Elephants, Rhinoceroses, Hippopotamuses and Tigers were also inhabitants of the country.

These critical-minded, skeptical, and well-respected investigators were stunned by their visit and absolutely convinced that the Frenchman's original observations had been accurate. At the conclusion of his study of the Abbeville site, Evans wrote that the handaxes

> found among the beds of gravel were evidently deposited at the same time with them—in fact the remains of a race of men who existed at the time when the deluge or whatever was the origin of these gravel beds took place.

The year 1859, then, was thus a landmark for studies of human prehistory. Darwin's concept of the origin of species by natural selection provided the theoretical foundation for understanding the antiquity of humankind. Boucher de Perthes' studies provided the actual evidence of human cultural change over a long period of time. Very rapidly these ideas and facts were finally being accepted by the skeptical scientific community.

Interestingly, Darwin's *Origin of Species* deliberately neglected consideration of humans in the overall scheme of biological evolution. Although he had accumulated evidence and arguments on the issue, he didn't deal in print with the ticklish subject of human evolution for over a decade. Instead Darwin first pursued a more neutral, less controversial tack—how organisms evolve, or how new species emerge over the course of time in the natural world. In 1859 he merely suggested that "much light will be thrown on the origin of man and his history." Darwin argued that a species could change or evolve over time simply by what he called "natural selection." All the sundry forces in the environment worked on the diversity of offspring produced by a population, weeding out less fit organisms and choosing others for survival and procreation, generation after generation. This was much the same process, he contended, as the "artificial selection" carried out by animal breeders, only it was the pressing requirements of the natural environment doing the selecting. (In artificial selection, humans choose traits in particular organisms according to various human needs or whims, whether it might be for a cow able to yield more milk or a dog able to retrieve badgers from underground tunnels—that is, the dachshund.) Darwin's concept of evolution provided a critical framework for the growing, unmistakable evidence of human antiquity—it allowed for both great depth of time *and* change in biology and culture over that time. Thus all the enigmatic, primitive-looking tools that had puzzled and stymied Europeans for decades were at last acknowledged as tangible signs of our remote, evolutionary past.

FILLING IN THE GAPS: SEARCHING FOR TOOLS AND FOSSILS

The next step was to find fossil evidence that the human species had passed through earlier, more primitive stages of development. The first fossils to be recognized as protohuman were those of Neandertals. Although Neandertals had previously been discovered on Gibraltar and at Engis, Belgium, the most famous discovery was in 1856, when a partial skeleton was found in the Neander(*thal*) Valley in north Germany. Considerable controversy raged about its significance (the anatomist who first described it called it an ancient human, while others dismissed it as a pathological specimen or an ape), but Neandertal fossils are now believed to represent an archaic form of *Homo sapiens* that existed in Europe and the Near East between about 130,000 and 35,000 years ago. With these human fossil discoveries, people were coming to grips with evidence not only that we had a distinct cultural prehistory, filled with strange and primitive tools unlike any known in modern or recent times, but also that we had a biological past, marked by creatures with distinctive similarities to ourselves, yet jarring, even startling, differences.

In 1887, former schoolteacher and Dutch army surgeon Eugene Dubois hypothesized that the earliest fossil protohumans would be found in southeast Asia, and he began searching, first in Sumatra and then in Java. As incredible luck would have it, he found the first evidence of a separate species of upright-walking humans, which he called *Pithecanthropus erectus* (now called *Homo erectus*), along the Solo River in Java, although not apparently associated with recognizable stone tools. This creature, with huge, beetling brow ridges and an estimated brain capacity that was only two-thirds the size of a modern human's, seemed to Dubois to be an ideal missing link between modern humans and an ape ancestry. Although Dubois's claims were generally ridiculed in his time (in old age he would call the fossils an extinct form of gibbon and hide the materials from the scientific world), it is now widely believed that *Homo erectus* is indeed an ancestor of modern humans.

The search was on for evidence of our Stone Age prehistory as well. From the middle to late nineteenth century, a large number of people took up the pick and shovel and (often with little scientific method) delved into the caves and rock shelters of Europe, uncovering a wealth of evidence of early human occupation: human and animal bones, hearths, strange tools of flint, ivory, and bone, and even occasional snippets of artwork such as bone engravings. One of the most dedicated of these self-appointed archaeologists was Edouard Lartet, a Frenchman who had left law in midlife to turn to paleontology, only to settle into archaeology for the rest of his seventy years. Working on his own and later with his friend Henry Christy, an English banker, Lartet discovered, excavated, studied, and wrote about a wealth of evidence of Stone Age occupation in France. Most important, Lartet and Christy recognized that the Stone Age wasn't just one slice of time that briefly preceded more recent human accomplishments, but internally it showed a long progression of changes over time. A single cave or rock shelter often showed a deep sequence of levels with different tools and even different forms of extinct animals in each one. In Lartet's scheme, diagnostic fossil animals served as the main criteria to identify different periods of time.

During the later nineteenth century, French prehistorian Gabriel de Mortillet began to classify Stone Age periods based on *technological* grounds. Sites with similar types of stone tools were classed together into such entities as Mousterian, Aurignacian, and Magdalenian, each named after an archaeological site in France that typified the technological stage. Although de Mortillet believed that these technological stages could be universally applied to the Stone Age record on a worldwide basis, later studies showed this was false, as not all regions show precisely the same technologies and the same sequence. Nevertheless his technological approach to understanding human cultural evolution is still a major emphasis in archaeology today.

THE EARLIEST HUMANS: MISSING LINKS AND FOSSIL FINDS

The supposed discovery of the Piltdown man in England in 1912 became a major red herring in human evolutionary studies for the following four decades. Lawyer and amateur archaeologist Charles Dawson claimed to have discovered parts of a cranium as well as the jaw and isolated teeth of a missing link. He claimed they had been found in an ancient river gravel deposit along with fossils of other extinct mammals, crude stone tools (called "eoliths," or dawn stones), and a pointed bone implement made of a fossil elephant legbone—in the rough shape of a British cricket bat!

Subsequent research in the early 1950s exposed the Piltdown *Eoanthropus* (dawn man) as a forgery; the cranium was a prehistoric modern human, *Homo sapiens*, probably from predynastic Egypt, and the jaw was that of a recent orangutan, with the cheek teeth filed down and all the bones and teeth stained or painted a dark brown color. The animal fossils were authentic but had been seeded there from other localities. The eoliths had also been introduced by an unknown culprit; in fact, most are now believed to be fractured naturally and not the product of human workmanship. Finally, the Piltdown cricket bat had been artificially shaped with a modern steel knife.

The identity of the forger (or forgers) has never been conclusively demonstrated, and to this date we don't know exactly who perpetrated this hoax or why. Unfortunately it did hinder human evolutionary studies somewhat in the early twentieth century and made it difficult to understand the real meaning of valid fossil discoveries being made. In the end, though, science was vindicated and clearly and undisputably revealed the fraud. Scientific and popular interest was fired up, and the search was on for human ancestors, for various missing links in our past.

In South Africa, anatomist Raymond Dart startled the anthropological world in 1925 by claiming to have found an even older missing link, which he named *Australopithecus africanus* (southern African ape-man). This was based upon the fossil skull of a small child that came from a limestone cave breccia at Taung in southern Africa. One of Dart's students had been given a block of rock from the cave. It was basically a chunk of limestone filled with angular rubble and bones of animals, especially baboons. Dart noticed part of an immature skull in the block, and he slowly freed the fossil from the stonelike matrix using a chisel and his wife's knitting needle. He perceived that the "Taung baby" was not a monkey or an ape, but a creature with more human affinities—a possible ancestor to ourselves.

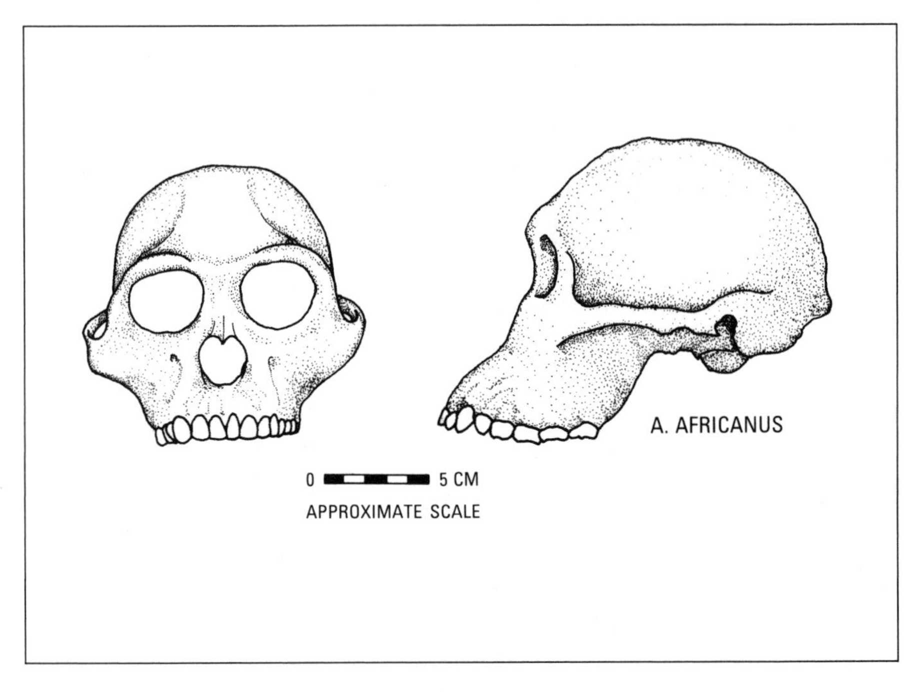

Australopithecus africanus. These fossils, between three and two million years old, come from South African cave deposits. This reconstruction is based on a specimen from Sterkfontein cave. No stone tools have been found in the deposits containing these hominid fossils.

Because of the Piltdown discovery (still believed at the time to be a valid fossil) and a scientific prejudice that the earliest human ancestors would come from Europe or Asia, at the time Dart's claims were not taken very seriously by much of the anthropological world. But by the 1930s and 1940s many more remains of *Australopithecus africanus* had been discovered at the caves of Sterkfontein and Makapansgat by Scottish paleontologist Robert Broom and his co-researchers. Since 1936 hundreds of fossils of A. *africanus* (first called *Plesianthropus transvaalensis)* have been found, representing at least fifty individuals. The layers these come from have been dated to between approximately 2.5 and 3.0 million years ago. Above these are layers with very different hominid fossils from a later period, which will be discussed in chapter 3. In addition, fossils from the nearby cave deposits of Swartkrans and Kromdraai showed that another bipedal protohuman also had existed. This one, though, had very large molar teeth and a skull designed for heavy chewing, including a massive crest of bone along the top of the skull (the sagittal crest) in males for the attachment of heavy chewing muscles. This creature was dubbed *Paranthropus* (now usually called *Australopithecus robustus*). All of these are now thought to be fossil relatives, whether or not they are in our direct ancestral line (it seems definite that the robust version, at least, is a side branch). But none of these seem to be the earliest stone tool maker in our past.

THE SEARCH FOR THE EARLIEST TOOLS

It is interesting that from the very beginning most studies of biological evolution tacitly accepted the idea that our early bipedal ancestors definitely, necessarily, *must* have had tools. When fossils of evident human ancestors were found in Europe (the Neanderthal sites) or in Java (the *Pithecanthropus* sites), tools were either associated at the site directly or were found in the region and taken to be connected somehow with the ancient human form. Even at the Piltdown site, stone artifacts and bones were apparently planted in the pit with the fossils. The expectations of the time were very clear: early humans had tools. In fact, if you were in doubt that what you had found was really "human," tools in effect clinched it: where you find humans you should find tools, and where you find tools you have found a human. So it is not surprising that when fossils of very early ancestors were found, there was a strong desire, even a drive, to find tools with them. But this can lead to some mistakes, particularly when dealing with very ancient fossils.

In fact, even the astute discoverer of the Taung baby, Raymond Dart, may have reasoned along such lines. As numbers of australopithecine fossils were recovered from other limestone deposits in South Africa, Dart began looking carefully at what else came out of these ancient caves. Finding tools alongside these fossils would no doubt convince the remaining skeptics of the humanness of these creatures. His pursuit over several decades indicates his strong conviction that an early human ancestor simply *must* have had tools.

At Makapansgat, a fossil site discovered soon after the announcement of the Taung baby, Raymond Dart once again found hominids among the bones. Although Dart originally put these into another group of australopithecines (A. *prometheus*, because he thought black staining on materials at the site showed hominid use of fire), most researchers now see no significant differences between these and the rest of the A. *africanus* fossils and put them in the same group.

Although he could not identify any definite stone artifacts, Dart did notice some remarkable, and very suspicious, patterns among the animal bones dug out of these caves. Many were fractured into strangely shaped pieces: broken limb bones and horns forming daggerlike objects; skull and other bone fragments making what seemed to be scoops and bowls; and jaw fragments forming sharp, jagged sawlike or knifelike objects. Dart deduced that these were very early tools predating the use of stone in our past, what he called an "osteodontokeratic" (bone-tooth-horn) industry. This fit in with a much larger theory of what these australopithecines, literally "southern ape-men," were like and what they signified about the early beginnings of human beings.

Dart envisioned *Australopithecus* using limb bones as clubs, antelope mandibles as saws, and jagged-edged broken bones as daggers. To Dart, many of the whole and fragmented animal bones found in the Makapansgat cave deposits represented a tool kit carried to the cave by his ape-men. Dart also argued that *Australopithecus africanus* was a skilled hunter—and homicidal (actually, *australopithecidal*). In fact, some of these ideas had been voiced very early on by Charles Darwin, in 1871, in *The Descent of Man*, as he considered why upright walking would have been advantageous or selected by evolution: "They would thus have been better able to defend themselves with stones and clubs, to attack their prey, or otherwise obtain food."

In his many publications Dart gave colorful emphasis to this notion, portraying our ancestors as bloodthirsty predators, as in this description of australopithecines from his article "The Predatory Transition from Ape to Man" in 1953:

Man's predecessors differed from living apes in being confirmed killers: carnivorous creatures, that seized living quarries by violence, battered them to death, tore apart their broken bodies, dismembered them limb from limb, slaking their ravenous thirst with the hot blood of victims and greedily devouring livid writhing flesh. Further, if Darwin's reasoning was correct, man's erect posture is the concrete expression of signal success in this type of life.

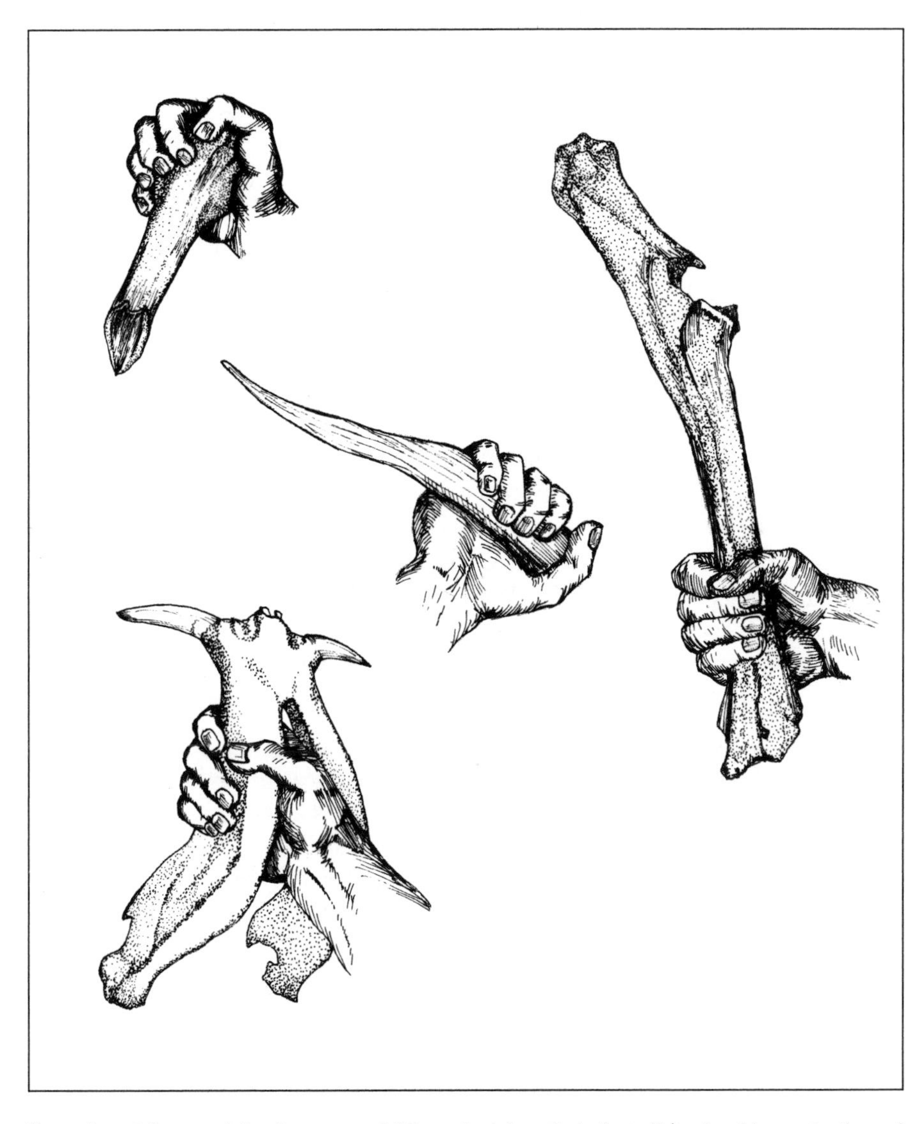

Examples of Raymond Dart's proposed "Osteodontokeratic Industry" (tools of bone, tooth, and horn).

In Dart's view, in our origins we were not only predators, but unbridled cannibals:

> The blood-bespattered, slaughter-gutted archives of human history from the earliest Egyptian and Sumerian records to the most recent atrocities of the Second World War accord with early universal cannibalism, with animal and human sacrificial practices or their substitute in formalized religions and with the world-wide scalping, head-hunting, body-mutilating and necrophilic practices of mankind in proclaiming this common bloodlust differentiator, this predaceous habit, this mark of Cain that separates man dietetically from his anthropoidal relatives and allies him rather with the deadliest of Carnivora.

Clearly this did not represent a totally detached, scientific view of things. Dart's concept of a killer ape in the human ancestry seems in part, an attempt to explain the distressing levels of aggressive behavior seen in modern humankind. It was nature, rather than nurture, he believed, that accounted for the darker side of human behavior. Dart cited several australopithecine fossils that he believed bore evidence of such aggressive tendencies: one with a fractured skull and another with a depressed fracture.

Dart's ideas were subsequently popularized by playwright and author Robert Ardrey in the book *African Genesis*, and the osteodontokeratic and killer ape hypotheses were chronicled in cinematic splendor in the opening scenes of the film *2001: A Space Odyssey*, written by Arthur C. Clarke and directed by Stanley Kubrick. In the beginning of this film a hairy ape-man (assisted by monolithic inspiration) invents the first tool: a bone weapon that is used to clobber members of a rival ape-man group in a primeval turf battle over a water hole in prehistoric Africa.

But the purported tools that Dart found were not all they were cracked up to be. More recent studies by C. K. Brain, the director of the Transvaal Museum in Pretoria, have shown that all the types of bone fracture that formed Dart's hypothetical tools can be produced by purely natural means, through carnivores and other forces breaking up bones after an animal dies. One of us (NT), in a collaborative project with Timothy White of the University of California, Berkeley, has conducted a worldwide analysis of pre–anatomically modern fossil hominids, funded by the Harry Frank Guggenheim Foundation of New York. The purpose was to assess anthropological claims of injury and trauma in the fossil record that might have been due to interpersonal aggression and violence. There was little if any direct fossil evidence from the South African australopithecine sites to support such interpretations. Most of the trauma to hominid skulls appeared to have been postmortem damage within these cave deposits, probably due to falling rocks or the pressures of overlying sediments. There was absolutely nothing that would stand up to cross-examination in a modern court of law as evidence of a predatory, weapon-wielding hominid.

This is not to say that early hominids couldn't have used bone tools or would not have had a certain level of aggressive behavior: this has been amply demonstrated in modern chimpanzees, among whom we see a low level of murder, infanticide, and cannibalism (but along with a great deal of cooperative and even altruistic behavior). But the early fossil hominid record does not, at present, demonstrate such aggressive tool-using behaviors. There is just no evidence of a human hand in the modification of these hominid bones.

Thus, Dart's osteodontokeratic industry died a natural death. It is ironic, however, that his sites seem to be just a little too early for the beginnings of recognizable tools. Some other researchers working simultaneously in another part of Africa had the luck—and the perspicacity—to dig in just the right place and the right time period to be able to push back our knowledge of human tool making very close to its recognizable beginning.

While many researchers were still looking toward Asia to find the earliest evidence for hominid tool makers, Louis and Mary Leakey were exploring sites in East Africa in search of such evidence. Louis (who had grown up in Kenya, the son of an English missionary) and Mary began working at Olduvai Gorge in Tanzania in the mid-1930s and continued there for the next three decades, discovering an unparalleled prehistoric sequence spanning almost two million years. Their studies eventually revealed a fantastic progression of changes in hominid evolution, stone technology, and environments in this part of the African Rift. The Leakeys recognized and named the most rudimentary types of stone tools known in the archaeological record, dubbed the Oldowan since they were found in the lowest and oldest levels of Olduvai Gorge.

Discovery of these Oldowan tools revealed an indisputable beginning for stone tool making. The artifacts are extremely simple but definitely manufactured, not just modified through use. Moreover they are not just a fluke: further research has dug up many more of these early tool sites. They have been found in Africa in sites spread over hundreds of miles and throughout hundreds of thousands of years of time, indicating some degree of cultural continuity and learning in maintaining these tool traditions. They show that some bell had gone off, that some

The grande dame of early Stone Age studies in Africa: Mary Leakey with one of her famous Dalmatian dogs. Her excavations and studies of the Stone Age sites at Olduvai Gorge established a foundation for many subsequent researchers. (© David L. Brill 1985)

important dependence on tools was under way, undoubtedly developed through intuition and trial-and-error and spread through culture. What's more, at Olduvai this technological breakthrough began almost two million years ago. Thus, by the middle of the twentieth century archaeologists were faced with a time depth for prehistory that was simply inconceivable one hundred years ago when the first prehistorians were digging their caves and rock shelters in Europe and puzzling over the stone artifacts they found. We had finally developed a better sense of the Stone Age: a technological past that was very long and quite complex, with many different stages. It was up to the growing discipline of archaeology to figure out this prehistoric record, but to do this archaeologists have also had to figure out *how* best to study this past.

ARCHAEOLOGY: COMING OF AGE

Mary Leakey pioneered in East Africa the archaeological approach of uncovering large horizontal areas of a discrete Paleolithic archaeological level (then called a living floor) to examine the spatial patterns of prehistoric materials on that ancient surface. Other archaeologists of that period would often dig tall, narrow trenches through a succession of sedimentary layers to see major differences in geology, paleontology, and technology over time. Mary Leakey's approach allowed prehistorians to start asking more pointed questions about the archaeological record—about the behaviors of early hominids, what they ate, how they lived, what they did at these sites—as opposed to what is often called the cultural-historical approach, or simply establishing what types of tools occur in different time periods in a particular area. Studies of other early Stone Age sites subsequently followed Leakey's methods: large-scale excavations were later carried out by J. Desmond Clark at Kalambo Falls in Zambia, F. Clark Howell at Isimila in Tanzania and Torralba and Ambrona in Spain, and Glynn Isaac at Olorgesailie and East Turkana in Kenya. In these and other recent studies, not only have the excavations been larger, but the archaeologists' questions have been broader. In recent decades archaeology has moved far beyond asking just what, where, and when, to inquire in greater depth into the why and how of our prehistoric past.

During the past few decades, archaeology has begun to look even more deliberately and doggedly for evidence of prehistoric behaviors. Along with this emphasis on a behavior-oriented approach to the prehistoric record, there has also been a surge in use of a self-conscious scientific approach in archaeology: archaeologists have been stressing the building and testing of scientific models that attempt to explain aspects of the past. Overall, recent archaeology is characterized by these two elements: emphasis on behavior and the use of models or hypotheses in exploring the past. Most modern archaeologists try to develop some firm sense of expectation of what happened in the past and why, then compare very closely what they actually find to what they had expected or predicted.

A major emphasis in this new behavioral approach has been to use modern analogues to help interpret the prehistoric past. We can't reconstruct the past in a vacuum: once we moved beyond mere description in archaeology, beyond saying what artifacts look like (how big they are or thin or symmetrical or whatever), we needed some means of understanding how they fit into the lives of our ancestors. The early artifacts lay in stony silence; they don't scream out their significance. (In fact, most nonarchaeologists would probably not recognize them as artifacts at all.) If we seek from them information about all sorts of interesting questions about ancient human lifeways, adaptation, cognitive levels, and so on, then we need some basis for formulating these ideas. We need some outside references, some living models for a dead past, to help us understand what the archaeological record might *mean*.

There are three main contemporary reference sources Stone Age archaeologists turn to regularly: hunter-gatherers, nonhuman primates, and experimentation.

To interpret our prehistory, archaeologists can get some basic ideas about what is entailed in living off the land from living peoples who still hunt, gather wild foods, and live a foraging, nomadic life-style. But these people are *not* prehistoric Stone Age people, and, most problematic, most no longer regularly use stone tools in their daily lives. Closely related primates, such as apes, also give valuable ideas about possible diets, social organizations, tool uses, and so forth in our far distant past, but they are not protohumans, nor do they make or use flaked stone tools in the wild. So experimentation has become a major avenue for looking into how stone tools were made and used and how they might have fit into the lives of our early ancestors.

It has been largely since the 1960s that experimental replication of Stone Age artifacts has gained wide acceptance in the archaeological world as a valid approach to understanding the complexities of prehistoric technologies. Experimental archaeology attempts to use the same raw materials, techniques, and strategies used by ancient stone-working peoples in the remote and recent prehistoric past to create modern facsimiles of stone hand axes, spearheads, arrowheads, and other tools.

AFTER 150 YEARS: PUTTING IT ALL TOGETHER

Today early Stone Age archaeological research includes extensive field reconnaissance and survey, meticulous excavation techniques, detailed recording with the assistance of computers, in-depth analysis of materials at both macroscopic and microscopic levels, and the use of modern analogues to help interpret the prehistoric past. It also usually involves close collaboration with a host of other scientists, including geologists and paleontologists.

The Leakeys' work at Olduvai Gorge established without a doubt a great antiquity for the prehistory of human tool making. Since their work began, scores of other scientists have concentrated their research in areas of Africa where the fossils and artifacts of ancient humans are found. These labors have uncovered important new fossils and archaeological sites and pushed the beginnings of stone tools back in time several hundred thousand years.

Glynn Isaac of the University of California, Berkeley, emerged as one of the leading figures in the archaeology of human origins in the 1970s and up until his death in 1985. Using the early Stone Age archae-

The late Glynn Isaac was one of the most innovative and stimulating figures in the archaeology of human origins. He is shown amid the excavated scatter of prehistoric stone artifacts and fossil bones at site 50 at Koobi Fora in northern Kenya, dated to about 1.5 million years ago.

ological sites at Koobi Fora in northern Kenya as a testing ground for many ideas about early hominid behavior, he emphasized a multidisciplinary approach to the prehistoric record and the testing of multiple hypotheses, a perspective widely used by many researchers in the field today.

Thus we appear to have identified a critical turning point in our past: that pivotal time in our evolution when flaked stone tools began to have a central place in our lives and adaptation. Though it is always possible that future discoveries might modify this picture somewhat, fill in and sharpen a few more details or add a few years to the story, we are now able to focus on an important beginning for humankind—the dawn of human technology.

CHAPTER 3

DAWN BREAKS: THE FIRST STONE TOOL MAKERS

The stream meanders sluggishly through the grassland plain, lined on either side by a gallery of huge fig and acacia trees as well as scrubby underbrush. Colobus monkeys feed in the lower canopies of the trees, and a troop of baboons slowly forages its way through the shrubs below. In a deep pool downstream a hippopotamus snorts, his nostrils, eyes, and ears barely above water level, and a needle-nosed crocodile slithers into the water from the steep bank. Upstream, the current can be heard as it rounds a bend, the inner curve of which contains a large gravel bar of stone cobbles. A group of bipeds is foraging along the river. One of the females notices the gravel bar.

She goes down the grassy slope to the edge of the stream, where half-buried lava cobbles are exposed in large numbers, their dull, gray-and-brown surfaces highlighted in patches from the sunlight streaming through the canopy. She looks around to be sure that there are no crocodiles in or near the water, nor any terrestrial predators nearby. Crouching down, she selects a cobble that has a smooth, uncracked surface and turns it to examine the one relatively thin edge.

Selecting an egg-shaped hammer stone with her right hand, she directs a hard, glancing blow against the thinner edge of the cobble held in her left hand. A sharp crack rings out over the stream as a large flake falls onto the sandy gravel bar. Turning the cobble over, she looks at the dark scar she has made in the cobble. She once again swings down the hammerstone on the cobble, aiming at this scar, and knocks off another flake. Continuing this way, she removes about 5 or 6 flakes from both sides of the cobble. She detects a faint, sulfurous smell produced from the impacts of rock upon rock.

Reaching down, she selects several of the larger flakes, feeling their edges with her thumb to test for sharpness. Carrying these in her hands along with the hammer and cobble, she climbs up the bank of the channel and rejoins her group before they move off.

> The skillful man is, within the function of his skill, a different integration, a different nervous and muscular and psychological organization. . . .
>
> **Bernard de Voto (1897–1955)**
> ***Across the Wide Missouri,* 1947**

THE DAWN'S EARLIEST LIGHT: THE FIRST ARCHAEOLOGICAL TRACES

The first hard evidence for the emergence of human technology is found in different geological contexts in two different regions: ancient river, delta, and lake margin sediments in East Africa's Rift valley system and the fossiliferous, usually cemented infillings of limestone caves in the Transvaal region of South Africa. As we saw in Chapter 2, both areas have yielded traces of bipedal australopithecine fossils prior to the first Stone Age findings. These same regions bear witness to the first flaked stone technologies, between 2.5 and 1.5 million years ago. (This period is often called the Plio-Pleistocene, referring to the time spanning the boundary between the Pliocene and the progressively cooler Pleistocene epochs of geological time, about 1.6 million years ago. There is so much continuity in evolutionary events from the later Pliocene into the earlier Pleistocene that this has become a convenient way to indicate this time span.)

Establishing exactly *who* the earliest tool makers were is not easy. Unlike the present age, when our species is the only hominid form in existence, during the late Pliocene (2.5 to 2.0 million years ago), when we see the first stone tools, there were at least two hominid lineages coexisting on the African continent: robust australopithecines and an evolving form of *Homo*.

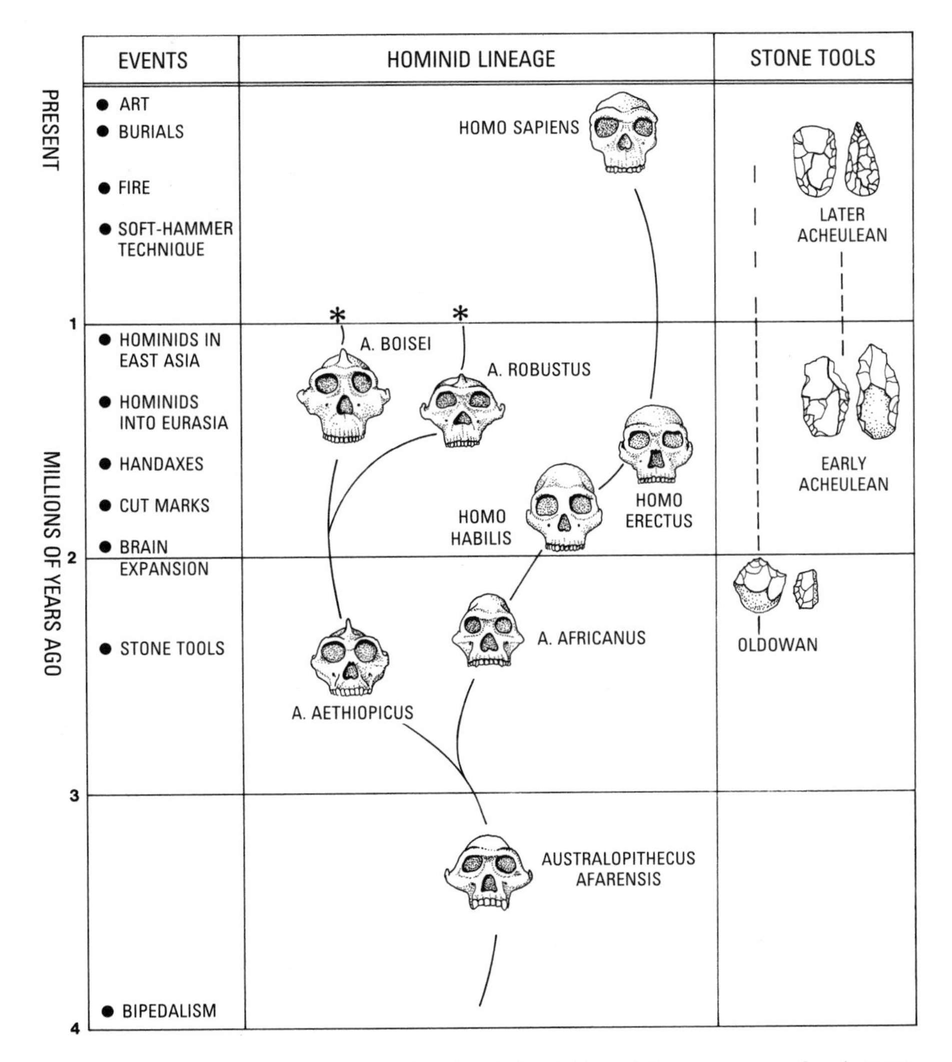

Chart showing possible evolutionary paths of early hominids and the emergence of early stone-age traditions.

1. *The robust australopithecines.* Fossils appearing approximately 2.6 million years ago and lasting over a million years suggest a branch of smallish-brained hominids with massive jaws and huge molar teeth for grinding food. One lineage is known in East Africa and another in South Africa. The adult males exhibit a keel, or sagittal crest of bone, running along the top of the skull for the attachment of powerful chewing muscles. Originally it was assumed that the body size of this lineage would have been very large as well, but more recent analysis has suggested that they may have been roughly the same body size as other contemporary

hominids. This means they had incredibly large teeth, both absolutely and relative to their size.

The discovery in 1985 of a 2.5-million-year-old cranium (dubbed the black skull because of its dark mineralization) on the western side of Lake Turkana demonstrated a greater antiquity for this robust lineage than some anthropologists had suspected.

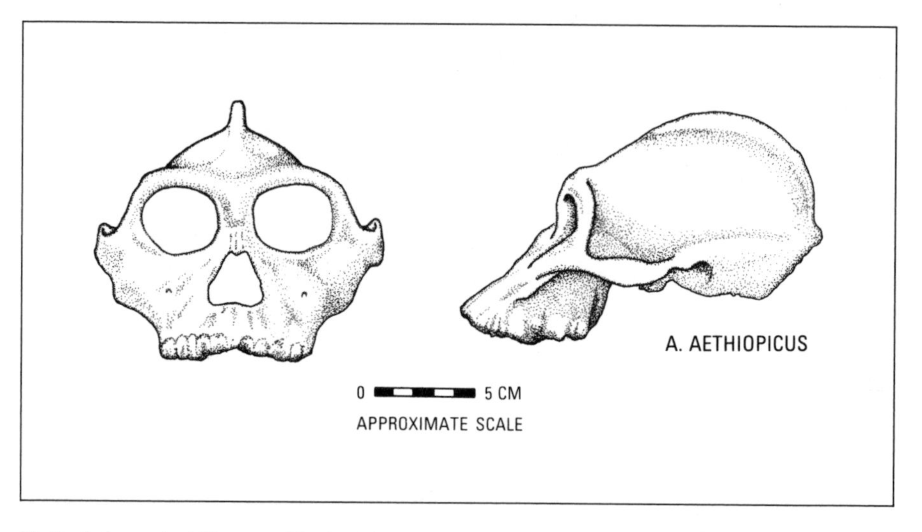

Skull of *Australopithicus aethiopicus* from West Turkana, Kenya, the "Black Skull," dated to approximately 2.5 million years ago.

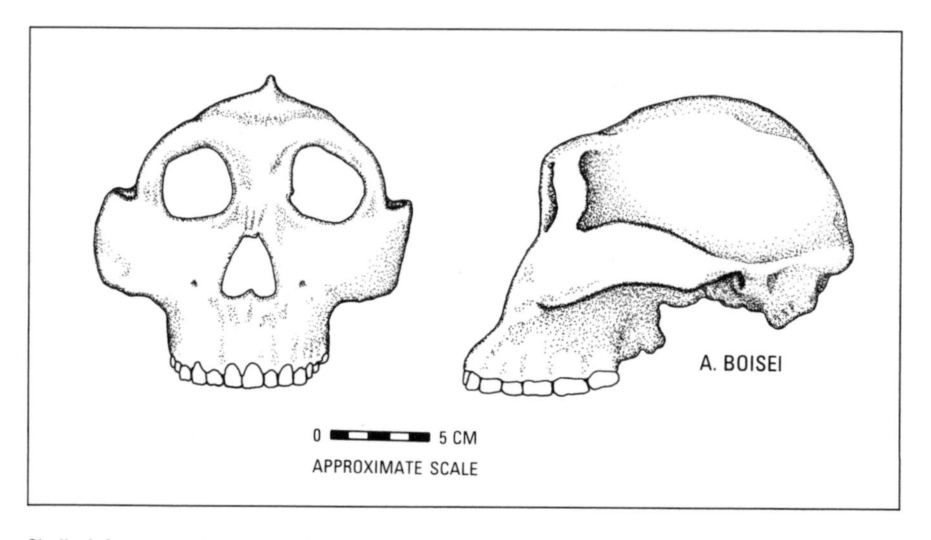

Skull of *Australopithicus boisei* from Olduvai Gorge, Tanzania, dated to approximately 1.7 million years ago.

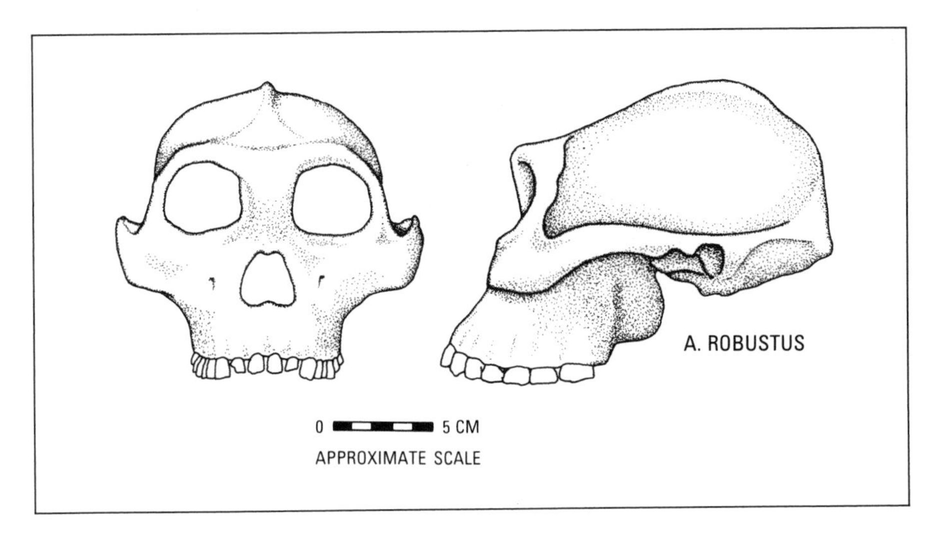

Skull of *Australopithicus robustus* from Swartkrans, South Africa, estimated to be about 1.5 million years old.

This very early, very robust form is now called *Australopithecus aethiopicus* by most anthropologists. By at least two million years ago, this lineage had evolved into *Australopithecus boisei* in East Africa. In South Africa a similar form is called *Australopithecus robustus* (sometimes called by its original genus name, *Paranthropus*), found in the limestone cave deposits at Swartkrans and Kromdraai, estimated to be about 1.5 to 2.0 million years ago. Each of these robust forms was present when stone artifacts start appearing in their areas, by about 2.4 million years in East Africa and 2.0 million years ago in South Africa. The cranial capacity of these forms ranges from about 400 to 550 cubic centimeters.

2. *Early Homo.* By at least 2.0 million years ago the distinctively larger-brained, smaller-toothed *Homo habilis* begins appearing in the fossil records of both East and South Africa, with brain sizes ranging from about 600 to 750 cubic centimeters. As we noted in chapter 2, the fossil record between 2.5 and 2.0 million years ago is still pretty sparse, but there is nevertheless fragmentary evidence suggesting to some that *Homo* evolved even earlier, by about 2.3 million years ago, at the Omo region of Ethiopia, perhaps still in relatively small numbers and limited populations. These forms may have evolved from A. *africanus* or a similar, yet undiscovered form. In fact, some argue that these fossils are so varied that more than one species of early *Homo* must be present

in East Africa, larger body sizes as well as smaller forms, some with smaller brains. Others argue that the differences in size are only between larger males and smaller females.

In any case, it seems that both the robust australopithecines and fossils assigned to early *Homo* are roughly contemporary with the earliest stone tools. And each of these branches continues as we see the archaeological record become more pervasive and conspicuous over time.

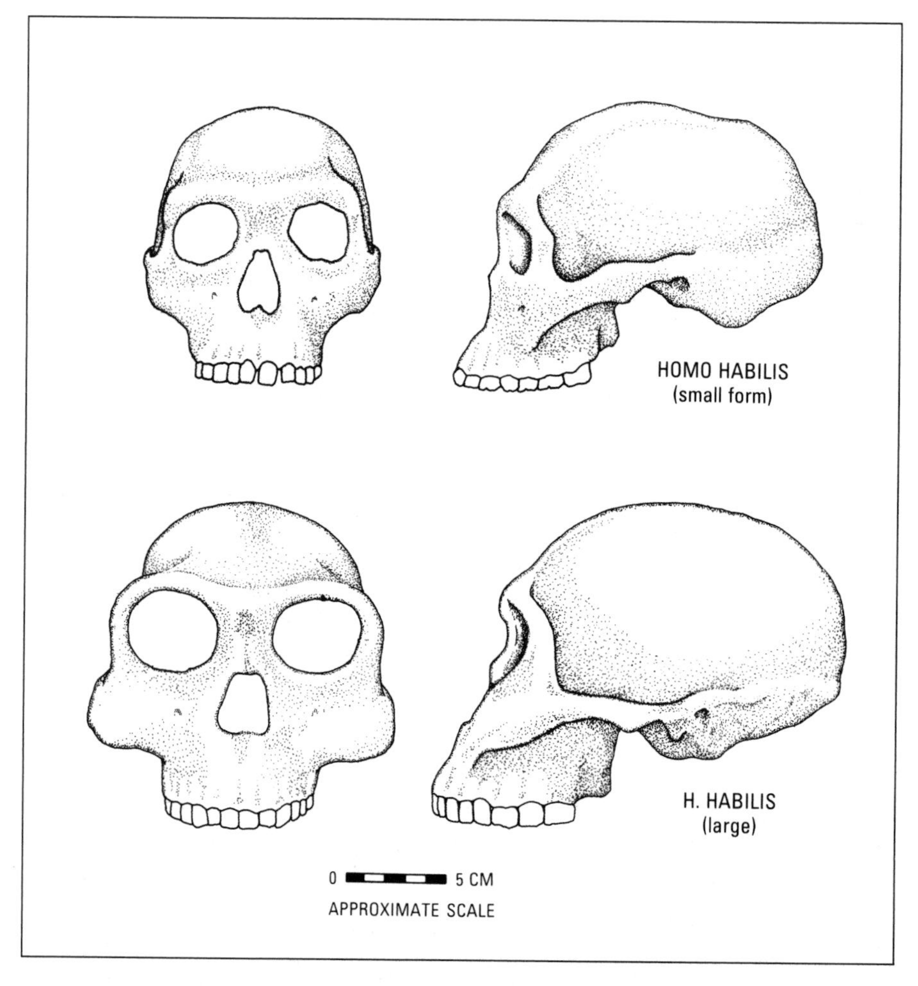

The first stone tool-makers? The first evidence of significant brain expansion in human evolution, seen in skulls attributed to *Homo habilis* between 2.0 and 1.7 million years ago. Both fossils are from East Turkana (Koobi Fora), Kenya. Some anthropologists believe the smaller and larger forms should be assigned to two different species of early *Homo*.

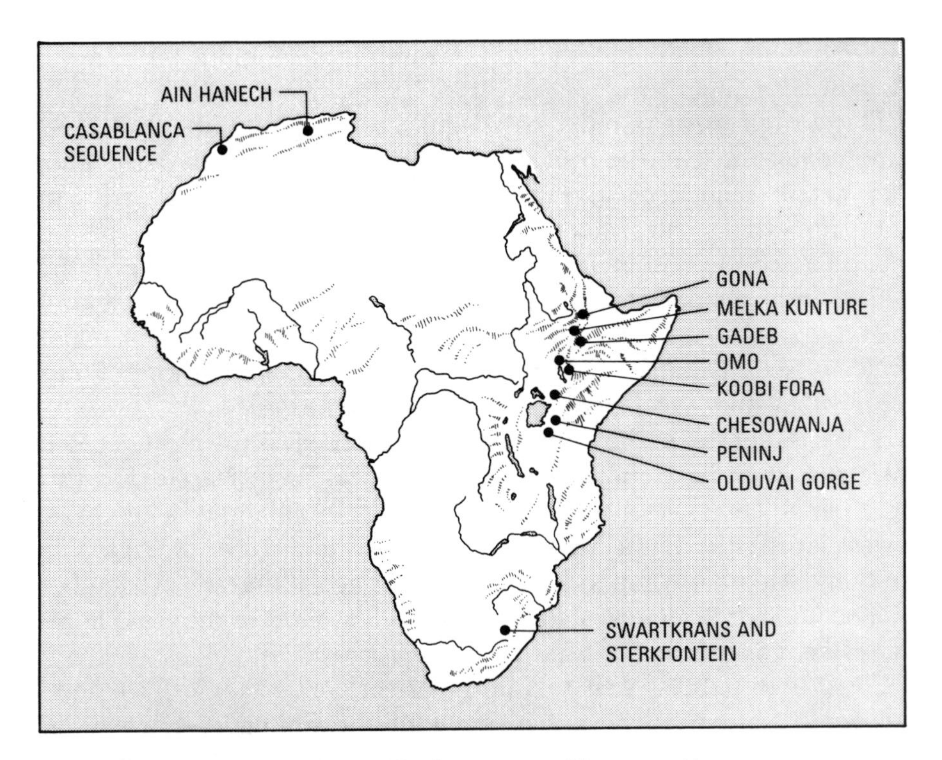

Early Stone Age sites in Africa believed to be over one million years old.

THE EAST AFRICAN RIFT VALLEY SITES

The Omo Region, Ethiopia

The Omo River is a major waterway that flows south from its headwaters in west-central Ethiopia through the semi-arid badlands of southwestern Ethiopia, emptying into Lake Turkana near the border between Ethiopia and Kenya. In its southern reaches the perennial Omo and its many small tributaries of gullies and sand channels, which flow only after heavy rains, have carved a valley through ancient geological deposits.

This area was investigated paleontologically by C. Arambourg in the early 1930s. It was then studied extensively in the late 1960s and early 1970s by an international team of scientists working at a number of localities. The team's holistic, multidisciplinary approach, incorporating dozens of scientists—geologists, paleontologists, and archaeologists—has been considered a model of paleoanthropological research, emulated by a number of more recent research projects in East Africa and elsewhere.

Due to the great number of volcanic ash deposits, the sediments exposed by the Omo constitute one of the longest well-dated Plio-Pleistocene sequences in East Africa, combining a long series of potassium-argon dates, documentation of the paleomagnetic record (periodic switches in magnetic north back and forth between the poles), and evidence of major changes in faunal evolution throughout the stratigraphic record (called biostratigraphy). There are four major cycles of sedimentation here that formed four geological formations, each with a story to tell about evolving animal forms.

The latest of these, the Shungura Formation, gives the most detailed information we have about animal evolution in East Africa during this time period. It ranges from approximately 3.3 million years ago in its lowest layers, the Basal Member (a member is a major stratigraphic unit) and Member A, up to 1.2 million years ago at the top, in Member L. The Omo archaeological localities as well as most of the Omo hominids have come from the Shungura Formation.

Fossil hominids, as well as a range of vertebrate fauna, have been recovered from nearly every member of the Shungura Formation. Most of the hominid remains consist of isolated teeth, jaw fragments, cranial fragments, and parts of the postcranial skeleton (the bones below the skull).

Though their fragmentary nature sometimes makes classification difficult, there are some definite patterns over time. The oldest hominid fossils (Usno and Shungura formations) may represent *A. afarensis* and are dated between 2.5 and 3.0 million years ago. Higher up are fossils that appear to be early *Homo*, starting about 2.3 million years ago, as well as the robust lineage, *A. aethiopicus* and then *A. boisei*, which may last here from about 2.5 to 1.2 million years ago. In the later deposits here, another even more evolved form of *Homo* (usually called *Homo erectus*) appears at about 1.4 million years ago.

This pattern of coexistence of *Homo* and robust australopithecines occurs at other sites in East and South Africa, over a span of more than a million years—two experiments in hominid evolution, only one of which continued over the long haul up to the present.

The earliest unambiguous, well-dated Paleolithic sites in the world have been located and excavated from Members E and F of the Omo stratigraphic section, dated to between 2.3 and 2.4 million years ago. These artifactual sites, unfortunately, do not themselves contain any fossil hominids, but both *Homo* and robust australopithecines have been found in sediments of this age nearby and elsewhere in the Omo region. And at these sites there are stone tools that are fascinating in their simplicity: small pebbles of milky white or yellowish vein quartz fractured into numerous flakes and fragments.

The Gona Sites of the Hadar Region, Ethiopia

The Hadar region is well known for its wealth of *Australopithecus afarensis* fossils, including Lucy and the First Family, discussed in the previous chapter. It is located in the Afar triangle, a depression sunken into the northeastern end of the African rift where a junction of three fault systems rupturing East Africa had divided it from the Arabian peninsula. From younger stratigraphic levels above the australopithecine horizons are some sites in ancient river floodplain and channel deposits near the Gona River, excavated by Hélène Roche of C.N.R.S. Paris and J. W. K. Harris and Sileshi Semaw of Rutgers University. The artifacts here consist primarily of flaked river cobbles of lava and the associated flakes and fragments. Fossil animal bone has sometimes been found with these artifacts.

The dating of the Hadar sediments, and especially these archaeological sites, has been somewhat problematic over the past two decades. Recent studies, however, indicate that these sites are approximately 2.4 million years old, about the same age as the Omo sites approximately 550 miles (800 kilometers) to the southwest.

The East Turkana Region, or Koobi Fora, Kenya

Just to the southeast of the Shungura Formation lies another major fossiliferous area, in northern Kenya on the east side of Lake Turkana (formerly known as Lake Rudolf). This area has been referred to alternately as East Rudolf, Koobi Fora, and most recently East Turkana (to associate it with yet distinguish it from West Turkana, the area on the other side of the lake).

Work by University of Utah geologists Frank Brown, Craig Feibel, and Thure Cerling indicates that this region, from the Omo through East and West Turkana, was part of a larger unified basin filled up with sediments during Plio-Pleistocene times, although each has its own local history. As might be expected, a sequence of members at East Turkana was laid down through much the same time span as the Shungura Formation, from over 4.3 million to less than 600,000 years ago. These have likewise produced a great deal of fossil bone, including that from hominids, as well as dozens of known archaeological sites investigated by Glynn Isaac, J. W. K. Harris, and others. Ashes allow these to be dated and correlated with the Omo and other deposits in East Africa.

Excavations at site 50 at Koobi Fora, Kenya, uncovered a well-preserved Oldowan site containing stone tools and fossil animal bones dating to about 1.5 million years ago.

Fieldwork was begun here in the late sixties by Richard and Meave Leakey, whose teams of researchers, including paleontologists, geologists, and archaeologists, have carried out studies in the region for over two decades now. Our first African fieldwork was at Koobi Fora with Glynn Isaac, where we worked during three major archaeological field seasons from 1977 through 1980.

Archaeological sites appear in the geological deposits here at about 1.9 million years ago, when the first relatively complete fossil skull of early *Homo* appears on the scene. The *Homo* lineage continues here, with the earliest *Homo erectus* fossils then appearing about 1.7 million years ago and continuing to about 1.3 million years ago. A. *boisei* is also contemporaneous during this time span in the same region. Stone artifacts here are made almost entirely out of lava cobbles, which could be found at the time in stream channels, just as they are today.

The West Turkana Region, Kenya

These deposits are located on the western side of northern Lake Turkana, less than thirty miles (fifty kilometers) from the East Turkana region. As part of the Omo-Turkana basin, the deposits here can be linked up with neighboring sediments just to the northeast at the Omo and to the east at East Turkana. Called the Nachukui Formation, they also contain many volcanic ashes indicating that they span almost precisely the same time range as the others. Here also are abundant animal fossils, including hominids, as well as archaeological sites, that help amplify and sharpen the picture of human evolution we get from the surrounding deposits.

A hominid resembling A. *afarensis* has been found at about 3.2–3.3 million years, but one of the most striking finds here is the skull of a very robust australopithecine at about 2.5 million years (A. *aethiopicus*). After this, a fossil attributed to *Homo habilis* appears, at around 2.0 million years, and a remarkable skeleton of *Homo erectus* at about 1.6 million years, to be discussed later. Artifacts here have been reported by Mzalendo Kibunjia and Hélène Roche starting at about 2.3 million years ago.

The Olduvai Gorge Sites, Beds I and II, Tanzania

The most famous of the early Stone Age localities in East Africa is Olduvai Gorge, a fifty-kilometer-long, hundred-meter-deep erosional gash through the Serengeti Plain in northern Tanzania. In modern times a small river here carved its valley through a thick sequence of sediments laid down during Plio-Pleistocene times, a sort of miniature Grand Canyon of prehistory. The work of the Leakeys at Olduvai Gorge laid the initial foundation for our understanding of hominid biological and technological developments in Africa during the Plio-Pleistocene. Added to this, the research of University of Illinois geologist Richard Hay and others has greatly enhanced our understanding of the ancient environments these hominids were exploiting.

In this period a small lake basin had formed here, nestled north of volcanoes and lava highlands. Most of the deposits represent either the lake bed itself, the margin of the lake where a large plain provided a rich habitat for many animals, or the streams feeding into the lake from higher ground at the edge of the basin. Layered throughout are various ash deposits that, as at the other East African sites, allow us to date and correlate the sediments as well as the fossils and artifacts they enclose.

Our window into the past here opens by about two million years ago,

The type site of Oldowan technology: Olduvai Gorge in northern Tanzania. Louis and Mary Leakey worked here for several decades, investigating prehistoric occurrences that spanned from almost two million years ago to recent times, an unprecedented record of human evolutionary change.

when a volcanic eruption spread layers of lava over the landscape. After this, by about 1.9 million years ago, the lake had formed and the sediments began to build up, incorporating the remains of all sorts of animals, including numerous hominids, as well as abundant archaeological sites, at one time the earliest ones known.

The first four layers, Beds I through IV, show evidence of early phases of human evolution and technology. In the lowest, earliest layer, Bed I, the remains of both *H. habilis* and *A. boisei* have been found, in addition to the earliest stone artifacts from Olduvai. These are flaked pieces of lava and quartz and are found at many localities exposed along this great gorge toward the base of the deposits. The bottom of Bed I, about ninety meters down, begins at about 1.85 million years ago and lasts for about 150,000 years, until about 1.7 million years ago. About thirty to fifty meters of sediment accumulated during this time period, the deepest part closer to the lake.

In the next layer, Bed II, between 1.7 and 1.2 million years ago, both of these hominid forms continue along with archaeological deposits until about 1.6 million years ago, when we see the last of *H. habilis* here. As we have seen elsewhere, A. *boisei* continues and by late Bed II is also accompanied by the evolved *Homo* form called *H. erectus* (which is found as late as Bed IV, between 600,000 and 800,000 years ago).

All of the East African evidence adds up to a remarkably consistent story: the emergence of stone tools by about 2.4 million years ago and the appearance of at least two major hominid branches by about the same time, one maintaining a relatively small australopithecine brain but developing a massive chewing apparatus, the other developing a larger brain but showing reduction in its chewing teeth.

THE SOUTH AFRICAN SITES

The Swartkrans Cave Deposits

The Transvaal region of South Africa, characterized by high veldt grasslands today, contains numerous limestone cave deposits, called breccias, bearing prehistoric fossil bones of many animal species. One of these localities, Swartkrans, has yielded hundreds of hominid fossils representing over ninety different individuals. The hominid-bearing layers have not been dated by radiometric means, as they have no associated volcanic deposits, but the fossilized mammals here can be correlated with dated forms in East Africa to get age estimates, and the paleomagnetic sequence can also be used to cross-check these.

The oldest layer here, Member 1, has the oldest and the most abundant hominids, estimated to span approximately 1.8 to 1.5 million years ago. Among the hominids in this level, robust australopithecines are far and away the most common. This is the South African variant called A. *robustus*. More than eighty individuals of this australopithecine species have been found in Member 1, compared with only a few representatives of *Homo*, thought to be *H. habilis* or perhaps even *H. erectus*.

The abundance of robust australopithecines in this cave deposit is really quite peculiar: they are by far the most abundant species found in these early levels. There are about as many baboons found in total, but these comprise four different species of animals. Why are these robust hominids so plentiful here? Furthermore, when we add the baboons and the hominids together, well over half the animals here are primates, a very unusual, even suspicious, pattern. Why so many primates here?

These questions intrigued C. K. Brain, who conducted studies of the cave for many years following its initial exploration in the late 1940s. In his fascinating book *The Hunters or the Hunted?* he gives persuasive evidence that many of the hominids found in the cave may have been dinner rather than the diner: they had probably been preyed upon by leopards, which dragged them up into trees near the cave entrances to eat them at leisure, undisturbed by covetous hyenas. One adolescent skull fragment, in fact, bears the telltale puncture marks of a leopard's canine teeth. The primate-rich diet of these predators very possibly also shows that they were raiding sleeping places of these primates within the cave, just as modern baboons regularly use caves and other sheltered areas as sleeping sites.

Relatively small numbers of simple stone tools have been found in the Swartkrans deposits from Member 1 through Member 3 (from 1.8 to about .5 million years ago). These are made primarily of flaked pieces of quartz and quartzite rocks. It is possible that many of these artifacts could have washed into the cave deposits from the surrounding landscape rather than been brought into the cave by hominids. In any case, these Paleolithic specimens represent the earliest known stone artifacts in southern Africa.

Besides the flaked stone materials, Brain has also identified several animal bone shaft fragments from Member 1 that appear to be polished and microscopically striated on one pointed end. He suggests that these were digging tools used by early hominids to uncover and uproot shallowly buried underground vegetable foods. If this interpretation is correct, these specimens would represent the first clear examples of bone tools in the archaeological record, supporting Raymond Dart's suggestion that bone was also employed as a material for tool use during the early Stone Age.

The Sterkfontein Cave Deposits

Across a valley and within sight of the Swartkrans locality lies the site of Sterkfontein. As already discussed, some of the lower levels in the cave deposits here (Member 4) produced partial remains of dozens of individuals of *Australopithecus africanus* as well as a wide range of other animals, estimated to be approximately 2.5 million years old. No definite stone tools have been discovered in association with these hominids, however.

By the time of the overlying Member 5, estimated to be approximately two million years old, *A. africanus* disappears from the scene here. In this period were found fossils of *Homo habilis* along with the first evidence of stone technology; a few hundred simple stone artifacts.

The cave of Sterkfontein in South Africa with Swartkrans across the valley in the distance. Excavations here have uncovered numerous fossils of *Australopithecus* and early *Homo.* Early stone tools believed to date from between 2.0 and 1.5 million years ago have been recovered from both sites.

Brain has also noted an intriguing, telltale pattern here, similar to that at Swartkrans: the lower level also has a preponderance of primates. Here baboons outnumber the hominids (*A. africanus*), even by a ratio of about 4 to 1. By the time *Homo* and stone artifacts appear in Member 5, however, baboons all but drop out of the deposits and most of the animals are antelopes and other grazing animals. Brain's tantalizing explanation for all of this is that, at first, various carnivores (such as leopards, saber-toothed cats, and hunting hyenas) controlled the cave, hunting the vulnerable primates seeking shelter there. But by the time of Member 5, Brain proposes, *Homo* had taken over the cave, keeping out the predators and using it as a lair. There, these hominids were able to eat various animals, some of which they may have hunted but others of which they had managed to pilfer from carnivore kills.

The South African evidence largely corroborates the pattern we have seen in East Africa: by about two million years ago we see the emergence of stone tools and the emergence of *Homo*. Compared with the East African sites, however, the evidence and settings here may not be as rich and diverse. But the deposits open up unusual windows into the past: these caves were special environments, which unselectively trapped or sucked in all sorts of debris. These underground caves may be giving us a unique opportunity to see the interplay between hominids and carnivores, which might have provided close competition during this phase of hominid evolution.

ARTIFACTS VERSUS NATURE-FACTS

If we were to spread out on a table examples of the first recognizable stone tools, most readers would probably be quite unimpressed and perhaps downright skeptical. How can we Paleolithic archaeologists argue with conviction that these nasty-looking bits and pieces of rock are the unambiguous product of intentional hominid craftsmanship? Some would probably say, "I can find stuff like that in my backyard." Our response would be that if you can *really* find stuff like that in your backyard, you should probably call an archaeologist, because you have located a Stone Age site.

How can we identify a stone artifact? What criteria do archaeologists use to discriminate between broken rocks shaped by nonhuman forces (called "nature-facts" or "geofacts") and those created by hominids? How do we make sense and order out of these constellations of fractured and battered stone artifacts at these early sites? This has always been a major challenge of Paleolithic archaeology, especially for the earlier time periods, and it is worth considering how prehistorians have approached this problem. Surprisingly, on close examination there is really little ambiguity. The basic features of human-manufactured stone artifacts were identified by attentive antiquarians in the nineteenth century, and subsequent research has shed even more light on their characteristics, which the archaeological novice can learn to recognize in a short period of time.

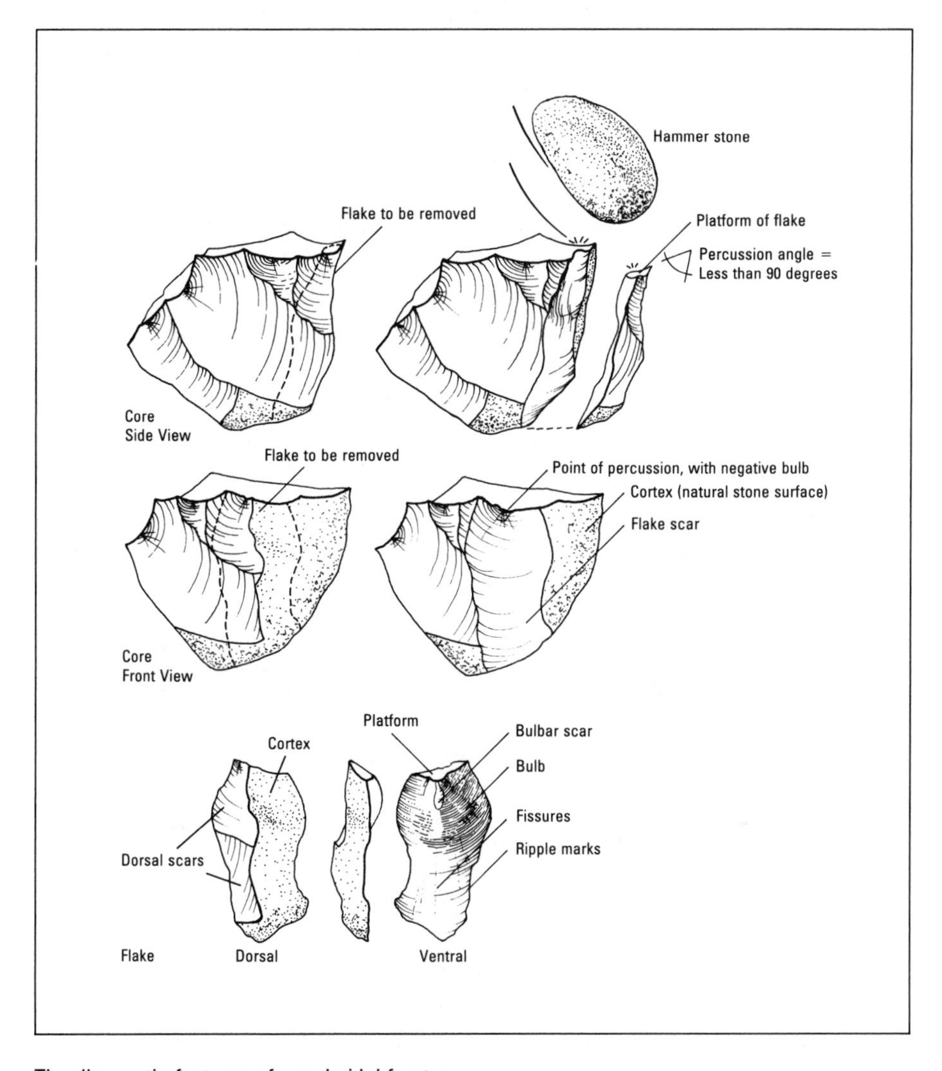

The diagnostic features of conchoidal fracture.

RECOGNIZING THE EARLIEST STONE ARTIFACTS

First of all, stone flaked by humans normally exhibits a breakage pattern that geologists call *conchoidal fracture.* This term originally meant that the fractured surface of the rock had smooth undulations reminiscent of the surface of a bivalve shell.

When a suitable stone breaks from percussion, as when a hominid strikes a rock with a stone hammer, there are two principal products: the *core*, the parent piece of rock from which fragments are struck, and the *flake*, one of the fragments detached from the core.

The flake may seem at first a paltry, featureless item, but in fact it has a number of very definite parts, all of which tell us something about how it was made. These include:

- the *striking platform*, the surface where the hammer struck the core and knocked off the flake;
- the *bulb (or semicone) of percussion*, a bulge or swelling just below the platform on the inside (core-side, or ventral, surface) of the flake, which represents the energy waves of the flaking blow as they spread into the core; sometimes a small, paper-thin *bulbar scar* (or *éraillure*) can also be seen on the bulb;
- *ripple marks*, wavelike undulations on the inner surface of the flake that radiate in a series of progressively larger arcs from the point of percussion, much as water ripples radiate in a circular pattern when a pebble is dropped in a pool;
- *fissures*, or very fine cracks on the inner surface, radiating from the point of percussion;
- the *dorsal surface* (the back, or outer, surface) of the flake, representing the outside of the core when the flake was made, which may show patches of cortex (the natural, weathered rind on the outside of a stone cobble) or scars called *dorsal scars* (concave surfaces showing where previous flakes had been removed).

The core, likewise, has features that can reveal information about its manufacture:

- the *cortex*, or outer rind of the cobble;
- *flake scars*, concavities showing the negative areas where flakes have been removed from the core. A complete flake scar should match up exactly with the inner surface of the flake that was removed. Within a scar on the core, then, we can often see mirror

images of features on the inner side of the flake, including a *negative bulb,* which matches up with the positive bulb of percussion on the flake, as well as ripples and fissures. As flaking proceeds, however, the earlier scars may be truncated or obliterated by later flake scars (rather like taking bites out of an apple: as you continue eating, the marks left by your first bites are chomped through and slowly disappear).

These features of conchoidal fractures can be best appreciated by the archaeologist who actually spends some time learning to make stone tools. As we shall see in the next chapter, one can examine these components immediately after hitting off a flake and then refit the flake back onto the core to see how the parts of each match up. The archaeologist can read such features in prehistoric artifacts and discover how hominids were making their tools, see their decision making, and recognize their successes and failures. Forensic scientists examine similar characteristics—ripples, striations—on broken glass to tell whether a window was struck from the outside or inside of a building.

In practice, conchoidal fractures can be seen most easily in some of the better raw materials for making flaked stone tools: hard, fine-grained rocks, such as flint or chert, in which individual rock crystals cannot be seen with the naked eye; or in amorphous rocks with no crystal structure, such as obsidian, a volcanic glass. In coarser-grained rocks, such as many quartzites and lavas, the patterns of fracture are still there, but their features are less obvious to the untrained eye.

Whenever we find a broken rock in prehistoric sediments showing these features of conchoidal fracture, can we then say that it has definitely been produced by deliberate hominid activity? Not always. Usually when rocks are fractured by natural means, they do not exhibit signs of conchoidal fracture, as when they break along internal flaws and lines of weakness, or through differential heating and cooling, or through weathering. But there are a few other forces in nature that can grossly mimic flaking done by humans, generally where high-energy geological forces act on rocks in the natural world: cliffs or waterfalls where rocks can drop an appreciable distance and fracture by percussion; beaches with a violent zone of crashing waves propelling rocks against each other at high velocities; and steep mountain ravines and alluvial fans spreading out from mountains where, especially during flash floods, rocks may be thrown against bedrock or each other with powerful force.

Another geological force that can fracture stone chonchoidally in higher latitudes is glaciation, when massive ice sheets rake over landmasses and subject rock to incredibly high pressure. Such forces can sometimes flake rock in a way that superficially resembles humanly induced flaking. (Many of the eoliths that were the focus of controversy around the turn of the century, previously discussed in chapter 2, had been produced by such forces.)

Archaeologists must recognize situations such as these to avoid confusing stones fractured by nature with those fractured by humans. Luckily, such situations are fairly easy to decipher in the geological record, and, moreover, any misleading stones are usually few and far between. (For instance, only a small percentage of rocks going over a waterfall fracture this way.) In addition, a telltale feature signals a geological origin: the angle of the fracture. To flake stone efficiently, humans need a sharp angle on the edge of the core (usually much less than a right angle, or ninety degrees). If the angle is less acute than this, it's difficult or impossible to initiate and control the fracture. So when prehistoric humans made stone tools, even in the early Stone Age, they were intentionally searching, however intuitively, for *acute* edge angles on cores where flakes could be readily detached. This angle can easily be observed and measured on the flakes that have been struck from a core; it tends to be around seventy to eighty degrees in human-made tools, whereas in stone that flakes conchoidally in geological circumstances, the angle averages close to a right angle, or ninety degrees.

In addition, human-made tools tend to exhibit the repeated removal of flakes in a variety of different directions, showing that the hominid turned the core this way and that, searching for those opportune acute edges. This can be seen on stone cores by examining their flake scars (and, to a lesser extent, on the flakes themselves). This pattern, present at even the earliest sites in the archaeological record, is normally *not* found in nonhuman contexts. Where stones have been fractured by geological forces, cores tend to have only a few bold flake scars, often struck from the same direction.

TYPOLOGICAL CLASSIFICATION: IMPOSING ORDER ON BROKEN STONES

Typology is, quite simply, the science of classification, of pigeonholing things in neat, orderly groups, an activity that is itself a modern human characteristic. (The word is actually derived from the Greek *typtein*, meaning "made by a blow," which is especially appropriate for stone tools.) Generally, things are classified on the basis of shared traits, whether among flowering plants, mammals, celestial bodies, volcanic rocks, cloud patterns, or stone tools. Moreover, typology usually descends through different levels, from general to more specific, so that something might be a plant, a flowering plant, a composite, and a dandelion. These types are essentially abstractions, archetypal models of similarity or difference that allow the classifier to assign a specimen to one group or another.

Typology is, for better and worse, deeply rooted in the past of Paleolithic archaeology, which emerged out of eighteenth-century antiquarianism into a more rigorous discipline only in the late nineteenth and early twentieth centuries. A major interest at that time was identifying the fundamental stages of humanity—the major phases of human biological and cultural evolution. This was accomplished primarily by classifying stone industries based upon the presence or absence of certain classes of stone artifacts, which became the type specimens (in French called the *fossiles directeurs*) of that industry.

If these tool industries could be found in stratigraphic sequences, as they could in the caves and rock shelter deposits of western Europe, then a chronology could be established for a region, which could be correlated with climatic events and human evolutionary stages. For this reason stone artifact typology became, and to a large extent remains, a dominant methodology in Paleolithic archaeology.

For the earliest Stone Age sites of East Africa, a major typological system for classifying stone artifacts was developed by Mary Leakey in the 1960s, based upon her excavations and analyses at Olduvai Gorge in Tanzania. These first Paleolithic assemblages had been labeled Oldowan by Louis and Mary Leakey, using Olduvai Gorge as the type site for this level of human technology.

STRONG, SILENT TYPES: EARLY STONE AGE ARTIFACTS

Mary Leakey's typological system has served as the basic model for the classification of African sites of between 2.5 and 1.5 million years ago, so it merits some attention here. The major categories of this typology are listed below.

Heavy-duty tools:

These are cores made of cobbles or chunks of rock from which fairly large flakes have been struck. In Leakey's view, these forms are called "tools" with the assumption that many of them were intentionally

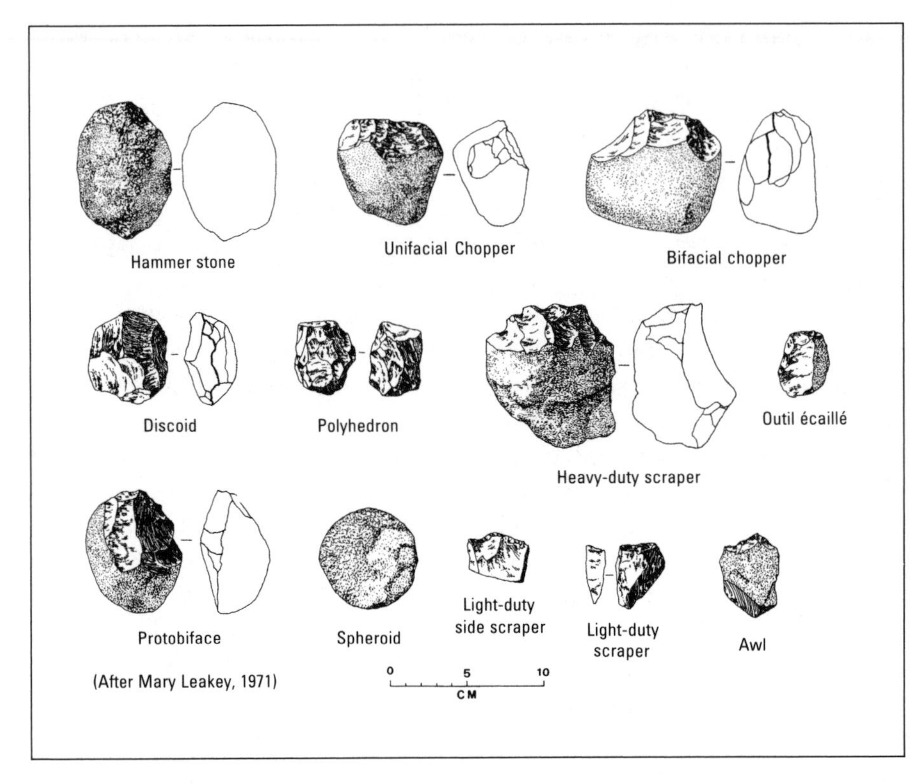

The dawn of human technology. Early examples of flaked and battered stone artifacts from Olduvai Gorge, characteristic of Oldowan technology. Conventional typological designations are noted.

shaped and used. They are classified into a number of different types, some of which are named after functions they are thought to have had (for example, a chopper, a scraper, or an awl). In other cases, they have been named after the overall shapes of the artifacts (for example, polyhedron, discoid, or spheroid).

Types of tools called "choppers" are among the most common forms identified in the earliest Stone Age of Africa. These are cores from which flakes have been removed along part of the perimeter of the cobble or chunk of rock. Mary Leakey divides these into "unifacial choppers," where flakes are only removed from one face, and "bifacial choppers," where flakes have been removed from both faces. They are also subdivided based upon the location and nature of the chopping edge: side choppers, end choppers, end and side choppers, pointed choppers, chisel-edged choppers, and so forth.

Another of these functional categories of artifact has been called the "core scraper" (or "heavy-duty scraper"). This is another large piece of rock that has been flaked along at least one side to produce a fairly steep, sturdy edge along a flat surface, suggesting a possible scraping function.

Among the types based on distinctive shapes, there are "polyhedrons" (flaked in several directions along multiple edges), "discoids" (flaked around the perimeter of a more or less circular core), and "subspheroids" and "spheroids" (flaked or battered into roughly, or very, spherical balls).

Light-duty Tools:

These constitute another class of tools made on smaller flakes and rock fragments, from which tiny flakes have been removed (called retouch) to provide a useful edge for some purpose or function. The specific types include "light-duty scrapers," "awls," and "burins" (engraving tools).

Utilized Pieces

This term is used for another class of artifact that has not been deliberately flaked into a specific form for a specific function but is thought to have been shaped as a *by-product* of its use for some task. These are usually flakes or chunks with some incidental damage or chipping of an edge, or specific types such as hammer stones and anvils, which become shaped and pitted through their use in stone tool making.

Débitage

Finally, a "waste," or debris, category, consisting of flakes and fragments. Aside from all of the so-called useful types mentioned above, assumed to be used or made deliberately for some purpose, this type allows that some artifacts seem to be more by-products of tool manufacture. Most show no specific retouch or damage that might give some hint that someone wanted to or actually did use them. These are put in the category of *débitage* (literally, "debris"), which includes simple whole and broken flakes and chunks that show no sign of modification after they had been struck from the core.

Mary Leakey then examined the relative proportions of these types to classify early Stone Age occurrences at Olduvai into different industries: the *Oldowan*, characterized by high proportions of choppers, discoids, and polyhedrons; and the *Developed Oldowan*, with higher proportions of light-duty tools such as scrapers as well as more spheroids and subspheroids.

GETTING BEYOND TYPECASTING

This system has served as a widely used artifact typology applied to the early archaeological record. A typology is important as a first approximation or description of the prehistoric material. We need somehow to be able to talk about our evidence, to communicate it to others, to be able to compare and contrast different sites throughout space and time, and ultimately to understand hominid behavior. For this we obviously need words, we need some names or labels to apply to things. Some of the shortcomings of this or any similar typological system have been alluded to here—typology can be arbitrary, it can assume function or intent without just cause, it can oversplit hairs, it can emphasize labels rather than contents—but we haven't really gotten to the crux of the problem.

More recent approaches to stone tools try to look at what the hominids seem to have been doing, such as the specific technological patterns they had followed in making their artifacts, and at the more elusive evidence of a tool's function. In the next chapter we will see how this is attempted and how much more we can learn—about the purposes and functions of stone tools and how they fit into the lives of early hominids—by avoiding typological thinking and looking at artifacts in a more lively, dynamic way. Once we learn to see and interpret these patterns, we can in a sense "read" the artifact and learn a lot about the actions, even about the decision making, of the hominid that produced it.

IDENTIFYING THE EARLIEST STONE AGE SITES

So it is usually extremely simple to distinguish between the artifact and the nature-fact. But archaeologists are faced with yet another crucial problem: how do we know when we have found a real, bona fide early archaeological "site," a place where some remnant of prehistoric activities has been preserved for us to study?

After more than a century of fieldwork and study of Stone Age archaeology, basic ground rules have been established for demonstrating the presence of early Paleolithic tool makers in a given area. These rules were formally stated more than a decade ago by Glynn Isaac.

1. We must see definite, unequivocal signs of human activity. First and foremost, we must establish that the prehistoric stone mate-

rials are, in fact, artifacts, showing manufacture or modification by humans rather than by natural forces.

a. We can sometimes find stone hammers with characteristic areas of battering, representing the actual instruments of percussion used by hominids to strike flakes off cores.
b. At sites that had been buried under very gentle circumstances (for instance, under silt that was blown in by the wind, or in a quiet, muddy pond), pieces of flaked stone can often be put back together to establish a blow-by-blow sequence of flake removals. This fitting together of flakes and cores has been called "conjoining."
c. The flaked rocks (and sometimes even unmodified rocks) at many Stone Age archaeological sites are often found outside their natural geological settings (for example, lots of flaked river cobbles found on a silty floodplain), implying that some outside force, namely hominids, carried them from one place to another.
d. Other materials are sometimes found with the flaked stone artifacts at an archaeological site, such as animal bones with cut marks made by stone tools, or, when we are very lucky, the fossils of the hominids themselves.

2. We must establish where in a geological sequence of deposits the artifacts are found. In order to determine the antiquity of the artifacts, they must be found in sealed, stratified geological deposits (that is, within natural layers of sediment). This way we can know that the stone tools were made at roughly the same time that the geological deposit was accumulating. Finding stone artifacts lying on the surface of the ground can give us an important clue as to where the artifacts are lying, but this is *not* proof enough that they, in fact, did come from that geological layer. They must also be uncovered from sealed deposits through excavation.
3. We must be able to get some sense, general or specific, of the age of the artifacts. This means that the sediments or their associated materials must be dated by reliable scientific methods. Ideally, more than one dating method can be used as a means of cross-checking. For the time period of the world's first known stone tools, between two and three million years ago, the most reliable dating technique is the potassium-argon method, which can be

applied to volcanic lava flows or ash deposits within the geological layers. (We are *not* dating the rocks that the artifacts are made from, for these could have come from rock formations much older than the tools themselves.) Such dates should be cross-checked with the paleomagnetic stratigraphic record, faunal correlations, or any other dating technique.

WHO WERE THE EARLY TOOL MAKERS?

With fossils attributed to both *Australopithecus* and *Homo* being found at the time of early Stone Age sites, can we identify the first tool makers? Traditionally, anthropologists have favored early *Homo* as the more likely candidate for the Oldowan tool maker. Recently, however, Randall Susman has also argued in favor of the robust australopithecine. He contends not only that these small-brained hominids made and used flaked stone tools, but that these creatures may even have *initiated* the first technologies. He cites the following lines of evidence to support his views:

1. The robust australopithecines tend to be more numerous than the gracile forms *(Homo)*, at the East and South African localities where early artifacts are found;
2. The human hand differs from that of the apes in having a much longer thumb, less curved finger bones, and broad, spatulate fingertips. It is capable of holding objects in a power grip (as one would hold a tennis racket) and also has a highly developed precision grip, where an object is deftly held between the thumb and one or more fingers (called opposability, as one might hold a razor blade for cutting paper).

   Hand bones attributed to *Homo habilis* at Olduvai Gorge suggest a creature with a power grip and a refined precision grip, which led anthropologists to infer that this hominid would have been an able tool maker.

   The hand bones found at the cave site of Swartkrans are also strikingly similar to a modern human hand and unlike that of a chimpanzee or even the earlier *Australopithecus afarensis*. Distinctive features are a flattened bone at the end of the fingers and a robust thumb bone with markings for a strong flexor pollicis longus muscle, a feature of the powerful mobile thumb in the modern human hand. Since most of the skull and tooth fossils from this cave represent *Australopithecus robustus*, Susman argued that most of the hand bones do as well.

3. Robust australopithecines show a slight increase in brain size over earlier forms of *Australopithecus* as well as over modern apes.

This is a fascinating argument, but to us it is not totally convincing. Some anthropologists have questioned whether the hand bones at Swartkrans can be definitively attributed to *Australopithecus* or *Homo*; furthermore, this hand morphology might be indicative of more habitual manipulatory behavior, but not necessarily Oldowan tool making. Although there is at present no way of establishing with certainty that the robust australopithecines were *not* making stone tools, the australopithecine legacy does not seem to be profoundly technological. We do not find tools with the earlier australopithecines, nor do we find tools contemporaneous with any of them until about the time *Homo* emerges in the fossil record (F. Clark Howell and Gen Suwa have both suggested that fragmentary fossils from about 2.3 million years ago at the Omo appear to represent early *Homo*.) By 2.0 million years ago, when the first relatively complete skulls of *Homo* are found, these show evidence of profound increase in cranial capacity as well as possible reorganization of brain structure (as seen in the casts or impressions of the brain's lobes that appear on the inside of the skull), in contrast with australopithecines, who show no such evidence for cerebral reorganization. Such extreme asymmetry is evident in modern humans, in whom it appears to be correlated with lateralization, between left and right hemispheres, of various functions of the brain. Such asymmetrical organization may also be correlated with preferential right-handedness (we will return to this question later). But australopithecines do not show such asymmetry in their brain. It seems that this evolved when the larger-brained hominid was doing something critical for its survival that depended upon and selected intelligence.

The huge cheek teeth and massive muscle and bone structure for heavy chewing would also suggest to us that the robust australopithecines were not relying heavily on tools as an aid in processing food resources. They appear to have been more specialized feeders than the more generalized *Homo* fossils, which would appear to be better candidates for a more habitual tool user using a broader range of food resources. Microscopic studies of the teeth of these creatures by Fred Grine of SUNY Stony Brook and Alan Walker of Johns Hopkins University suggest that their diet included tough, gritty foods that left many scratches and pits on the teeth from heavy crushing and grinding.

At Swartkrans, in levels containing large numbers of *A. robustus* fossils, only a few hundred stone artifacts are found. This scarcity is particularly interesting considering that these deposits may represent sleeping quarters for these hominids for tens of thousands of years and also that a huge quantity of sediment was excavated from this site. It would seem that if these robust australopithecines were sleeping here and used stone tools, they did not habitually carry these tools around and discard them near their sleeping places in any quantity.

Furthermore, when the genus *Australopithecus* went extinct about one million years ago, the genus *Homo* continued without a pause—and so did stone technology. We can say with confidence, then, that *Homo* definitely was a stone tool maker and, in our view, probably the dominant stone tool maker. But could the robust australopithecines also have made some of the stone tools? It is certainly very possible that these creatures were casual stone tool makers and users, but we find it unlikely that they were responsible for the dense concentrations at Oldowan sites of large numbers of stone tools with large animal bones that suggest more carnivorous forms of behavior. The relatively derived *Homo* morphology makes this lineage seem a more likely candidate for a creature who is beginning to rely upon technology to an increasing extent in its adaptation.

HOMAGE TO OUR ROBUST COUSINS

Still, it might be appropriate to pay some tribute to this other major lineage of hominids, the robust australopithecines. With the characteristic pride or, at least, wonder we sometimes feel in the accomplishments of our own lineage, it is easy to forget that each of the robust australopithecine groups (in East Africa and in South Africa) was remarkably successful for over a million years.

This is even more impressive considering that they carried this off while coexisting with our lineage, the evolving species of *Homo*, in the same environments. It may even be that these large-toothed creatures were *so* successful that they forced our ancestors into new food niches that increasingly required the use of stone tools. They went off on their own fork in the road and lived and adapted to its consequences for at least 1.5 million years. We are still rambling down our pathway, but if we look over our shoulders at our relatives, we should realize there are no guarantees of continued survival.

The robust australopithecines may have had different niches, different foods that they would spend their days looking for and consuming, but they did overlap with *Homo* in time and space. They most probably were aware of the larger-brained forms, possibly saw them frequently, possibly even watched them carry out some activities on occasion.

Perhaps they sometimes used some of the same food resources, and conceivably interacted with them in various ways from time to time. Whether these close encounters were hostile or benign we have no way to tell so far, and we probably never will.

Our ancestors, the early *Homo* creatures, surely were aware of these massive-jawed bipeds as well. They may have had some special perception of them due to the strangeness of their gait. Anyone who has spent time on the African savanna, whose senses have been honed to screen sights and sounds for signs of approaching predators and other dangers, knows how striking, how absolutely unmistakable, is the sight of a hominid walking in the distance. Perhaps our ancestors would have strained their eyes at such an approach on the horizon, picking up subtle signals that told them which biped was nearing and the appropriate action to take.

The robust australopithecines are, alas, gone, one more grand evolutionary experiment that came to an end. It would have been fascinating to have met up with them. Musings of this sort, while driving to the field in Koobi Fora a number of years ago, prompted the writing of this song by one of us (NT). It is a lament of A. *boisei* (originally called *Zinjanthropus* by the Leakeys when they found the first East African robust australopithecine fossil in 1959 at the FLK site in Bed I of Olduvai Gorge):

Zinjanthropus Blues

I got the Zinjanthropus blues
My name's been abused
They call me dumb, they call me thick
My tools don't even do the trick, oh no
I just can't take it no more
I may just do myself in
Down on the FLK floor

I got the Zinjanthropus blues
My name's been abused
From the soles of my feet to my sagittal crest
I know my lineage is the best, uh-huh
Don't gimme no big-brained buffoon
'Cause under that dome
He's just an aberrant baboon

I got the Zinjanthropus blues
Across the savannas
It's already the early Pleistocene

And I ain't seen no bananas, oh no
I never had a jolly time eatin' seeds
Tubers and roots,
They never satisfied my needs

I got the Zinjanthropus blues
I'm really bad news
I'm strong as a lion, got jaws like a croc
I'm smart as a wildebeest, may take a walk over you
I pity the fool who hassles Mr. Z
I'll give you thirty-two diastemas*, instantaneously

I got the Zinjanthropus blues
My syntax is confused
I've mastered five nouns, but I'm weak on my verbs
You can't think of meat while you're browsing for herbs, oh no
It just doesn't make any sense
To contemplate extinction, please
Someone invent a future tense.

SUMMARY

By at least 2.0 to 2.5 million years ago, some hominids seem to have made a decisive adaptive shift leading to the species we are today. As we can see from the variety of australopithecines that have ever lived, being a bipedal hominid in and of itself was not necessarily good insurance for survival in the changing world of the Pleistocene. It was perhaps a necessary condition but not entirely sufficient to permit the unprecedented, rapid radiation the human lineage has made all over the world. We had to undergo other critical changes *after* we became bipedal to direct us down this unusual road we have taken.

The robust australopithecines apparently concentrated on foods that were very tough and probably weren't very nutritious overall, so that they likely had to eat large quantities of them. This diet was probably supplemented with other foods, such as invertebrates, eggs, small birds and mammals they might encounter, but their cranial and dental anatomy as well as tooth-wear studies suggests that tough, gritty vegetable foods were a principal part of their diet. To survive on this, to process a lot of tough, fibrous plants, nature selected for larger-jawed, more powerful chewers, these creatures developing over hundreds of thou-

* A diastema is an anatomical term for a gap between teeth.

sands of generations into tremendous masticating machines, the hominid food processors of the Plio-Pleistocene.

Meanwhile, our probable ancestors started down an entirely different route. We never developed such a specialized dietary adaptation, although some shift in our diet, as we shall see in chapter 5, was probably quite important to our survival and ultimate adaptation. We seem to have made another beginning in our evolutionary life, between 2.0 and 2.5 million years ago, when we began an adaptation stressing intelligence, with a higher degree of deliberate planning and controlled manipulation of our environment. The emergence of stone tools and the genus *Homo* signal this event.

These seemingly insignificant-looking stone tools, we feel, played a key role in this new adaptive strategy: technology is a primary means to manipulate objects for some particular purpose, *and* the successful use of technology places a higher premium on the intelligence that guides it. An in-depth examination of the anatomical, cognitive, and adaptive consequences of a profound reliance on stone tools is essential to our understanding of the emergence of the human condition.

CHAPTER 4

FASHIONING OUR FUTURE: THE MAKING OF EARLY STONE TOOLS

Lend me the stone strength of the past and I will lend you
The wings of the future, for I have them.

Robinson Jeffers (1887–1962)
"To the Rock That Will Be a Cornerstone of the House"

MAKING SILENT STONES SPEAK

In the last chapter we witnessed the emergence of flaked stone tools, a profound landmark in our evolution. For the first time we have access to direct, tangible signs—beyond biology—of the manipulatory behavior of early hominids, of what they were doing to adapt and survive.

But the stones they used do not speak for themselves. As we have seen, there are some real limitations to some of the traditional approaches to dealing with early stone tools. In particular, an exclusively typological approach, assigning a name or label to each artifact and putting it back on the shelf, can thwart any genuine understanding of what stone tools signify. We need to find out what they can tell us about the lives, adaptation, and capabilities of these early tool-making hominids.

A STONE AGE MANUAL: OLDOWAN SECRETS TO SUCCESS

Imagine yourself as an early hominid tool maker two million years ago, entering a new geographical area you are not familiar with. You have to learn to recognize and locate not only edible foods, but also raw materials for stone tools. Most rocks found in nature are *not* suitable for stone tools. So our tool-making ancestors had to become competent field geologists in order to locate sources of premium raw material in their habitats and to discriminate between rocks that were superior and those that were unsuitable for tool making.

What were the qualities that Oldowan hominids were looking for in a rock? And how did these creatures turn a rock into a tool? To answer these questions, we must not only examine prehistoric artifacts, but put ourselves in our ancestors' place and try to duplicate their methods through the techniques of experimental archaeology.

A NEW SENSE OF PROBLEM: ACCENT ON BEHAVIOR

In 1977, when we were graduate students at the University of California, Berkeley, Glynn Isaac invited us to participate in the Koobi Fora Research Project in northern Kenya. This project, codirected by Isaac with Richard Leakey (then director of the National Museums of Kenya), focused on the rich paleoanthropological evidence coming out of sediments on the northeastern side of Lake Turkana. For the next several years we became intensely involved with this project, which incorporated scientists studying early hominid paleontology, geology, invertebrate and vertebrate paleontology, fossil plant pollen, and archaeology.

The Koobi Fora study area had already supplied an incredible amount of hominid fossils from one to three million years ago, as well as numerous archaeological sites dating between one and two million years ago. The archaeological program, under the directorship of Isaac with J.W.K. Harris, had excavated over twenty early Stone Age sites there since 1970, providing a large, diverse collection of evidence ripe for further problem-oriented research.

Our research was a part of what Isaac called the second round of investigations here, all trying in diverse ways to probe deeper into the *meanings* of the archaeological sites. The excavated sites, analyzed by Isaac and Harris, had provided a good overall sense of what kinds of evidence they bore; the problem now was, just how far could we interpret patterns of early hominid behavior?

For the next several years Isaac continued to study the distribution and density of stone artifacts and fossil animal bones at one geological horizon that could be traced for long distances, trying to see not only the large megasites, but also the scatter of smaller sites in between, which together might tell us something about how hominids were using the overall environment. J.W.K. Harris continued excavations and studies of archaeological sites on the Karari Escarpment, most dating to about 1.5 million years ago. Henry Bunn studied the nature of the animal bones found at the sites to learn what they might tell us about early hominid subsistence and diet. Ellen Kroll studied the spatial distribution of prehistoric materials at these early Stone Age sites and their potential for inferring early hominid behavior. Jeanne Sept and Anne Vincent investigated in modern, analogous environments what types of plant foods early hominids might have exploited. Zefe Kaufulu studied the fine-scale geology at the archaeological sites.

One of us (KS) explored how early Stone Age sites could have formed and been modified through hominid actions as well as non-hominid, or "natural," forces (for instance, geological agents such as floodwaters). An important part of this study was creating experimental sites made up of replicated stone artifacts and modern animal bones and examining what types of spatial patterns were produced by such activity as stoneworking and animal butchery. These experiments were set out in a variety of environmental settings and monitored over a number of years, to see how they became buried and to compare their ultimate patterns with the prehistoric sites.

The other one of us (NT) carried out an experimental archaeological program to find out how the range of early stone artifacts at Koobi Fora (and other early Stone Age localities) could have been made and how they could have been used in antiquity. Our two research projects dovetailed in such a way that we often worked together on experimental projects, each assisting the other.

WORKING AND LIVING IN THE FIELD

The spectacular drive north from Nairobi through the Rift Valley in Land Rovers normally took two full days, passing through river and lake valleys lined with volcanoes, equatorial forests with browsing elephants, lava wastelands, and deserts with occasional nomads and their camel trains and spending the first night on folding cots hastily set up as darkness encroached. The first glimpse of Lake Turkana (called the Jade Sea) after the hard drive was always a welcome sight.

At Koobi Fora, the archaeological field season normally lasted from September through December. As the majority of the archaeological sites were many miles inland to the east of Lake Turkana, we would normally live in tent camps of five to ten archaeologists and geologists and a field crew of ten to twenty African employees of National Museums of Kenya, many of them veteran excavators with many years' experience working on projects with the Leakeys in Kenya or Tanzania.

On a typical day we would rise at 5:30 A.M., shake out our clothes and our shoes for stowaway scorpions before getting dressed, and have a quick breakfast of maize porridge and coffee or tea before setting off for the site, where we would start excavating by 6:30 or 7:00. To dig through the hardened, one- to two-million-year-old sediments, we would normally use small chisel-ended tools (made by driving a large nail into a wooden handle carved from a branch and flattening the nail head to make a sharp edge), dental picks, and brushes to uncover and clean sediment from the fossils and stone artifacts we would find. As we slowly dug down through the ancient deposits, each fossil or artifact would be given a catalog number, plotted within the grid of one-meter squares that we had strung out over the site, and then leveled in with surveying equipment. This ensured that we would know the three-dimensional position for each prehistoric piece for later computer analysis of the spatial layout of the excavated materials.

Bones uncovered were usually coated with a thin layer of chemical preservative to strengthen them. The sediment removed during excavation was swept up and passed through a fine screen in order to catch up any tiny artifacts or bones that might have been missed. By 1:00 P.M. we would return to camp for lunch, typically a hash of canned corned beef and rice. After lunch and a brief rest, we would either continue excavation at the site, work on excavated materials at the main project tent, or delve into various experimental archaeological projects. The afternoon work would continue until 5:30 or 6:00 P.M., with a tea break to keep us going.

As the sun tended to set quickly so close to the equator, at dusk we would wash up at our tents—normally the ration was half a bucket of water per person. Water was a precious commodity obtained by our crew every few days from a well in the highlands or, if we had had sufficient rain in the area, from the sandy beds of ephemeral streams where we could dig holes to meet the water table and haul up slightly silty but perfectly safe drinking water.

We would normally convene before dinner for a drink of Lemon Squash laced with a capful of rum. Sitting on folding chairs in the dry sand channel under the spreading branches of acacia trees, we would discuss the day's progress and plan future work. After dinner we would normally retire to our tents, where we could listen to music tapes, catch up on our field notes, or even read a novel by lantern light, before falling asleep by 9:30 or so.

Every weekend or two the team would drive to the main camp established by Richard Leakey on the shore of Lake Turkana, where crocodiles cruised and hippos waded offshore, and the prints of lions were often seen on the beach in the morning. We could hand-wash our clothes by the lake and lay them on the coarse beach grasses to dry in the sun. We stayed in thatch-roofed stone buildings (bandas) and net-fished for Tilapia fish for the evening dinner. We would often have impromptu seminars there led by one of the scientists on a selected topic, normally with lively debate among participants.

At the end of the field season, all excavated fossils and artifacts were wrapped and packed and taken back to the National Museum in Nairobi for curation along with all the field notes on the excavations. There we were able to continue analyzing the excavated stones and get down even more seriously to the critical matter of comparing them to the experiments we had conducted and trying to understand what these early stone artifacts actually *mean*.

DESIGNING THE TOOL REPLICATION EXPERIMENTS

How does one go about understanding early stone tools? Early on in this study, it became clear that it would be necessary to set up a feedback between analysis of prehistoric artifacts from the archaeological sites on the one hand and experimental archaeology on the other. So there was a lot of going back and forth between studying the actual excavated artifacts, doing experimental studies of how they were made, seeing the prehistoric artifacts in a new light, doing further experiments, and so on.

It was first necessary to get a good sense of the problem: What was the nature of the artifacts being dug up at these early Stone Age sites in East Africa? What really went into their making? Neatly stored on shelves in the archaeology wing of the National Museums of Kenya in Nairobi were all the materials uncovered from the Koobi Fora sites as well as those excavated by Mary Leakey at Olduvai Gorge. These provided a good basis for designing an experimental program. Before conducting any major replicative experiments, a thorough technological examination of the excavated artifacts from Koobi Fora was the first order of business.

In order to understand how the hominids were making their tools, what sorts of ideas and intentions went into them, obviously we needed to understand what they were doing to their cores—that is, how they were flaking them. Cores and retouched pieces from these sites were carefully scrutinized for several specific features:

1. what the hominid chose to work with—that is, the original form of the artifact, whether a cobble or a flake;
2. the overall complexity of the flaking, whether the piece was worked on one side or face (unifacial), on two sides (bifacial), or on many sides (polyfacial);
3. how extensive the flaking was—whether the hominid worked around the whole circumference of the piece or just on some portion of it;
4. the overall morphology (size and shape) of the resultant core or retouched piece.

The combination of these four features can give an excellent idea of some of the decisions hominids made in their tool making and perhaps of what they were after.

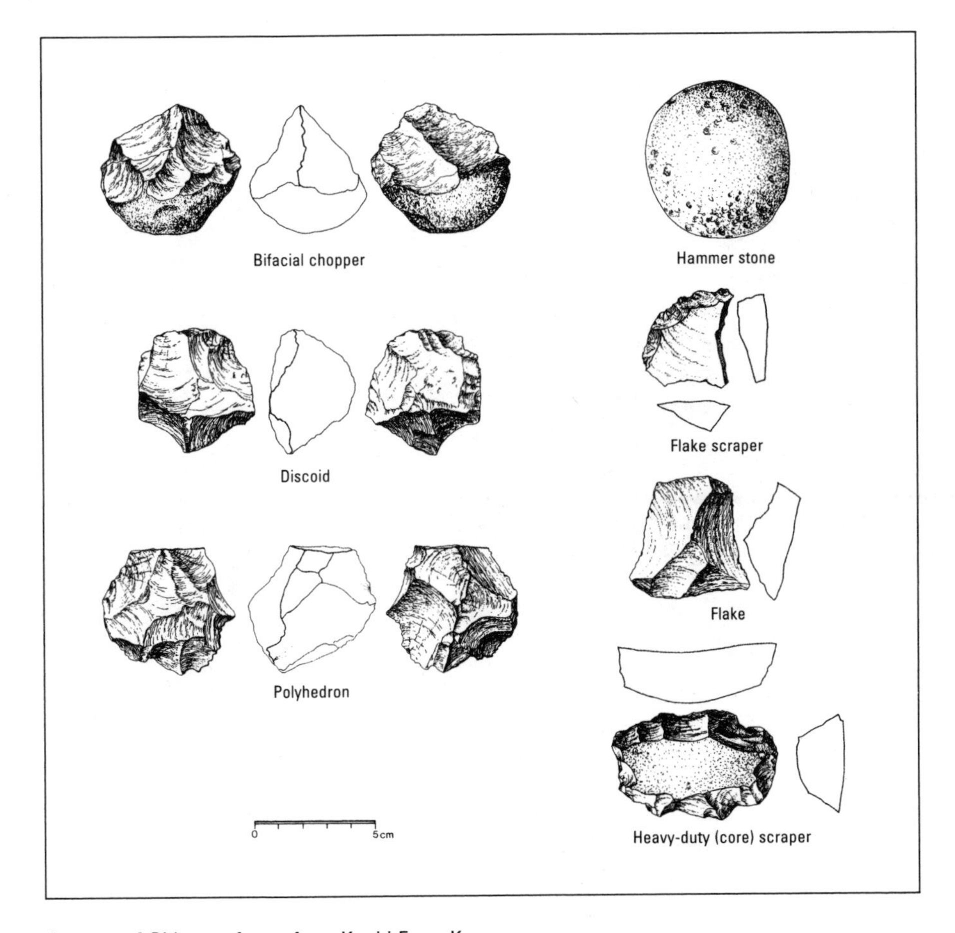

A range of Oldowan forms from Koobi Fora, Kenya.

We looked for signs of the techniques the hominids had used, which can be deciphered in the flaked artifacts themselves. Also, battered or pitted stones that might have been used as hammers or anvils were examined for additional clues. With these preliminary observations, we moved to the experimental program.

The Koobi Fora study area in many ways proved to be an ideal place to conduct experiments into early Stone Age artifact manufacture, as virtually all major raw materials used in antiquity there are still available in similar sizes and shapes within the channels and gravel bars of modern streams (which flow down from volcanic highlands bordering the sediments to the north and east, as in the Plio-Pleistocene). In addition, geological exposures in various parts of the Koobi Fora region

Rolling stones: cobbles of lava and other types of rocks that were used in prehistoric times can be found in modern stream beds at Koobi Fora in Kenya, as they could in early Stone Age times. These stones become increasingly rounded by transport during floods, producing a smooth weathering surface (cortex).

preserve some of the ancient river gravel deposits, showing generally what sorts of raw materials were available to the early hominids in different areas.

The first experiments into Oldowan stone artifact manufacture were intentionally informal: they were to see what flaking techniques and strategies led to what sorts of artifacts. During the first field season, literally thousands of experiments were conducted to see how typical Oldowan artifacts could be produced and which flaking techniques were most suitable. We conducted most of these experiments at our campsite, creating an impressive early stone tool workshop just outside the front door flap of our field tent. We intentionally carried out stone tool-making experiments at specially designated areas away from the archaeological sites so that there would be no confusion in the future as to which artifacts were ancient or which were modern.

Some Departments of Anthropology now offer laboratories or workshops that include practical stonework, and some field schools specifically teach a course in primitive technology. We had been trained in stonework first at the University of California and then at a flint-knapping field school led by Jeffrey Flenniken at Washington State University. For those interested in learning the secrets of stone technology, we strongly recommend working with people who have mastered these skills rather than learning entirely on one's own. Even Oldowan hominids would have probably relied on learning from more experienced tool makers in their group, if only by observation.

Since flaked stone technology produces very sharp fragments of rock, we find it advisable to observe some simple rules of caution: our students wear a good pair of gloves to protect the hands from cuts, goggles or glasses to protect the eyes from flying shards, and footwear to keep sharp flakes from puncturing the soles when least suspected. Stoneworkers often flake some distance apart so that their shrapnel doesn't fly in one another's faces. (In our own experiments, however, we often chose *not* to wear protective clothing, except eye protection, in order to get a sense of the range of injuries early hominids might have incurred from flaking. Also, the spatial patterns of flaked stone debris tend to be different when not wearing such protective clothing.)

Another danger that stoneworkers are now aware of is from flaking indoors: this produces surprisingly huge quantities of angular silica dust. Many of the gunflint makers of the nineteenth century, who would work indoors for long hours their entire professional lives, died at early ages from what was then called consumption. This was almost certainly silicosis, a lung disease brought on through long-term inhalation of silica dust, a danger with modern sandblasters as well. For this reason it is advisable to flake outdoors or in well-ventilated areas to minimize such risk. And, needless to say, due caution should be taken about open cuts and threats of infection. Although such cautionary measures were certainly not taken by prehistoric tool makers, it is useful to remember that natural selection was constantly removing the hapless ones from the evolving populations!

After a first round of experiments in trying to create early Stone Age forms, back in Nairobi we inspected the prehistoric collections a second time in greater detail with a better appreciation of the probable ways these prehistoric artifacts could have been produced. Also, we could see if there were any artifacts whose mode of production could not easily be explained.

Producing an Oldowan chopper core by hard-hammer percussion and the resultant scatter of flakes and fragments. Both the core and the hammer are made of volcanic lava.

A range of characteristic Oldowan forms produced by experimental replication from lava cobbles. *From top left:* hammer stone, unifacial chopper, bifacial chopper, polyhedron, core scraper, discoid. *From bottom left:* flake scraper and six flakes.

Our second round of tool-making experiments was much more controlled and detailed than the first: it turned out that there were essentially about twenty major ways to produce the whole range of flaked artifacts at Koobi Fora. We conducted over one hundred experiments, carefully documenting the size and shape of the cobble or flake each step along the way. This was an exercise in patience: we numbered sequentially as each whole flake was struck off, and then we collected all the flaking debris in a sheet and sieved it through an excavation screen (5-mm mesh) to retrieve the very small flakes and fragments. All these materials were measured and analyzed and the data ultimately entered into a computer. This painstaking detail was worth it: it gave a very firm basis of comparison between the experimental tool making and the early archaeological sites.

Finally, six of the prehistoric Koobi Fora Oldowan sites and one early Acheulean site from between 2.0 and 1.2 million years ago were analyzed in exhaustive (and exhausting) detail, and these data were also entered into a computer. These sites were selected to sample a range of early stone technologies through time as well as in different geological contexts (delta, river floodplain, river channel). These sites were then compared directly to the experiments, to look in detail at the actual process of early hominid tool making.

EXPERIMENTAL TOOL MAKING RESULTS AND THEIR IMPLICATIONS

THE EARLIEST STONE AGE TECHNIQUES

When discussing stone technology, Paleolithic archaeologists often separate *techniques*, the physical forces applied to the rocks, from *methods*, or strategies in the mind of the tool maker. Techniques and methods became much more sophisticated as human evolution proceeded, but in the time period between 2.5 and 1.5 million years ago, four major techniques appear to have been used by early Stone Age hominids.

Hard-Hammer Percussion

An extremely simple technique, this is the most common one at the earliest archaeological sites. It consists of hitting a rock with another. The one rock becomes the core (after flakes are knocked off), and the other is the hard hammer. (This is distinguished from a technique not seen until later in the Stone Age, soft-hammer percussion, in which the hammer is much softer, more yielding and resilient than the stone core, and produces quite different results; often wood or bone is used for this.)

Choosing a hammer stone requires some thought: the most desirable choice is usually a cobble that has a nicely rounded surface (to knock against the core) on at least one end or side and is fairly smooth over most of its surface, so it doesn't dig into the hand while flaking. Many times, the preferred hammers are slightly oblong or egg-shaped. The size of the hammer stone varies: for a pound-size core, the hammer is about the same size; for flaking a small cobble, the hammer stone is somewhat larger than the core; and for very large cores, the hammer tends to be smaller in order to be wielded in the hammering hand. Also, it's very important to avoid cobbles with obvious flaws or cracks in them, as they can shatter and cut the hammering hand. (All of these precepts were gained *not* from the armchair, but from the direct experience of experimental archaeology.)

EARLY STONE AGE TECHNIQUES

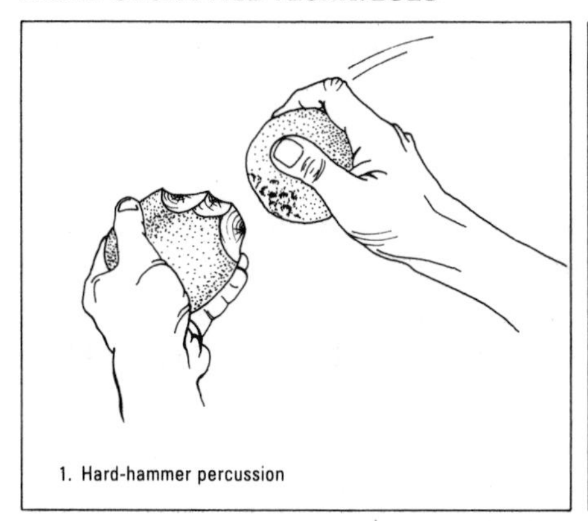

1. Hard-hammer percussion

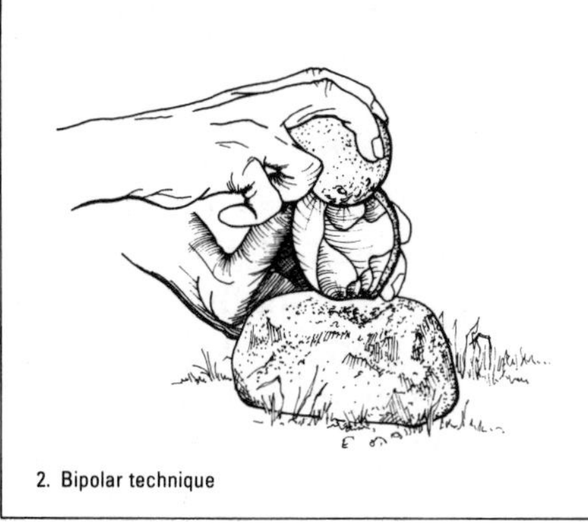

2. Bipolar technique

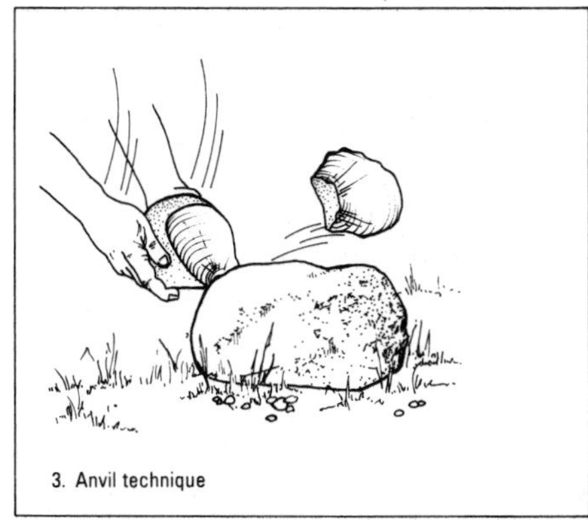

3. Anvil technique

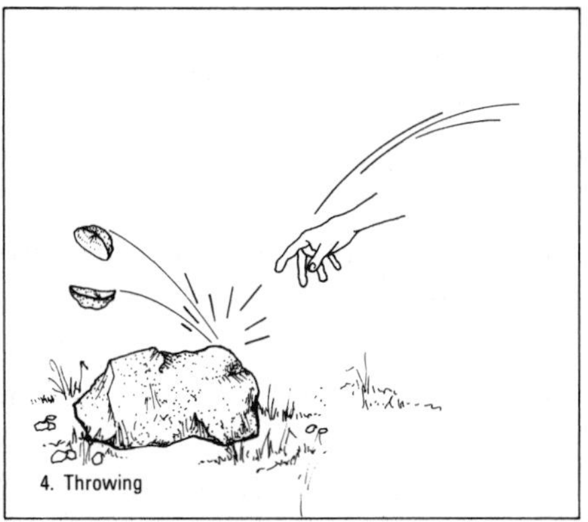

4. Throwing

The major techniques of flaking stone that would have been available to the earliest stone tool–making hominids. Each produces characteristic fracture patterns and by-products.

With this technique the hammer is usually held in your preferred hand (whichever one you might use to hold a hammer when hitting a nail) and the core in the other (for a right-handed person, it almost always works best to hold the hammer in the right hand and the core in the left). Before swinging into action, it's important to line up the core for the flaking blow. As we indicated above, a sharp angle is critical for success. Normally you try to choose some area of the core with an overhang, where the angle formed by two sides coming together is much less than ninety degrees. This angled edge of the core is held so that it projects out from the hand, with a flat surface facing upward (this is where the hammer will strike). Then, holding the hammer firmly with the other hand, with a rounded surface projecting outward and downward, you strike with a hard, glancing blow fairly near the edge of the core. The flake is removed from the undersurface of the core. With practice this procedure produces flakes fairly reliably and predictably.

Bipolar Technique

This is also seen in some of the early archaeological sites. It also entails hitting the core with the hammer, but in this case the core is first rested or braced upon another rock, an anvil. Flakes are normally knocked off by the hammer, but sometimes the force of the core against the underlying anvil also detaches flakes from the anvil end of the core (hence, bipolar). This technique produces pitted hammers and anvils as well as small, distinctive core forms (called "outils écaillés," meaning scaled tools, or simply "bipolar cores.") This technique often tends to be used for flaking very small pieces of stone or for rock types such as quartz that fracture unpredictably and could break up in the hand.

Anvil Technique

In this technique the core plays the active part, as it is held in one or both hands and swung down forcefully against a stationary, usually fairly large anvil. This technique is sometimes the choice when the core is very large, so that it is hard to hold in one hand and needs a very strong blow to initiate fracture. It is also a more dangerous technique, as flakes are removed from the upper surface of the core, toward the body and face.

Throwing

A final technique that could have been used by early Stone Age tool makers is an extremely simple one: throwing the core or cobble against a stone anvil in order to initiate fracture. This technique can split a cobble neatly in half with a pronounced area of crushing at the point of impact. As we will see shortly, we have seen this develop spontaneously in experiments teaching a bonobo (pygmy chimpanzee) to make stone tools.

Based on experimentation and examination of the prehistoric artifacts, we found that virtually all the major types of Oldowan stone artifacts produced by early hominids at Koobi Fora sites could be produced simply by hand-held, hard-hammer percussion. The unmistakable fracture patterns on all artifacts—cores, retouched pieces, flakes, and fragments—indicate this. This is not too surprising, since the overwhelming rock type used was a basalt lava, which requires appreciable force from a hammer to produce fracture. Even though smaller cobbles of quartz, chert, and chalcedony were occasionally used as raw materials, there is little if any evidence of bipolar technique at these sites in the early Stone Age. The battered stones found at Koobi Fora sites, normally also of lava, are consistent with their use as stone hammers.

At some other Oldowan sites, notably at Olduvai Gorge, there is more evidence of bipolar technique in addition to the standard hard-

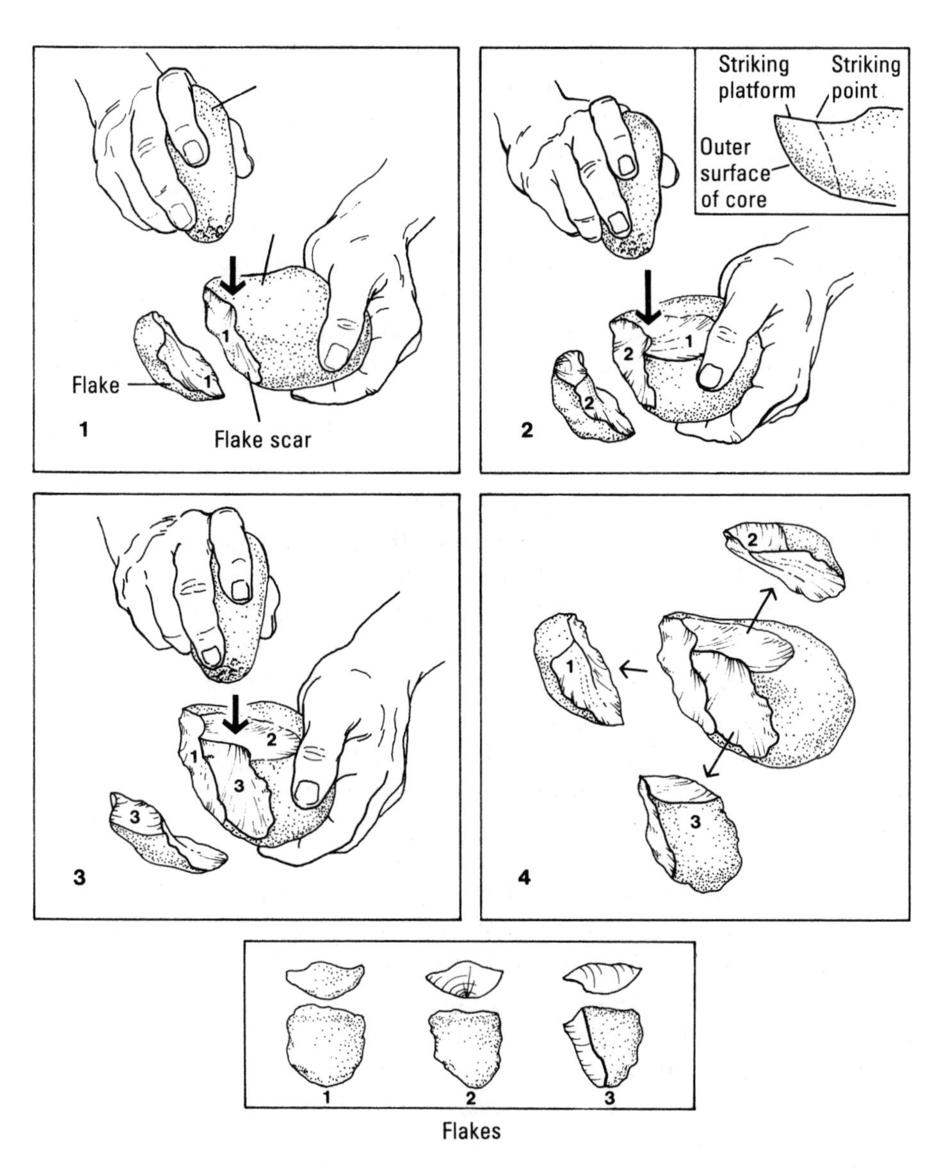

The production of a simple Oldowan chopper core and the resultant flakes.

hammer percussion. We can see this in pitted rocks at a number of the Oldowan sites, which suggest their use as hammers and anvils for bipolar flaking, and in typical bipolar cores often found (Mary Leakey's "outils") usually made from quartz or blocky chert. Still, hard-hammer percussion appears to have been the favored technique overall.

HOW SELECTIVE WERE EARLY HOMINID TOOL MAKERS?

How many miles would a hominid go for a good rock? In the recent prehistoric past, there is evidence that hunter-gatherers were sometimes very selective about rocks they used as raw materials for their tools. Within the past forty thousand years or so, they periodically took special care to select extremely high-quality raw materials, occasionally transporting or exchanging them ultimately over very long distances, tens or even hundreds of miles from their geological sources. But what about hominids at the dawn of human technology—how choosy were they? As we learned in our own tool-making experiments, there are several qualities of a good rock that early hominids would have looked for.

First of all, the rock should be relatively hard and consolidated. Many rocks, such as most limestones, siltstones, and sandstones, can be too soft or unconsolidated to flake well and to hold an edge while being used. These were rarely flaked in the Stone Age. Harder rocks, such as those that are high in quartz content (called silica, or SiO_2) and whose minerals are cemented tightly together, are better suited. (Sandstone is made of grains of hard quartz, but the individual grains are normally not cemented together very firmly, making it a poor flaking material that tends to crumble.)

Second, the rock should be relatively fine-grained, meaning that the individual crystals in the rock should be fairly small, often invisible to the naked eye (what geologists call cryptocrystalline.) Since rock tends to break around crystals, coarser-grained rock will have more irregular flaked surfaces and edges, and flaking will be harder to control. Fracture in fine-grained rocks is typically conchoidal in nature, producing characteristic bulbs of percussion, ripples, and fissures.

Third, the rock should fracture equally well in any direction when force is applied. In geological terms this is called an isotropic material. Rocks with preferential cleavage planes, such as shale, slate, mica, and schist, or with internal crystal faces, such as calcite and feldspar, will break easily along some planes but not others, making it difficult or impossible to control their fracture or, sometimes, to produce sharp edges.

Finally, the rock should be fresh inside (unweathered by chemical alteration) and unflawed; there ideally should be no inherent fracture lines in the material (often apparent as hairline cracks on the outer surface, or cortex, of a rock).

Igneous Rocks

Igneous rocks are formed from a hot, molten state (magma). Those suitable for flaking have normally formed at or near the earth's surface and have cooled quickly, so that large crystals have not had time to develop before the rock hardened. These include:

- *lavas*, rocks that have been produced from molten magma flows. These rocks are subclassified based upon their chemistry and mineralogy, into such forms as basalts, andesites, trachytes, phonolites, and rhyolites. Lavas suitable for flaked stone tools tend to be fine-grained and homogeneous without large crystals or vesicles (holes in the rock from gas bubbles forming in the molten magma before it cools). Many lavas tend to be gray or black in color, but shades of brown, red, and green are also common.
- *ignimbrites*, rocks produced by volcanic ash, where the ash particles have become fused, or vitrified, by the intense heat. These rocks often originate from hot clouds of ash *(nuées ardentes)* that are ejected from volcanoes (such as the famous eruption of Mount Pelée in Martinique, which destroyed the nearby town and most of its occupants). These rocks are sometimes called "welded tuffs." These rocks tend to be gray, brown, red, or green.
- *obsidian*, a volcanic rock that also forms from a molten state but has cooled so quickly that crystals have not had a chance to form. It is a volcanic *glass*, an amorphous rock whose cutting edge, when flaked, can theoretically be as thin as one molecule, so that it is incredibly sharp (it has even been used in modern hospital surgeries on several occasions). Obsidian is usually black, but it can also be brown, red, gray, or green. Obsidian is normally found in massive formations that represent ancient flows within volcanic hills or mountains. In such volcanic areas, obsidian is usually a minority of the types of volcanic rocks present.

Sedimentary Rocks

Sedimentary rocks are formed from particulate matter (such as clays, silts, and sands) that has settled and become cemented over time, or through various minerals crystallizing or congealing in solution (such as limestone or dolomite). To be suitable for flaking, such rocks should be hard and fine-grained and are usually made up for the most part of very small quartz crystals. These include:

- *cryptocrystalline siliceous rocks*, which have quartz grains (silica, or SiO_2, crystals) that are so small they cannot normally be seen with the naked eye or even an optical microscope. Classification of these rocks is somewhat complicated by different systems of geologists and gemologists (as many of these rocks are classified as semi-precious gems). For archaeological purposes they are sometimes lumped together as cryptocrystalline silicas, or CCS. These rocks may form in tabular bands in geological strata, as nodules in chalk or limestone deposits, or as fillings to cavities in other rocks. Most notable among them are

1. *flints*, by far the best-known of the rocks used in prehistory, usually dark-colored rocks composed of a high proportion of silica. In European terminology, flint normally derives from within chalk deposits.
2. *cherts*, similar to flints, but often lighter in color and less pure in silica content, and often found within limestone deposits.
3. *jaspers*, forms of cryptocrystalline silica that tend to be red, brown, or yellow in color.
4. *agates*, cryptocrystalline rocks usually banded with different colors.
5. *chalcedonies*, cryptocrystalline rocks usually translucent, with a waxy luster.

- *opal*, which is a form of noncrystalline rock (amorphous, like glass) formed from silica and water. It tends to have a waxy, iridescent luster.
- *quartz*, formed when silica (SiO_2) makes large crystalline rock, and often suitable for flaking as well. Two major varieties of quartz are normally recognized:

1. *vein quartz*, a milky-white or yellow type of quartz, often found filling the cracks of other rock formations. Fracture is often irregular, so vein quartz artifacts are notoriously difficult to analyze by archaeologists. In the gem world, a reddish variety is called rose quartz and a violet one amethyst.
2. *quartz crystal*, large geometric crystals of quartz, usually of a

colorless, transparent nature, which were occasionally flaked in antiquity. The fracture pattern is conchoidal and resembles broken glass.

- *silicified limestone*, formed when silica from groundwater permeates limestones, making them harder and more amenable to controlled flaking.
- *silicified sandstone and siltstone*, formed when silica cements relatively unconsolidated rocks such as sandstones (called orthoquartzite when silicified without metamorphism) and siltstone (called silcrete in its silicified form).

Metamorphic Rocks

Metamorphic rocks began as sedimentary or igneous rocks but have been changed through heat and/or pressure into another type of rock. These include:

- *quartzites*, sandstones with grains cemented so firmly in the rock that fracture tends to go through the sand grains rather than around them. Those suitable for flaking tend to be highly cemented and have a matrix of relatively fine sand grains with a sugary texture. These rocks tend to form in thick, massive geological formations.
- *indurated shale*, shales that have been metamorphosed into a rock that has lost most of its tendency to fracture along certain planes and becomes more easily flaked into artifacts.

In some geographical areas only one major raw material suitable for making stone artifacts may be available; in others there may be a number of suitable rock types from different sources, and in these areas the hominids had more choice of which rocks they could use. Raw materials suitable for stone tools may be found in two different contexts. They may be in their *primary* context (in their original place of geological formation), such as lava or obsidian flows, flint nodules in a chalk deposit, or cliffs of quartzite. Alternatively they may be in a *secondary* context (moved by natural forces from their original geological setting), such as cobbles in a river gravel or beach deposit, or rocks lying on a desert surface.

Although prehistoric hominids would not have classified rocks as modern geologists do, even the earliest archaeological record indicates that Oldowan hominids had learned to identify such rock sources and distinguish good from poor materials with some expertise. Over time, by trial and error they determined which types of rocks were amenable for flaking and which were not. Almost certainly they developed a well-defined mental map of their landscape, remembering the precise location of these sources (even chimpanzees in West Africa show some development of mental maps of the rocks they use for nut-cracking activities).

How far would hominids travel for good rock? Not very far by recent human standards (but as we will see later, a fair distance for a modern ape). Rocks for early stone tools in the Koobi Fora region were found primarily in beds of streams flowing out of volcanic highlands: very large cobbles and boulders could be found far upstream near the volcanics, while farther downstream smaller cobbles and pebbles could be found. Even farther downstream, around sites closer to modern Lake Turkana, there were really no stream gravels available worth flaking: the nearest cobbles of appreciable size were at least several miles upstream and must have been carried by hominids to the sites. But possible rock sources were relatively nearby most of the sites studied in the region, which are often situated along or near river beds containing cobbles large enough to flake.

Hominids at Koobi Fora appear to have been fairly unselective in the sense that they seemed to choose rocks in the general frequency in which they could be found in the ancient river gravels. These streams contained cobbles and pebbles composed mainly of lava, and at all of the Koobi Fora sites lava is the predominant rock used. Usually 95 percent or more of artifacts are made of this volcanic rock, mostly the black basalt ubiquitous in the region. The raw materials found less commonly in the ancient stream beds, such as ignimbrite, chert, chalcedony, and quartz (which tend to be easier to flake and produce sharp edges), are likewise relatively rare at these sites.

Nonetheless, these hominids appear to have had a good eye for good rocks with sound flaking qualities: rarely does one see evidence of vesicular lava (full of bubble cavities) being used or of rocks with serious weathering flaws in them. By the time the hominids arrived at the sites to flake their rocks and/or use their tools, inferior rocks had already been rejected, probably left back at the stream bars where they had originally picked up their stones.

At Olduvai more selectivity may be seen in the early Stone Age. Quartz was obtained from bedrock outcrops to the north of the ancient lake basin, while lava cobbles were generally to be found in alluvial fans and streams farther away, draining down from volcanic highlands southeast of the lake. Usually, some proportion of quartzes and lavas are found at each site, showing that these localities were focal points where hominids concentrated stones they had collected from more than one source on the landscape. The maximum distances of rock transport to most of the earliest Olduvai Gorge sites are usually about two or three miles, but, as we will see below, it appears that early hominids were habitually carrying stone around here.

In the earliest period (Bed I) at Olduvai at about 1.8 million years ago, lava was the most common rock type for stone artifacts, but by Bed II about 1.5 million years ago quartz was becoming the predominant material. How much of this was due to hominid selectivity, and how much to hominid land-use patterns is not clear. For example, if by Bed II, hominids tended to forage to the north of the lake basin, they would primarily have had access to the quartz outcrops there and would have been farther away from lava gravels to the south. But they still carried this quartz for several miles as they moved around the landscape.

The general pattern in this phase of hominid evolution seems to indicate moderate transport distances and some, though not an overwhelming, amount of selectivity for particular rock types. Strong selectivity and long carrying distances appear to increase through time in the Paleolithic and become especially pronounced with anatomically modern humans during the past forty thousand years.

ASSESSING THE COMPLEXITY OF HOMINID PLANNING

Does Oldowan stone technology represent merely a means to an end, devised when some immediate need arose? Or did hominids plan ahead?

The simplest, most unsophisticated technological response to a need would occur when some circumstance developed in the life of an early Stone Age hominid (or a group of hominids) that required a stone tool. The tool maker would travel to the nearest source of usable stone, a river with a gravel bar of cobbles or a bedrock outcrop, and gather some rocks. He or she would carry the stones to where they were required, flake them, use them, and discard them. All the tool-making and tool-using behaviors would be centered around the immediate, pressing need of the moment.

If this were so, we might expect that *all* stages of making the tools could be found at the site where they were used and finally discarded. All the artifacts produced—the rocks that were taken to the site and flaked, producing the usable tool as well as other incidental stone debris—would be left behind at the site after the task was completed. Theoretically the stone artifacts at the site could be fit back together with the flaked cobbles left behind if undisturbed by major geological forces.

But this is definitely *not* the pattern we detect in the earliest Stone Age. Every site we have examined shows that cores were almost always first flaked elsewhere, sometimes extensively, *before* they were brought to a site. At each site we are missing critical parts of the entire technological story: we find that most cores are missing many of the flakes removed from them, in particular, flakes removed early in the flaking process (which are very distinctive, since they should show the outside rind, or cortex, of the cobble). Likewise we often find flakes of distinctive raw materials but no corresponding cores of the same rock. Altogether this suggests that we are seeing *not* a simple story of the most expedient technology, but a more complicated one.

Work by the Koobi Fora research team reconstructing (or conjoining) flakes and cores at various sites gives us a glimpse of how the hominids were transporting cores around from place to place. At one site in particular (with the cumbersome name of FxJj 50), from 1.5 million years ago at Koobi Fora, our group was able to reconstruct about 15 percent of the artifacts into separate core sets. These show a complicated mix of artifacts: almost all the cores had been previously flaked somewhere else, then flaked further at the site; then some of these were discarded there, but most were then *removed* from the area where they had been worked. This suggests a lot of carrying around of flaked stone by these early hominids.

In fact, we can now document in some cases the distances over which hominids transported rocks from their sources to the Oldowan sites where they left them. These distances range from only a few tens of meters (with the site in the immediate vicinity of the rock source) to several kilometers (with the rock source some distance from the site). Again, these longer distances appear to indicate systematic carrying of stone by hominids beyond anything seen so far among nonhuman primates and suggest intelligent behaviors involving foresight and planning evolved in tandem with the earliest stone technologies.

This is a much more complicated pattern than many would have suspected from this remote period of time. It bespeaks to us an elevated degree of planning among these early hominids than is presently seen among modern nonhuman primates. This is a fascinating pattern with serious implications, which we will look into more deeply later.

CORES AS MENTAL TEMPLATES?

Some other patterns revealed by artifacts do not mean what they seem to. Did hominids set out to make such types of tools as choppers, discoids, polyhedrons, and core scrapers in the early Stone Age? Did they have these artifact forms firmly in mind when they struck one rock with another? Throughout the study of Oldowan industries, there has often been an implicit or explicit assumption that the cores or core tools, rather than the flakes, were the principal implements and also the intentional target pieces of the hominid tool makers. This is what American anthropologist James Deetz has called, in later humans, mental templates: prescribed and highly standardized ways of producing artifact forms, with a clear idea in the mind of the tool maker what the end product should be.

It is worth noting that as archaeology emerged in the seventeenth and eighteenth centuries out of an art-historical pursuit, there may have been some subtle but very real reasons why Oldowan core tools became the center of attention. Why study Michelangelo's rubble heap of marble chips if you can study his *David* or *Pietà?* If stone cores are considered (however intuitively) as a form of sculpture, shaped by the removal of smaller pieces of rock, then these core forms are almost certainly going to become the principal object of study.

Our studies in stone artifact manufacture, as well as our experiments with novice modern stoneworkers with no prior knowledge of archaeology or stone technology, have shown that most of the Oldowan core tool forms do not, in fact, have to represent shapes deliberately planned by early hominids. The majority of these core forms can be produced simply by the production of sharp-edged flakes from lumps of stone. So a great deal of the variation seen in these forms at Oldowan sites can be explained by viewing them as *by-products* of sharp flake manufacture. But why so many different shapes? Experiments show the answer to this is simple: The starting point for the artifact—the size, shape, and type of stone used for flaking—has a profound effect on the shape of the final core produced, even with no intent or mental template on the part of the tool maker.

Thus the principal aim of early stone tool making may have been the flakes (previously thought to be mere by-products) rather than the core tools (previously thought to be the tools). This reassessment of early stone artifacts puts the emphasis on the technological processes the hominids went through rather than the various products they created. Many of the core tools identified in the earliest Stone Age may be figments of the archaeologists' imagination: they may be nothing more than by-products of making the flakes. Why the hominids would have wanted these flakes will be addressed in the next chapter, when we look at experiments in using various early stone artifacts.

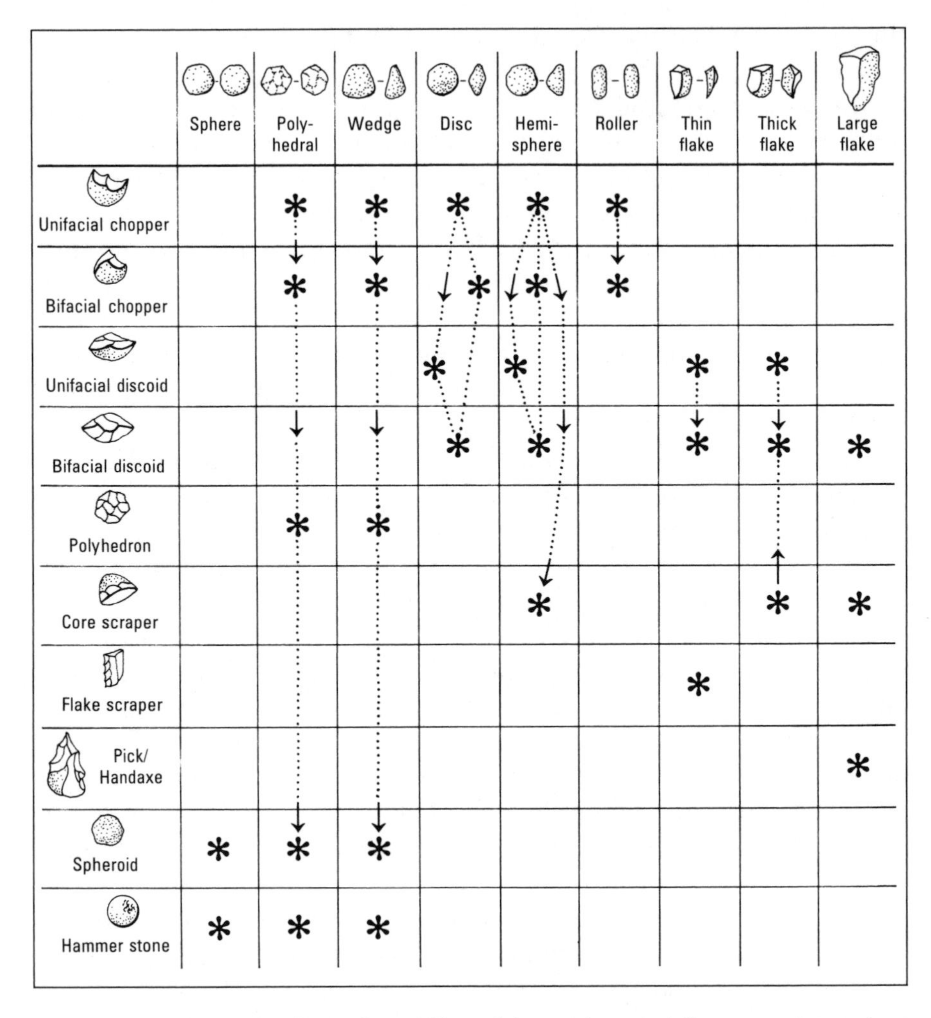

Diagram showing how the shape of a cobble or flake can have an influence on the resultant core or retouched form and how one type can grade into another.

A CASE STUDY: THE RIDDLE OF THE BOLAS STONES

One especially enigmatic artifact form that prevails throughout much of the Stone Age in the Old World is a globular, battered piece of stone called a "spheroid," "stone ball," or "bolas stone." Examples made out of quartz are found in large quantities at some early Stone Age sites in East and southern Africa. In other areas, such as North Africa, the Middle East, and China, similar forms are sometimes made of a dense limestone.

For over a century prehistorians have been impressed by these artifacts because of their striking symmetry and their recurrent presence at many prehistoric sites from the early Stone Age, from 1.8 million years ago up to the time of Neanderthals, about 40,000 years ago. Intuitively these forms appear to exhibit a great deal of predetermination in their manufacture. Furthermore, the ordeal of making them, the time that must have gone into creating them, requiring repeated and prolonged battering against another object, suggested that these objects must have served some valued purpose.

A thespian production set in the Ice Age, performed for the French Prehistoric Society in the early twentieth century. Note the use of three spheroids as bolas stones, a common interpretation of the function of these artifacts.

Thus these puzzling stone balls have often been assumed to be a major class of tools, used through much of the Stone Age. What they were tools *for* has been more problematic. The most common interpretation was that these were bolas stones, tied to thongs in twos or threes and used in hunting to trip up and bring down game in a manner like that of the gauchos and indigenous peoples of the Pampas of Argentina. Other interpretations for these artifacts have included club heads, thrown missiles, bone-breaking tools, and plant-processing tools.

We designed an experiment to see if there might be a simpler explanation for these common early Stone Age forms. This was done in Zambia in central Africa, where angular chunks of quartz are the most common raw material available in many areas.

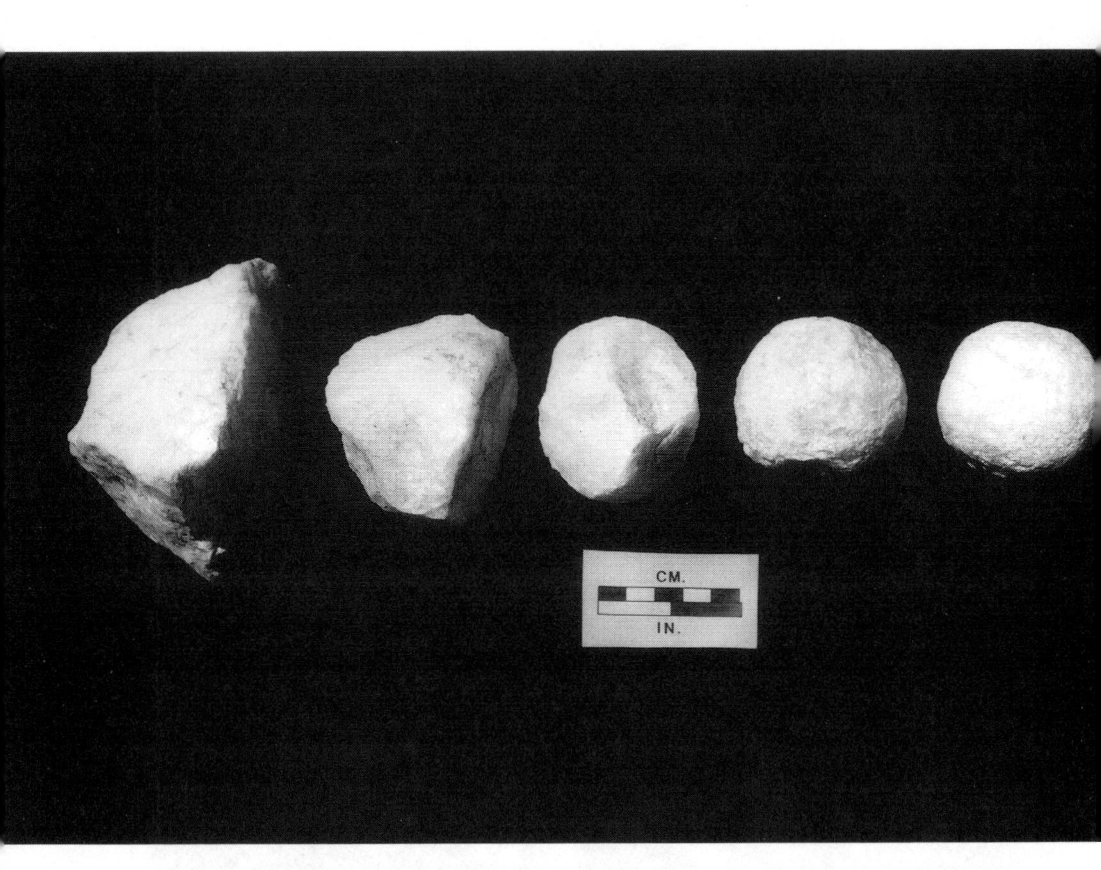

Spheroid production. *From the left:* an unmodified quartz chunk; a flaked polyhedral core; a polyhedral core used as a hammer for one hour; a hammer used for two hours; a hammer used for four hours, becoming a "spheroid"/"stone ball"/"bolas stone." Such forms can be produced simply by their use as hammers, without predetermination or intent.

We found that after approximately four hours of percussion these quartz hammers assumed a remarkably spherical shape *without any necessary intent or predetermination.* After four hours of use as a hammer stone, an angular chunk of quartz was transformed into a virtually perfect spheroid, just by being used as a hammer stone. Quartz is much more friable than most other types of stone used in flaking and tends gradually to develop a round shape through prolonged battering. From lower Bed I through Bed II of Olduvai Gorge, the use of quartz for tools increases sharply over time, and along with this, the incidence of spheroids as a tool type increases dramatically. This appears to be the result of repeated use of quartz as hammers for making tools.

That does not, however, mean that these forms are without interest in Stone Age studies; in fact, this finding makes them very interesting. Since the frequency of these types tends to rise throughout the early Stone Age, it seems likely that early tool-making hominids were *carrying* their hammer stones more habitually and/or that sites where such hammers were used were *reoccupied* on a more frequent basis. In either case they were apparently using their hammers over and over again and producing these heavily battered and spherical forms as by-products.

This type of experimental approach shows how a simpler, reductionist explanation can sometimes be forwarded for what appears to be a mysterious archaeological phenomenon. Although it does not prove that these objects could not have been used for some of the other activities suggested in the past, we have demonstrated that it is possible to arrive at these forms simply from the process of flaking stone, especially when more brittle, friable rocks are used as hammer stones.

ARTIFACTS AS REFLECTIONS OF INTELLIGENCE

What do these patterns of stone artifacts tell us about the cognitive abilities and intelligence of Oldowan tool makers? Although the *products* of Oldowan technology are quite simple, the *processes* required in the hominid mind to produce these forms show a degree of complexity and sophistication: in other words, *skill.* We feel that these early stone tool–making hominids had evolved, by 2.4 to 1.5 million years ago, to an important new level of intelligence and cognitive operations not present in earlier hominids or in modern nonhuman primates, including the highly intelligent apes.

It is clear to us that efficient flaking of stone requires a strong intuitive knowledge of three-dimensional geometry as well as sophisticated motor skills in order to detach flakes from cores. This is a skill that can take a number of hours to master, even for a modern human. As previously mentioned, to flake stone efficiently, one is constantly searching for acute angles on core edges from which to detach flakes. A carefully controlled, sharp, glancing blow from the hammer stone to the core is required to initiate fracture. This level of cognitive operational skill and motor control has not been documented in nonhuman primates in the wild.

OLDOWAN HOMINIDS AS BIPEDAL CHIMPANZEES?

Recently several researchers, such as primatologist William McGrew of the University of Sterling and archaeologist Thomas Wynn of the University of Colorado, have posed an intriguing question: Could the cognitive processes of *Homo habilis* essentially be on a par with those of modern apes and the differences that exist be due primarily to ecological, dietary, and technological differences rather than differences in intelligence levels?

Wynn has applied an interesting methodology used by cognitive scientists to examine the developing complexity of cognitive processes in modern human children. Using criteria proposed by the Swiss psychologist Jean Piaget, he analyzed Oldowan tools using the geometric concepts of proximity, separation, and order. Wynn concludes that Oldowan tool makers exhibited preoperational intelligence, similar to that of modern chimpanzees and very young children, and that these early hominids did not exhibit any cognitive complexity above that of the African apes.

Although we would agree that the *products* of Oldowan technology do not show many clear signs of deliberate or stylistic design, we would argue that the technological *processes* that produced these forms probably *do* exhibit higher cognitive capabilities than are observed among modern apes.

Is there another, more direct way to test how the thought processes of Oldowan tool makers were similar to or different from those of modern apes? Although anthropologists can compare endocranial casts (casts of the inner surface of the skull case, which preserve the gross morphology of the brain) of apes, modern humans, and fossil hominids, such anatomical information cannot show us the *actual capabilities* of these animals. And since modern African apes are not known to produce flaked stone artifacts in the wild (the principal behavioral manifestation left behind by Oldowan hominids), comparisons of the material cultures of these different creatures is difficult as well.

There is, however, another approach we can take to attempt meaningful comparisons between the cognitive abilities of early tool-making hominids and those of modern apes. If we could teach modern apes the basics of flaking stone tools, give them a reason to do it, and then allow them to develop their own expertise and abilities, we could then compare apes and early hominids in terms of their tool-making abilities. We know that apes are smart and that they can be taught a variety of complex tasks. But we wanted to see, given only the very basics of stone tool making, how far a modern ape could take it.

PAN THE TOOL MAKER

The ideal opportunity to study the technological capabilities of a modern African ape presented itself in 1990 at a conference in Portugal on tools, language, and cognition, organized by the Wenner-Gren Foundation for Anthropological Research. Psychologist and primatologist E. Sue Savage-Rumbaugh presented research results of communication studies with apes. We had been wanting for several years to collaborate with primate researchers on comparative studies of ape and hominid technology, and finally we met someone who also thought this would be a great idea.

Now we would have the opportunity to see if a bonobo, one of the two species of chimpanzees, could learn to make stone tools and, if so, how its actual tool-making procedures and the artifacts it might make would compare with the earliest tools produced by our hominid ancestors. Basically we hoped, if the bonobo did indeed get involved in making stone tools, to be able to see if its tools really differed very much from those the Oldowan hominids were making. Or were the earliest stone tool makers significantly advanced beyond what an ape is capable of?

Our van heavily laden with stone cobbles, we drove from Indiana University to the Language Research Center in Atlanta, Georgia, operated jointly by Georgia State University and the Yerkes Regional Primate Research Laboratory of Emory University. Here, chimpanzees and other primates are raised in an incredibly nurturing environment with lots of interplay with other members of their species as well as human companionship and tutelage. The research focuses primarily on symbolic communication studies. Our collaborators were psychologists E. Sue Savage-Rumbaugh, Duane Rumbaugh, and Rose Sevcik, as well as Kanzi, a 160-pound bonobo, often called a pygmy chimpanzee *(Pan paniscus)*.

In the 1970s Australian archaeologist Richard Wright demonstrated that an orangutan (that lived in a zoo in Bristol, England) could be taught to make a stone flake and use it as a tool to cut through a cord to open a box containing food. This was a fascinating start, but the orangutan was not really faced with many of the problems and decisions that early tool makers had to contend with (how to hold and manipulate the core, how to search for angles, how to orient the core in relation to the hammer, where to flake next, and so on). The core in this early experiment had been preshaped with acute angles all around it and was strapped to a platform so that the ape just had to learn to hit an edge and remove the flake. Nor was this research envisioned as a long-term project to see how the ape might develop its tool making over time.

Using Wright's study as a foundation for further research, we wanted to look directly at behaviors analogous to what was going on in early tool making, to compare what Kanzi did *and* what he made with what early hominid tool makers were doing and making. Immediately we saw that Kanzi was going to be an excellent pupil. He was highly intelligent, focused in his studies, and resourceful.

The first day we started by showing him that stone tools are pretty useful things: a stone flake could be used to cut a cord and open a box containing a treat (of Kanzi's choice—a bunch of grapes, a piece of watermelon, a cold juice drink, and so forth). By the end of the first day Kanzi was using flakes that we had made and cutting readily into box after box, developing a true appreciation for stone tools. At the end of the second day he had become an excellent judge of stone knives: given a choice of five different pieces of stone to cut into his box, he could choose the sharpest one nine times out of ten. He was also making casual attempts to hit rocks together to make a tool on his own.

Kanzi is principally involved in communication studies, so he could only engage in these experiments from time to time, spaced apart sometimes by weeks. Within about a month of starting this study Kanzi was striking his first flakes off cores by hard-hammer percussion.

Since he was first initiated into the mysteries of stone tools, Kanzi has been given the latitude to develop his own flaking style, with occasional demonstrations to show how a modern human would do it. He was shown merely the fundamental principle of hard-hammer percus-

Pan the toolmaker. Kanzi, a ten-year-old bonobo (*Pan paniscus,* or "pygmy chimp"), flakes a flint core with a stone hammer to produce usable flakes. His stone artifacts can be compared directly with those produced by early Stone Age hominids to gauge levels of cognitive complexity.

sion—hitting one rock against another—as well as how useful these stone flakes can be. He immediately was interested in having these flakes to use, and this has become his motivation to try to make them himself. Over time Kanzi has been developing his tool-making skill and gradually getting more adept and precise at flaking stone. He was not conditioned or shaped in his flaking of stone. He is motivated by his desire to cut his way into his box and has been experimenting with rocks on his own to see how he can best get usable cutting tools from them.

Several months after the start of this study, Kanzi even devised his own technique for making flakes: purely on his own he developed the innovation of *throwing* his stones on a hard tile floor to fracture them and produce cutting edges. (Curiously, he had never been very interested in throwing until this technological advance of his, although some of his siblings have shown a penchant for throwing things. But now he seemed to find a purpose to it.) Within a few throws he usually was able to make a sharp flake, which he promptly used to get into his box.

After this innovation he seemed to have made his own connection between the force of the throw and successful flaking. He now carried

Pan the tool-user. Kanzi uses a flake that he struck off to cut through a cord in order to open the box (at right) to obtain a fruit reward.

this association to his hard-hammer percussion work, which before this time had been somewhat lackluster but now became much more forceful. Working outside again, without a hard floor to throw against, he became more proficient at hard-hammer percussion. Within about nine months of the start of this study he was removing several flakes from cores, producing objects that we can begin to compare with those in the early archaeological record.

Although Kanzi is still continuing to improve his tool-making abilities, his present level of expertise is significantly below that seen in Oldowan hominids. His core forms are strikingly similar to the natural eoliths produced by geological forces, which confused prehistorians around the turn of the century. He still doesn't show the understanding of flaking angles that Oldowan hominids had: Kanzi bashes and crunches the edges of cores with his hammer stone rather than using highly controlled and forceful blows that we can see in the early Stone Age artifacts. Recently, throwing has become his favorite technique.

We are enormously impressed with Kanzi's tool-making development and his innovations, and we're anxious to see how far he will take all this. Future questions we will ask include whether other apes at the laboratory might learn tool making from him and whether he will start to carry tools around for use at activity sites out in the research center's surrounding forest. (We are starting to put reward boxes outside in the woods, away from the rocks stored at the lab.) But despite his advances, the products he is now making are fascinating in their *contrast* with ones at early archaeological sites. This suggests to us that there may be an even earlier stage of stone technology in the archaeological record, perhaps rare and difficult for the archaeologist to identify. It will perhaps take a sharper eye and more time in the field to identify such sites, which would probably be earlier than 2.5 million years ago.

Moreover, Kanzi's progress so far as a tool maker suggests to us that early Oldowan hominids may exhibit a much greater cognitive understanding of the principles and mechanics of tool making than modern apes seem to be able to develop. This indicates something important about our hard-wiring, the size and complexity of our brain and its connections to the motor control system, at this stage in our evolution. We feel that these hominids probably had surpassed modern apes and probably their australopithecine ancestors in their ability to modify stones.

In addition to displaying enhanced skill, early hominids show a significant behavior in transporting stone, occasionally over distances of several miles, and sometimes building great concentrations of artifacts and animal bones in certain places. Whether they did this all at one go or, more likely, in several trips, this suggests more foresight and planning with regard to carrying and collecting than is seen in nonhuman primates today. Chimpanzees in the wild rarely carry tools for more than one hundred yards or so, and they usually fashion a tool just prior to using it.

These early hominids, however, seem to be doing something quite different: they seem to have been planning and acting in greater anticipation of projected events and future needs.

DEXTEROUS EARLY HOMINIDS?

In modern human populations, approximately 90 percent of people are dominantly right-handed (dextral) and about 10 percent are dominantly left-handed (sinistral). This is a very unusual pattern, and it is unique to humans. In the rest of the animal world, including nonhuman primates, the breakdown of handedness (or pawedness) tends to be about 50 percent left-handed and 50 percent right-handed. (Among some primates there may be slight populational asymmetries in handedness for certain activities such as hanging from and feeding from trees, but these do not come close to the 9:1 ratio seen in modern humans.) It would seem probable that this preference for right-handedness in modern humans has a strong genetic component, and its absence among the apes would suggest that this trait became established in the human lineage after the split with the African apes.

Why should the human lineage have become so dominantly right-handed? No one knows for sure. It has been suggested that our extreme asymmetry in handedness may be correlated with an increased specialization (lateralization) in the hominid brain during the course of human evolution. Although there is increasing evidence to suggest that many animals appear to exhibit some lateralization in brain specialization, it seems to be on a much smaller scale than in human beings. In *Homo sapiens* the left hemisphere (which controls the dominant right hand) has become more specialized for such activities as time sequencing and language, and the right hemisphere has become more involved with spatial perception. With the development of manual skill during the rise of stone technology, perhaps natural selection favored populations that tended to use the same dominant hand to facilitate learning by observation (imagine a right-handed person trying to learn to play guitar by watching a left-handed person).

When did preferential right-handedness become a characteristic among hominid populations? During the past century and a half there has been a great deal of speculation, as this characteristic was discovered in human evolution. While doing research on the early stone tools

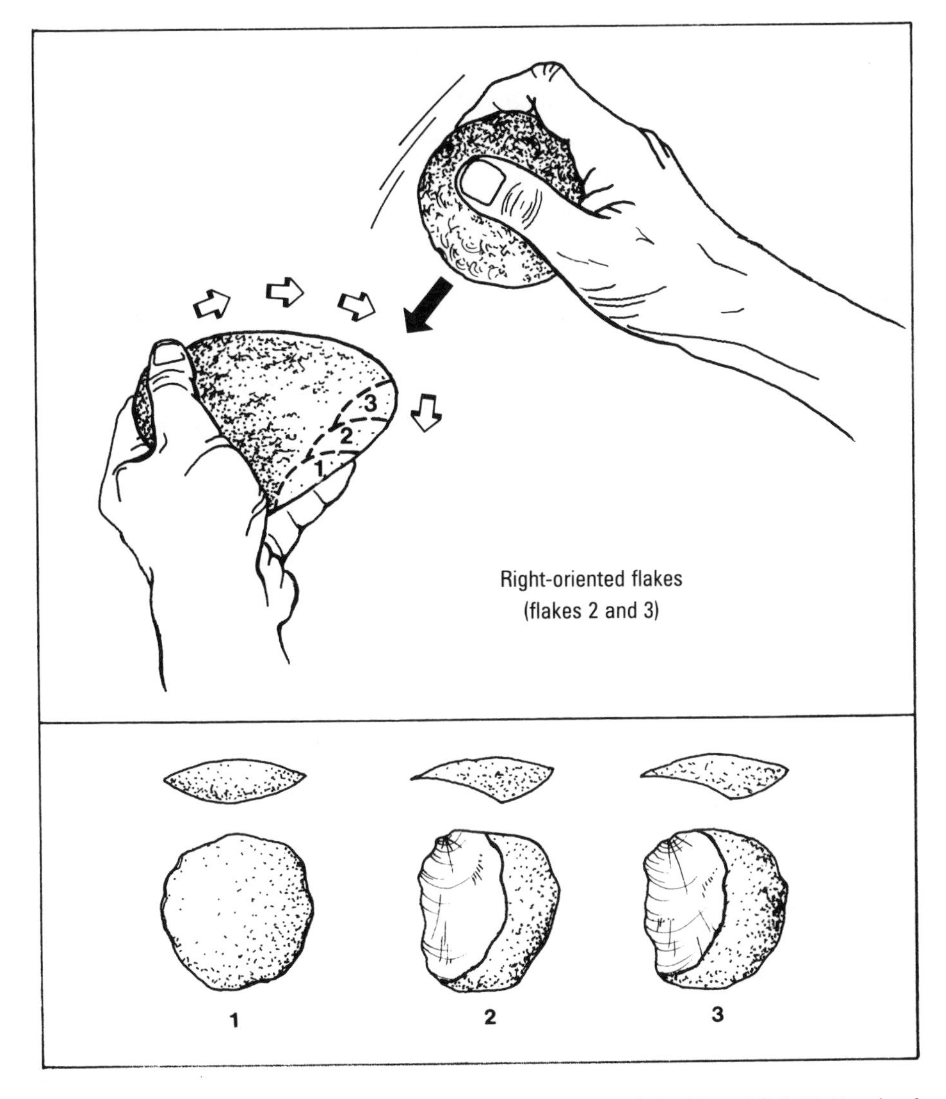

Production of "right-oriented flakes" (flakes with cortex only on their right side). A 56:44 ratio of "right-oriented" to "left-oriented" flakes was produced by the right-handed experimenter (left-handed people do the mirror reverse of this). Early hominids at Koobi Fora, Kenya, produced a similar 57:43 ratio, suggesting that they were preferentially right-handed, which could have important implications concerning the evolution of the human brain.

from Koobi Fora, an unexpected pattern was discovered that may shed light on this question.

Let's look at what a right-handed tool maker typically does during hard-hammer percussion. As outlined above, a right-handed individual normally holds the hammer stone in the dominant right hand (which gives more precision and power to the flaking blows and lessens the chance of hitting one's fingers) and the core to be flaked in the more passive left hand. The left hand essentially acts like a vise to securely grasp the core during repeated blows from the hammer stone, orienting the core properly for each successive impact.

Now, what effect does this setup have upon the flaking process? If a sequence of flakes is removed from one face of a core, there is a tendency for the left hand holding the core to rotate it in a clockwise direction as the flakes are removed. One hits off a flake, rotates the cobble a little, and strikes off another to the right of the first, rotates it slightly again, and flakes again, and so forth. If the core is made on a cobble or thick cortical flake, we can see this clockwise rotational bias by examining the flakes that have been produced. Successive flakes tend to have part of a flake scar on the left (where the previous flake had been struck off) and part of the cobble's cortex on the right. Thus, large samples of these flakes can tell us something about handedness: whether the cobble was being rotated in this way, as would a right-handed person, or whether it was being turned by a left-handed person, in the opposite hand and producing the opposite pattern. Experiments show that right-handed tool makers produce significantly more of these right-oriented flakes and left-handed people more of the left-oriented flakes. In our experiments, (we being right-handed), a ratio of 57:43 of right-oriented to left-oriented flakes was produced.

This is an experimental result that can be applied directly to early Stone Age artifacts. So far, every site we've examined from the early Stone Age, including those at Koobi Fora dated from about 1.9 to 1.5 million years ago, shows exactly the same pattern. Thus it appears that by the time of early toolmaking in the archaeological record, these ancestral hominid populations may have already become preferentially right-handed. For whatever reason or reasons, right-handedness seems to be an ancient trait in humans.

TECHNOLOGY AS AN EVOLUTIONARY ACCELERATOR

In the 1960s anthropologist Sherwood Washburn of the University of California suggested a "biocultural feedback" model of human evolution. In this model a feedback loop developed between human cultural and genetic evolution: as culture became more adaptive and advantageous to the species, genes that support its development were selected, which allowed for enhanced cultural abilities, with genes supporting these cultural developments again selected, and so on.

A similar concept of the evolutionary process called "gene-culture coevolution" has been suggested by sociobiologists Charles J. Lumsden and Edward O. Wilson, first in *Genes, Mind and Culture* (1981) and later developed more fully in *Promethean Fire* (1983). The authors contend that the origins of the modern human mind and intellect are due to the complex interaction of biology and cultural behavior. In their model, humans have evolved not through genetic change alone, nor through cultural developments, but through a unique interaction we have developed between genes *and* culture, shaping what we have become. The human mind and intelligence and the culture it both allowed and shaped have played a major role in the final evolution of our species. Both Washburn's and Lumsden and Wilson's models help explain why intelligence would have been selected so strongly since this gene-culture interaction began and thus why there has been such a rapid expansion of the human brain in the past two million years of human evolution.

A critical aspect of this reverberating effect between genes and culture, particularly in its early phases, would have been technology. This likely included not just making tools, but conceptualizing possible needs for them, anticipating more concretely possible times and places they might be needed, creating and maintaining more complex mental maps of stone tool and food resources in their environment, and amplifying this new conceptual realm by using tools to make other tools. Tools would have been very important in accelerating the tempo of our evolution, not just because of what they did or accomplished for early hominids, but for emphasizing thought, cultural sharing of information, and planning for a future more removed from their immediate present.

THE INVENTION OF FLAKED STONE TECHNOLOGY: SPECULATION

How did the very earliest stone technologies evolve? No matter how far we try to push back the visible archaeological record, its very beginnings will probably always remain hidden and undiscovered. So in the absence of real evidence, here are some conjectures about how this whole enterprise might have gotten under way.

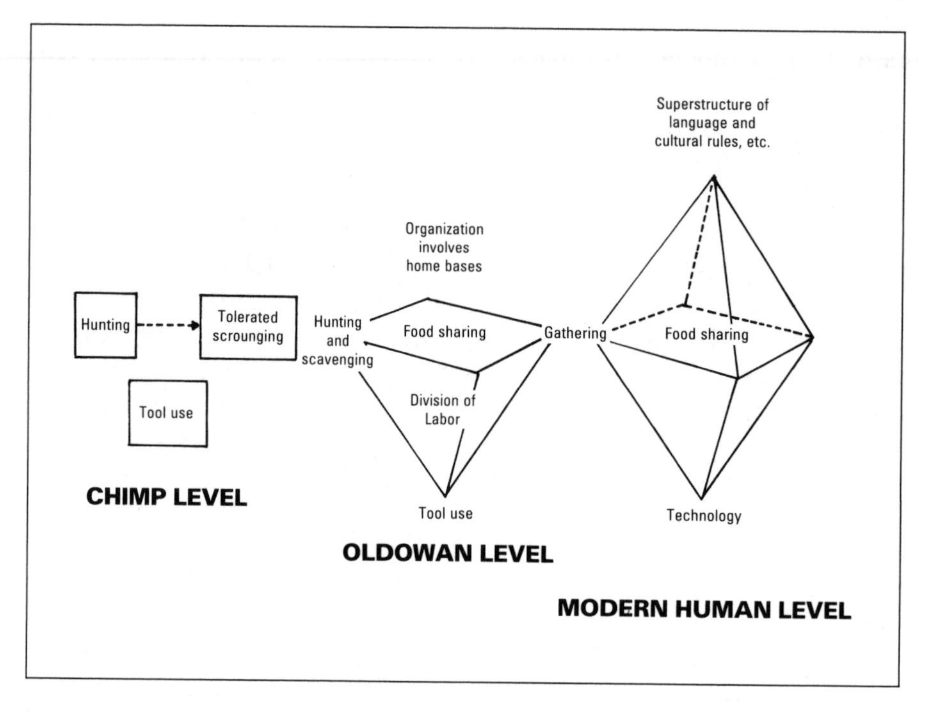

Isaac's model of increased sophistication during the course of human evolution.

THROWING

Throwing stones as offensive or defensive missiles, or in display behavior, is well documented among wild chimpanzees. Many people in recent nonindustrial societies have also been known for their accuracy, range, and effectiveness in throwing rocks. So an ability and propensity for throwing may be an ancient trait in our lineage.

Is there any way that throwing could lead to stone tool making? If we postulate a stage of early hominid evolution in which populations developed the technique of hunting small animals (such as rodents, small antelope, hyrax, birds, lizards, and snakes) with rocks (a behavior not seen among nonhuman primates today), it is likely that sooner or later some of these rocks would miss their target and strike other rocks or outcrops, producing sharp-edged fractured pieces. (In this scenario the sharp rocks might be found in the general vicinity of the killed prey as well, which could have spurred a connection between the available tool and the cutting task.)

BIPOLAR NUTCRACKING

As we have seen, some groups of chimpanzees in the wild have developed a cultural tradition of cracking nuts with stone hammers. We could not exclude ancient hominids from a similar technology. Various hard-shelled nuts, fruits, and seeds (such as legumes), commonly available seasonally, develop a hard outer rind or coating, and hammering would help hominids exploit these resources.

Since stone hammers and anvils often break spontaneously, a flaked stone technology could have developed as a by-product. And since a shattering stone hammer can cut the hand, a hominid could have been duly impressed with the sharpness of fragments. It would have then been necessary, though, for the hominid to explore how this new invention could be applied to other possible tasks in the daily round of activities, because it does not seem to fit in with the immediate task of extracting nuts or seeds.

BONE BREAKING

Another bashing activity in which early hominids might have been involved is bone breaking. If, in opportunistic hunting or scavenging, hominids were using stones to break open bones for their nutritious, fatty marrow contents, broken hammers could ensue just as at the nutcracking sites. Here sharp fragments could conceivably have been used for cutting off any relict meat, cartilage, or fat still attached to the bone.

EVALUATING THE POSSIBILITIES

Which of these scenarios is most likely? At our present stage of knowledge it is difficult to say. If hominids had actually gone through a nut-cracking or bone-breaking stage before habitual flaked stone technologies, there may be evidence that will someday be detected in the archaeological record in times *before* we find flaked stone: namely, pitted hammers and anvils or broken bones (ideally found in association with rocks that could have served as percussive instruments) that show evidence of having been fractured with hammer stones.

If hominids habitually or even occasionally transported rocks away from their geological sources, archaeologists might be able to identify these erratics, or "manuports," in the sedimentological record and infer that hominids were the likely source of such transport. Unless we could document a clear, consistent pattern of transport in some area, however, this conclusion would remain fairly tentative. It is also possible that the earliest flaked stone technologies developed directly out of a response to process animal carcasses effectively, in which case there was likely no earlier stone-using behavior that served as a direct precursor.

But even if the inception of stone tool making remains shrouded in mystery, the early archaeological record we do have at hand provides us with a great deal of information about early hominid behaviors. And experiments have afforded us a new understanding of how they were made and some measure of the cognitive processes of their hominid creators.

These early artifacts signal a new threshold in hominid intellectual development, one that separates developing Oldowan hominids from the apes—not just in locomotion, but in how they were thinking and planning their lives, which sets them forth on an ever-spiraling trajectory involving enhanced culture, elaborating technology, and rapid brain development. By 2.0 to 2.5 million years ago, the dawn of human technology had illuminated a new pathway for our ancestors to follow. But why were these stone tools so important in our adaptation?

CHAPTER 5

THE ROLE OF ROCK: USES OF EARLY STONE TOOLS

The wildebeest is grazing mindlessly on the edge of the water hole, away from the rest of the herd in the afternoon. From downwind the pack of hunting dogs approaches, using the cover of the tall grass and shrubs to conceal their small piebald figures from the eyes of their intended prey. In a flash they converge.

The hominids are foraging along the river when an adolescent female hears a low-pitched bellow. She climbs a nearby fig tree and sees the dust; flailing legs; small, darting carnivore bodies; and, finally, red flesh. She excitedly gives a food call and gestures in the direction of the kill. Several other hominids, adult and immature, climb the same tree, perching on adjacent limbs to view the carnage. One drops down, picks up a few cobbles with one hand and a large fallen branch with the other, and starts running in the direction of the kill. Others follow suit, until there is an irregular phalanx of adult males, and adolescent males and females, with adult females and their infants and juveniles bringing up the rear.

The hunting dogs are distracted from their kill by the flapping wings of several vultures that have been patiently watching and waiting several meters from the kill. The source of their disturbance is the group of advancing bipeds, who are brandishing sticks and branches and begin to throw stones, yelling and whooping aggressively. The dogs bristle and snarl their bloodied muzzles but are finally chased away by the hominid interlopers; a well-placed pitch of a lava rock strikes one of the retreaters in the flank, accelerating his speed of departure with an angry yelp.

The dead wildebeest stares vacantly, its tongue clenched between its teeth. The body cavity has been ripped open by the hunting dogs and most of the internal organs consumed. Otherwise the carcass is still fairly intact. A strong smell of blood, viscera, fresh meat, and musk fills the air, which buzzes with an ever-increasing number of flies attracted to the kill.

The hominids surround the wildebeest and with excited vocalizations scan the landscape to make sure there are no hyenas or lions nearby to steal their new spoil. A number of them squat down with rocks in their hands and begin to crack them together.

Several large lava flakes are struck off a core, and the sharpest one, with a smoothly curved, convex edge, is selected as a butchery knife. As one hominid grasps the rear ankle of the wildebeest and pulls the hind limb taut, another starts cutting through the skin and muscle of the leg, exposing red meat and, underneath, the white bone of the joint. Through the parting flesh he glimpses the round end of the thighbone partially hidden in the circular socket of the pelvis. The grating sound of a stone knife edge against fresh bone can be heard. After numerous cuts at this joint, the hind limb is wrenched free from the body.

Two adult females are working at the other end of the carcass. One pulls the left forelimb taut while the other cuts into and around the shoulder area, which easily removes the entire forelimb from the body with the shoulder blade attached.

Soon the entire carcass has been portioned into easily carried units and is being hauled off on the heads, in the arms, or over the shoulders of the hominids as they make their way toward the shady grove of trees along the tributary that flows into the larger river. This grove is well-known to the hominids. Many half-buried stone artifacts and fragmented animal bones are scattered on the ground. The grove is open enough for the hominids to see any approaching pack of hyenas or pride of lions that might be hungry enough to challenge them for their booty, in which case they will make a hasty retreat up the trees with as much of the carcass as possible. Today, however, there is no contest.

Under the shade of the acacia and fig trees, several of them take their lava cores and strike off some more flakes. Then they start cutting the flesh off the individual limb bones, clenching the meat between their teeth and cutting with a stone flake where the muscle attaches to the bone.

There is much bickering, for there are twenty-odd hominids and only nine dismembered portions of the wildebeest: the two hind limbs, the two forelimbs, the head, the pelvis, the two sides of the rib cage, and the backbone, all still covered with meat. The dominant members of the group have first access to prime parts such as the limbs and begrudgingly share portions with other individuals who sit nearby in anticipation and occasionally abscond with a chunk if attention is diverted. Rarely is there an actual melée, however, unless food has been very scarce and hunger overcomes the established social hierarchy.

Consumption of the wildebeest proceeds rapidly and with much clamor, with satisfied grunts and belches. Mothers occasionally give pieces of meat to their weanlings, who are not yet old enough to use the stone tools.

One individual places a femur against a large root protruding from the base of the huge fig tree. She repeatedly strikes it sharply along the shaft with a rock until long cracks form in the surface of the bone and it finally splits open, breaking into many fragments and splinters and exposing the greasy marrow inside. This is scooped out with the index finger as far as it will reach, then fished out with a twig and swallowed. Finally the ends are smashed, exposing the red, spongy bone, which is sucked of its fatty contents. Some walk down to the river, crouch down, and drink from the cool flowing water.

Soon only scatters of broken bone fragments and discarded stones are left, some of them covered with drying blood, meat, and fat. Ants begin to swarm over these. After a brief rest in the shade, with some grooming among adults and play and wrestling among the young, they begin to rise. Slowly, one by one, the hominids resume their foraging, halfheartedly, as no strong pangs of hunger gnaw at their insides. They spread out in open search of anything edible—fruits, berries, shoots, roots, insects, lizards, small mammals, birds' eggs, and such fare—along the strip of forest on their side of the river. As the afternoon progresses they begin to make their way to a cluster of fig trees that they climb, building their sleeping nests in the sturdier boughs.

The shadows lengthen on the plain as the reddish-orange sun sets behind the shimmering volcanic ridges on the horizon. As twilight turns quickly to night, the nearly full moon rises over the eastern highlands, splashing pale light on the reclining figures in the trees as it beams through the gently swaying foliage.

DISCOVERING PURPOSE IN THE STONE AGE

> He has invented and is able to use various weapons, tools, traps, etc., with which he defends himself, kills or catches prey, and otherwise obtains food. . . . When primeval man first used flint-stones for any purpose, he would have accidentally splintered them, and then would have used the sharp fragments.
>
> **Charles Darwin,**
> ***Descent of Man,* 1871**

When young Stamford was informing Dr. John H. Watson, recently returned from the second Afghan war, of a possible companion with whom to share lodgings, one Sherlock Holmes, Stamford expressed a certain amount of anxiety about bringing them together.

> "It is not easy to express the inexpressible," he answered with a laugh. "Holmes is a little too scientific for my tastes—it approaches to cold-bloodedness. . . . He appears to have a passion for definite and exact knowledge."
>
> "Very right, too."
>
> "Yes, but it may be pushed to excess. When it comes to beating the subjects in the dissecting-rooms with a stick, it is certainly taking rather a bizarre shape."
>
> "Beating the subjects!"
>
> "Yes; to verify how far bruises may be produced after death. I saw him at it with my own eyes."
>
> **Arthur Conan Doyle,**
> ***A Study in Scarlet,* 1887**

With the forensic sciences, experimental approaches to archaeological problems can sometimes take unusual, even bizarre, turns. In their attempts to match up processes with products, forensic scientists and experimental archaeologists have developed remarkably similar aims and methods.

Both attempt to reconstruct human behavior, and both advocate an experimental approach in order to understand the relationships between certain types of human behavior and their material by-products. In fact, as archaeologists, we have had frequent contacts with forensic scientists in order to see which of their methods and techniques might help us in studying the early Stone Age record.

What were early hominids actually *doing* with the stone tools they produced? What was the adaptive significance of these early technologies in the course of human evolution? These are problems that have perplexed prehistorians for a century and a half. To answer these questions is to explain why flaked stone technologies were invented in the first place, why they became popular among some early hominid groups, and why they spread throughout Africa and subsequently most of the Old World during the next two million years. This chapter will focus upon attempts to determine the function of stone tools through a variety of different approaches: the examination of stone tool use among modern human hunter-gatherers and chimpanzees, experimental use of stone tools, bone modification studies, stone artifact use-wear studies, and examination of organic residues on stone artifacts.

This Far Side cartoon by Gary Larson aptly illustrates the problems inherent in determining the functions of stone tools. (The Far Side © Universal Press Syndicate. Reprinted with permission. All rights reserved)

"So what's this? I asked for a *hammer*! A hammer! *This* is a crescent wrench! . . . Well, maybe it's a hammer. . . . Damn these stone tools."

ETHNOGRAPHIC AND ETHOLOGICAL MODELS

To see how simple tools can be used in living off the land before the use of metals, we can examine the technologies of chimpanzees (the only nonhumans to display any significant use of technology) and modern hunter-gatherers. From these we can compile a list of possible uses for stone technology in the early Stone Age.

As we have seen, some chimpanzee populations use stone and wooden hammers and anvils, stone or wooden missiles, sticks to capture termites, and leaf sponges. But other chimpanzee groups do not appear to use tools at all, so it would appear that the chimpanzees' adaptation as a species does not necessarily depend upon the use of such tools but rather allows some groups to exploit certain resources, such as termites, ants, or hard-shelled nuts in areas where these tools are used.

Among modern hunter-gatherers, such as the San of the Kalahari Desert and the aboriginal tribes of Australia, we can see stone tools being used today or at least until relatively recent times. During the last several centuries, the !Kung San and related groups have replaced many of their tools that were made of stone with metal implements, first through trade with iron-producing, Bantu-speaking societies and subsequently from contact with colonial and postcolonial European technologies.

We have constructed a list of possible uses for early stone tools, based upon what is known of chimpanzees on the one hand and modern or recent human hunter-gatherers, such as the !Kung San of the Kalahari Desert, the Hadza of Tanzania, the Pygmies of the Congo basin, and the Australian aborigines, on the other hand. For the hunter-gatherers, this includes what they have been known to use within the last one hundred years or so, before metal finally replaced many elements of their tool kit.

ACTIVITY	CHIMPANZEES	HUNTER-GATHERERS
1. Nut Cracking	S	S, A
2. Woodworking	0	A
3. Digging	0	A*
4. Animal Butchery	0	A
5. Weaponry	0, S (casual)	A*
6. Hide Working	none	A

Key: 0 = without stone tools; S = unmodified stones; A = modified stone artifacts; * = stone tools used to make other implements that are used for the activity.

Interestingly, recent hunter-gatherers usually do *not* use stone tools per se to gather most edible plants. Nevertheless, other simple tools such as digging sticks may be used to harvest underground vegetable resources—roots, tubers, and corms—and containers made out of natural fibers and materials are often used to help collect vegetable resources such as nuts, fruits, berries, and the like.

We can see from this list that chimpanzees show a very limited use of stones as tools and these are not deliberately modified in any way (though their nut-cracking stones become pitted or broken over time through use). Aside from plant gathering, recent hunter-gatherers used stone tools in most of the other crucial activities of their daily lives: in processing some foods (cracking nuts and cutting meat), in preparing skins for clothing and carrying devices, as hunting and defensive weapons, and in making a variety of other important tools such as spears and arrows or digging sticks for finding food and water.

POSSIBLE TASKS OF EARLY HOMINIDS

Now let us consider different activities that early hominids may conceivably have been pursuing and explore the possible use of tools made out of stone and other materials in various aspects of these tasks. For each of the activities here, we will first explore the importance of the particular function and look at analogous activities among nonhuman species as well as modern humans. Also we include results of our experimentation and consider the possible archaeological traces of such activities.

NUT CRACKING

A variety of hard-shelled nuts and fruits are found today in many trees in the woodlands of Africa. These hard-shelled nuts and fruits pose a special hazard to animals feeding on them as they may break teeth or wear them down very rapidly, which could lead to serious infections or at least to a reduction in their efficiency in feeding. Some primates, such as baboons, are able to use their dentition for cracking some of these hard-shelled species. Two modern primates, chimpanzees and modern humans (such as African hunter-gatherers), have solved this problem with very hard nuts and fruits by using stone or wooden hammers and anvils to do the work for them.

In our own experimentation, we have used stone hammers and an-

Nut cracking: a stone hammer and anvil are used to crack hard mongongo nuts from Zambia. Hard-shelled and tough-skinned vegetable foods could have been processed easily and rapidly by early hominids with an unmodified stone technology.

vils to crack many species of edible nuts. Even the toughest-shelled items, such as the highly prized African mongongo nuts from central Africa (which yield almost as much protein and fat as wild meat), can be opened easily with well-directed blows. Through prolonged use, the stones used as hammers and anvils start to develop small dimples or pits, the amount of time this takes depending on the hardness of the stone and the nut.

The prehistoric visibility for this behavior would be the presence of such pitted hammers or anvils (also created by chimps and hunter-gatherers today), often found under the trees of the most productive groves for nutting. We believe it is very likely that such nut cracking (or fruit splitting) could have preceded the earliest flaked stone technology, whether or not it was a cause of it. It would be a small step to transfer this technology to other activities, such as cracking long bones of animals for their marrow content.

At Olduvai Gorge, Mary Leakey has found objects identified as anvils at sites in the lower two levels, Beds I and II (the time of the earliest hominids and stone tools there). In the lower, earlier Bed I, she found these battered and pitted rocks to be rough blocks or broken cobbles of stone, but by Bed II they seem to have been fashioned into more circular shapes. At present, though, the prevailing interpretation is that these pitted stones are anvils used in the bipolar flaking technique, in which a stone to be flaked is rested on a stone anvil and shattered with a hammer stone. The attrition from this flaking causes pitting on both the hammer stone and the anvil. But there is nothing to rule out the cracking of plant foods on these as well, if hard-to-open, edible species of nuts were available in the vicinity in prehistoric times.

Interestingly, the mechanics of splitting rocks by the bipolar technique and that of cracking nuts are almost identical. As we mentioned earlier, it is possible, though by no means demonstrable, that the technology of bashing nuts, also known among chimpanzees, was later transferred by early hominids to stone flaking as well.

Experimenting with a range of digging tools: a sharpened stick, a broken ostrich leg bone, and a gazelle horn. Such non-stone materials make excellent digging implements that could have been used by early hominids for obtaining underground plant resources or burrowing animals, digging for water at a river bend, or freeing stone raw material eroding out of the ground.

DIGGING

Many animals have evolved ways of digging for underground resources, such as hidden vegetable foods or water, by developing special anatomical features. Bush pigs use their snouts and tusks for rooting up underground morsels, elephants use their feet to dig for water, and aardvarks use their robust claws to help open up termite and ant nests.

Among nonhuman primates, the baboons are the most prolific diggers, particularly when they search for edible roots and tubers lying at shallow depths below the ground surface. They use their fingers as probes to loosen earth and their hands to scoop it away and expose the vegetation.

Modern hunter-gatherers, such as the San of the Kalahari or the aborigines of Australia, use wooden digging sticks to exploit a wide range of deeply buried underground vegetable materials, to excavate burrowing animals, to expose partially buried blocks of stone to be used as raw material, and to dig for water.

Edible roots and tubers dug up in Tanzania. The use of a simple digging stick would have greatly facilitated early hominid exploitation of these plentiful and important resources, and stone or wooden scrapers could have helped clean the grit from their exteriors.

The earliest archaeological traces of wooden digging devices are, not unexpectedly, quite recent, since this highly perishable raw material tends not to survive in the archaeological record. At Gwisho Springs, Zambia, however, amazingly well-preserved plant remains were found in waterlogged levels of a hot spring's deposit, dated to around thirty-five hundred years ago. There, archaeologists Brian Fagan and Francis Van Noten found preserved wooden digging sticks that are virtually indistinguishable from those used by the San.

Of course, materials other than wood also could have been used for digging. What is needed is a stick- or shovellike device made out of some hard substance, which might help pierce and break up or scoop out earth: simple broken shafts of long bones naturally or artificially shaped to a point at an end; pointed natural rocks, including stalactites and stalagmites; juvenile elephant tusks; bark or wood; animal scapulae (shoulder blades); and so on. Experiments digging with antelope horns proved very successful, especially for piercing and breaking up the earth, and other animal horns could also serve quite well.

Many of these materials are quite perishable, and many would not require much modification, so they would not be very visible archaeologically unless heavily worn and polished at one end. We have already mentioned such pointed bone fragments found at the caves of Swartkrans and Sterkfontein, which C. K. Brain has proposed as early bone tools used for underground vegetation.

WOODWORKING

For early tool-using hominids, wood would have been one of the most abundant raw materials available in the grassland-woodland environments of prehistoric Africa. Wood was probably difficult to modify into many desirable tool shapes with bare hands and teeth, so the use of simple stone tools would have been a great asset.

Wild chimpanzees have a limited set of wooden tools in their technological repertoire, mainly termite-hunting sticks, nut-cracking hammers or anvils, or branches used as clubs or missiles. The termite-hunting sticks are carefully modified to a desired shape with the hands, and the nut-cracking tools are occasionally shortened to a desired length by stepping on a branch. Chimpanzees also commonly use branches for the sleeping nests they build for overnight stays in relative safety high up in trees. They commonly twine live branches into roughly woven nests (in a sense, primitive basketry).

Among hunter-gatherers, wood is one of the most popular materials for tools, and it is used for very diverse purposes. In addition to sticks used for digging, these tools include spears and arrows, atlatls (spear throwers), various containers (cups, bowls, trays) of wood or even bark, and probes for termites or bone marrow. In addition, shelters are often fashioned out of saplings and branches.

For a sturdy digging stick or spear, most branches that have naturally fallen from trees are too old, brittle, and weak to serve very effectively; ideally a fresh tree limb is much better. There are many steps in working wood, and we have tried each of these with the early types of stone tools that we have made.

Chopping a tree sapling with a lava chopper to make a wooden shaft. This task takes only a few minutes with such a simple tool.

Chopping

To sever a branch from a hardwood tree such as an acacia or grewia, a heavy, relatively sharp-edged stone tool works best. In our experiments a stone tool weighing at least four hundred grams (almost a pound) and with a working edge of under eighty degrees tended to be the best for such heavy-duty chopping. Large flakes or cores are excellent tools in this regard. Once a sizable branch has been chopped from a tree (say, one inch thick, with a length of five feet), it is ready for shaping into whatever form is desired. Primary shaping may be done with heavy cores or flakes by chopping, but finer work is usually done with scrapers.

With a stone flake, a wooden branch is slowly honed to a sharp point by scraping. Such an implement could have been used as a formidable spear, digging stick, or skewer for carrying meat. Final shaping was done by grinding the point against a rough rock. The use of one tool to make another tool is one more human characteristic.

Scraping

To shape and form the branch into a usable tool—say, a digging stick with a stout-pointed tip—it's best to scrape it down with a fairly sharp edge, either an unmodified flake, a flake scraper, or a large core form such as a steep scraper.

Hollowing

A variety of stone tools can be used to hollow out a large piece of wood, which can serve as a container or water collector. Especially useful are heavy, pointed forms, used as gouges, and smaller retouched flakes, to scrape away the inner wood.

Finishing

To smooth and even out the final surface of a spear or digging stick, a technique that works very well is to grind it back and forth against a piece of gritty stone, such as an outcrop of sandstone bedrock.

Although wood generally does not preserve well in the ancient archaeological record, we can find indirect clues of the use of wood in various ways. Sometimes a tool seems to have been made to mount on a handle (in recent prehistory and in modern times usually made out of wood or bone). We don't see any good evidence of this, however, until much later, only within the last hundred thousand or so years. Another sign is a distinctive alteration, seen through a microscope, of the edge of stone artifacts used to work wood. On 1.5-million-year-old flakes from Koobi Fora, Lawrence Keeley from the University of Illinois identified signs that suggested they had been used for scraping wood. Making what, we don't know—possibly some other tool.

HIDE WORKING

The skins of animals would have offered tremendous technological possibilities for prehistoric peoples. Many hunter-gatherer groups make extensive use of cured animal skins for clothing, blankets, bedding, carrying devices, water containers, and shelters. But this is not universal: even in recent times some groups, such as some Australian aborigines, made little or no use of such animal skins (although they had a wide range of other materials they used for clothing and containers).

The physical evidence for hide working in early prehistoric times tends to be indirect, since animal skins rarely survive in the prehistoric archaeological record and only under very special circumstances. In the African grasslands today, one sometimes comes across the remnants of large animal carcasses still bearing dried skin, especially in the area of the rib cage. It is conceivable that early Stone Age hominids made use of such materials for carrying devices or other activities, but such impromptu use would leave little evidence for the archaeologist. Remains of animal skins made into clothing or containers have sometimes been found in environments such as salt mines, anaerobic bogs, frozen Siberian tombs, the high Alps, and very dry cave deposits from the last several thousand years, but none go back farther than this. Farther back, approximately twenty-five thousand years ago, there are some specimens of Ice Age art that suggest parka-like clothing, and bone needles from that time also imply the stitching of some material, in all probability animal skins.

But in the very remote Stone Age past, our primary evidence for hide working comes from Lower Paleolithic sites in Europe, the earliest about three hundred thousand years ago, where stone artifact edges show the telltale microscopic wear pattern of highly rounded, rough, and pitted edges with scratches perpendicular to the tool edge, indicating a scraping motion. In the earliest Stone Age, we have no direct evidence of the working of hides. Once a skin is cut away from an animal's body, it can be worked with stone tools, ideally flakes or retouched flake scrapers, to remove adhering meat and fat, which will normally make the skin go rancid. It is usually easier to remove adhering tissues by staking the skin on the ground with the hair side down

An animal skin is scraped of its relict meat and fat with a stone scraper. The sun-dried skin is stiff rawhide and could be used as a carrying device or as roofing material for a simple hut. If the hide was cured with animal brains or other substances, it could then be softened to serve as clothing.

and allowing it to dry in the sun. A sun-dried skin (rawhide) can serve as a stiff carrying device, but to make a softer, more pliable material that can be used as clothing, usually some chemical tanning agent (such as brains, urine, ash, and water) must be used. Whether Oldowan hominids had such technologies is unknown, but it seems more likely that a strong reliance upon clothing for warmth was especially correlated to hominid movements into more temperate parts of the world later in time.

ANIMAL BUTCHERY

The rapid, efficient exploitation of relatively fresh animal carcasses through scavenging or hunting would have been an important addition to the early hominid diet. Stone tools would have provided a critical ability to rapidly disarticulate and consume such animal foodstuffs.

In butchering animals, modern hunters normally will first slit the hide on an animal, remove the major organs from the body cavity (intestines, stomach, heart, lungs, trachea), and then remove the skin before major disarticulation takes place. There is no certainty that early hominids, if faced with a relatively complete carcass, would have done the same; they may have disarticulated major body parts with the skin still on. At times they may have even "eaten as they went" from the whole carcass, more or less as modern carnivores do, but in this case with the assistance of stone tools.

For most aspects of butchery, fairly sharp cutting tools are required. In each of the butcheries we have done (testing different Oldowan artifacts), one tool proved superior for slitting the skin of the animal: a simple, sharp flake. In fact, a small lava flake proved adequate even when presented with a positively enormous hide-slitting job (as we'll see shortly). We have had the opportunity to test out stone artifacts in butchering a variety of wild and domestic species that have died of natural causes or have been slaughtered by traditional techniques for human consumption. These include elephant, oryx, topi, zebra, wildebeest, cow, horse, sheep, goat, pig, and deer. Although the primary activity consists of cutting, different phases entail slightly different motions with the tool. Also, animals can differ somewhat in the thickness and toughness of their skin and coarseness of their hair, calling for sharper edges or more force in the cutting action.

Hide Cutting and Gutting

Hide cutting is a crucial first step in consuming a large mammal, whose skin tends to be tough, often covered with thick, coarse hair, and very dirty and gritty as well. Considering the anatomy of early Stone Age hominids, it seems likely they would not have been very efficient at tearing through the skin of a larger mammal with their bare teeth and brute strength (even though they were probably considerably stronger

The role of rock: slitting the hide of a wildebeest with a lava flake. This photo demonstrates what we think is the principal reason for the emergence of flaked stone tools: rapid, efficient processing of animal carcasses for transport and consumption with the use of flaked stone cutting edges. (All animals used in these experiments in this research had died of natural causes.)

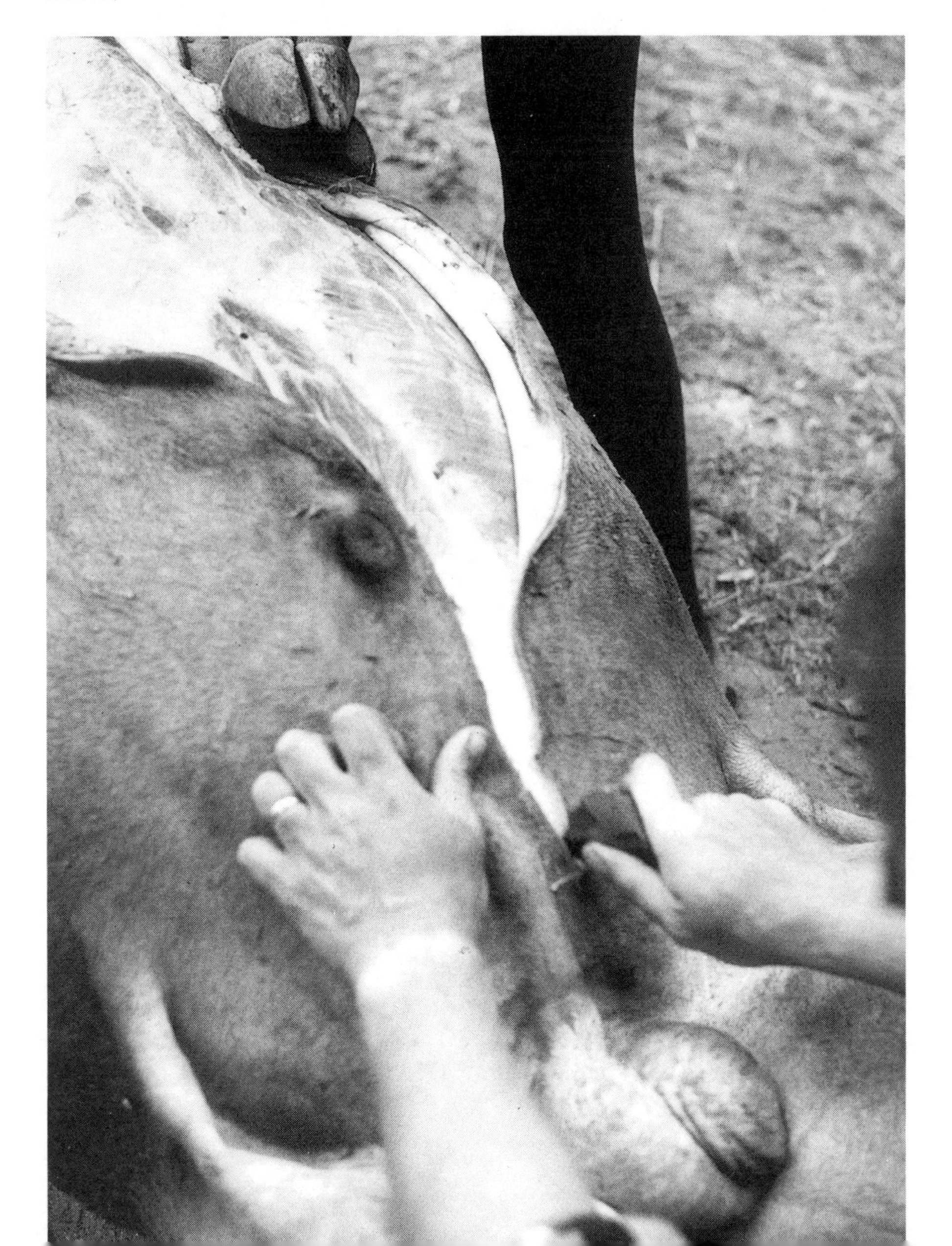

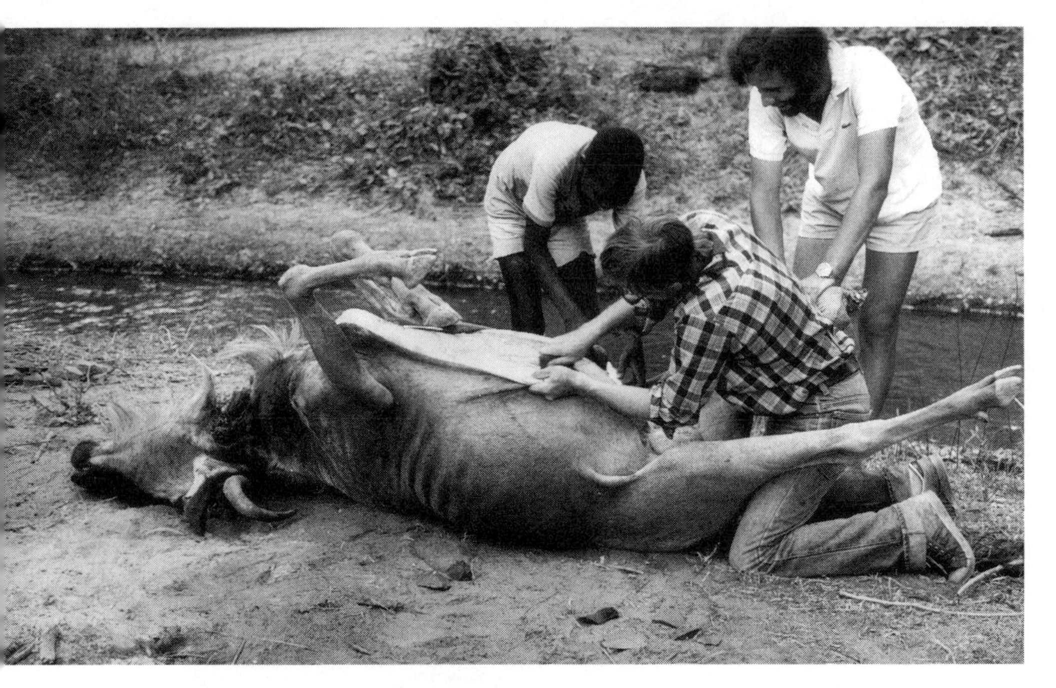

Using stone tools to butcher a wildebeest that died of natural causes.

than most modern humans). When chimpanzees manage to hunt and kill small animals such as monkeys, bush pigs, or small antelope, they will tenaciously munch through the entire carcass—skin, bones, and all (as some people do with chicken wings). They do not hunt larger animals, but if they could and did, they would be faced with problems of a different magnitude. Armed with a simple Oldowan flake, however, hide slitting and removing viscera are easily accomplished.

Filleting (Defleshing)

For this function (removing chunks of meat or muscle units from the bones of the animal), a sharp cutting edge is again favored and a simple, unmodified, sharp flake proves to be a superior tool. A sharply angled retouched flake can serve equally well, as long as the edge is not too steep. A flake with one relatively long sharp edge, straight or even slightly convex, serves as an excellent cutting tool. (With practice one learns to choose a flake with a relatively blunt side opposite the cutting edge, so that the tool doesn't also slice into the meat of the hand! One can also blunt this edge by flaking it steeply or else holding it in a wad of skin or leather.) The flake, when used either in a slicing or sawing

Cutting the meat away from a leg bone with a stone knife. Notice the cut marks produced on the shaft of the femur. Such marks have been identified on fossil animal bones from early Stone Age sites showing that hominids processed animal carcasses, but how they obtained such body parts, and in what condition they were obtained, are hotly debated issues in archaeology today.

motion, cuts very easily through the flesh and also severs the ligaments attaching muscles to the bone. The edge dulls in five or ten minutes of use, so it needs to be replaced or resharpened by removing small flakes to create a scraper. Such a retouched edge is normally not quite as sharp as an unmodified flake edge, but it tends to last longer.

Dismembering

This activity (removing parts of the animal carcass as individual units) usually entails disjointing the animal where different bones are held together by muscles, ligaments, tendons, or cartilage. For this function, flakes again won the prize among Oldowan artifacts, though for heavier-duty cutting and chopping at tough joints a sharp-angled core, such as a chopper or discoid, was useful as well.

How to Carve an Elephant

Are the possibilities endless for these cutting tools? How far might hominids have taken this technology? A recurrent pattern at some of the world's oldest archaeological sites is the presence of carcasses of "megafauna" (elephants, hippopotamuses) along with simple stone artifacts. For instance, two archaeological sites from Olduvai Gorge in Tanzania have yielded elephant skeletons (an *Elephas recki* in Bed I and a *Deinotherium*, an extinct form with downward curving, digging tusks, in lower Bed II). At each of these sites, dated between 1.7 and 1.5 million years ago, an array of Oldowan artifacts was found. At Koobi Fora in northern Kenya as well, parts of a hippopotamus carcass dating to 1.9 million years ago were found with simple stone tools.

Two major questions have emanated from such evidence. First of all, is there a *causal* relationship between these giant animal bones and these crude stone artifacts, or are these coincidental associations, places where humans discarded artifacts and these large mammals died independently, perhaps near the edge of a drinking spot along a watercourse or under the cool shade of a stand of trees? Second, if there could be cause or meaning connecting these stone artifacts and these pachyderms, what possible role could the stones have played? How could the world's simplest stone tools be used to process meat from animals weighing many thousands of pounds, with skins that can be over an inch thick?

We have had two opportunities to put stone artifacts to the ultimate test—to butcher elephants (which had died of natural causes). Somewhat daunted, we approached our task equipped with simple lava and flint flakes and cores, which looked more and more paltry as we got closer to the impressive body. Initially, the sight of a twelve-thousand-pound animal carcass the size of a Winnebago can be quite intimidating—where do you start? We had never seen a field manual on pachyderm butchery, and they aren't like smaller animals: you cannot move the body around (for instance, flip it over to get a better vantage) without heavy power machinery. You have to play the carcass where it lies. In fact, once the upper side has been filleted, most of the lower half remains untouched and almost inaccessible unless you dismantle the skeleton of the animal, a positively arduous task.

Butchering the world's largest terrestrial mammal with the world's simplest flaked stone technology: using a stone flake, Kathy Schick and Ray Dezzani cut through the one-inch-thick hide of an African elephant that died of natural causes. Most scavengers cannot or will not attack a fresh elephant carcass, so early Stone Age hominids might have moved into an "open niche" through the exploitation of such resources. To get a sense of the task required, imagine cutting through a car tire with a razor blade.

Despite the success of our tools in dozens of other butcheries, we were not really sure they were up to this task. We were amazed, however, as a small lava flake sliced through the steel gray skin, about one inch thick, exposing enormous quantities of rich, red elephant meat inside. After breaching this critical barrier, removing flesh proved to be reasonably simple, although the enormous bones and muscles of these animals have very tough, thick tendons and ligaments, another challenge met successfully by our stone tools.

No beast too large. The use of early types of stone tools in the experimental butchery of an elephant that died of natural causes. Such carcasses could have provided occasional huge "bonanzas" of meat to hominid groups that possessed flaked stone cutting edges. Nicholas Toth defleshes the carcass while Jack Fisher, now at Montana State University, loads meat into canisters for weighing.

Throughout these and many other butcheries, our tools soon became strewn around the carcass, as we used one for slitting here, another for filleting over there, and another for hacking at a tough muscle attachment. It was always simple enough to grab another tool or knock another flake off a nearby core. It was easy to see how artifacts would become lost and engulfed in the task, to be left behind with any animal remains ultimately abandoned.

Although we feel that the butchery of such pachyderms was probably a rare event in the early Stone Age, and probably resulted from scavenging rather than hunting, such experiments demonstrate that the simplest stone technologies can be used to process even the largest terrestrial mammals. Cut marks from stone tools have been found on some *Elephas* bones at Olduvai, which indicate that stone tools had some causal relationship with the carcass. Since modern scavengers normally do not eat a dead elephant until it has decomposed for several days, such carcasses may have provided occasional bonanzas for early Stone Age hominids, at least until chased away by larger scavenging social carnivores.

Simple Flakes, Powerful Tools

In many considerations of Oldowan tools, the flake has been treated simply as waste, something removed in order to make the presumed core tool. But our experiments in butchery drive home an important realization: with a simple flake a hominid could open up a whole world of possibilities.

BONE BREAKING

In our own culture, consumption of significant amounts of animal fats has become anathema to most people, due to strong association with the growing problems of obesity, cholesterol buildup and heart disease, and various cancers. But this is not the case among modern hunter-gatherers, nor was it likely the case in our prehistoric past.

In *The Paleolithic Prescription*, authors Eaton, Shostak, and Konner constrast our genetic adaptation to a hunter-gatherer diet in our Paleolithic past with the gross excesses, abuses, and strange new aspects of life, diet, and health in modern society. Among modern African hunter-gatherers, animal fats are considered a highly prized food. Their typical diet is basically high in complex carbohydrates, fiber, moderate in proteins, but low in fat (wild game meat contains about one-seventh the total fat of supermarket beef and several times more polyunsaturated fat). In fact, fat is in fairly short supply among those living off the land. The vital addition to the diet of bone marrow, which is predominantly fat, can help balance the protein-to-fat ratio (a minimum fat intake must be maintained in order to process protein properly) and serve as a great source of energy as well.

It is likely that finding marrow through scavenging or hunting was important for our hominid ancestors at this time. Nonhuman predators and scavengers, when feasting on an animal carcass, must decide instinctively whether the benefit of crunching a long bone to get into the inner marrow is worth the cost of serious potential damage and wear to the teeth. Lions, for instance, will often leave unbroken the larger bones of a kill, while hyenas, nature's ultimate bone-breaking machines, tend to be much more efficient at fragmenting bones, swallowing them, and digesting their fat contents and either regurgitating or passing the bone fragments through their digestive tract. Each carnivorous species has its limit: a maximum size and strength above which bones become difficult or impossible to crack.

For early stone-wielding hominids, however, these biological constraints on bone breaking could easily have been bypassed. In our experiments—breaking open mammalian bones to expose the inner marrow contents—a simple, fist-size cobble usually did the trick within a few blows, even on a large antelope bone. It's usually best to strike the bone in the midshaft area where it's weakest, bracing it against a hard, fairly nonyielding anvil such as a rock or wooden stump. Usually an incipient fracture grows and often spirals along much of the length of the bone shaft, and the bone can be twisted apart. The inner reaches of the marrow cavity can be accessed either through more blows along the shaft or by probing with a stick or bone splinter. An elephant limb bone, however, presents a major challenge: we found it necessary to throw a huge boulder repeatedly against an elephant thighbone before it finally cracked and exposed the oily, runny marrow. There did not seem to be a definite advantage offered by a flaked cobble or core tool over an unmodified cobble in cracking these bones, but many cores and core tools could certainly have been used.

OFFENSE AND DEFENSE

Hominids would likely have had ample need for some forms of defense and offense on the African savanna, whether against the attacks of various predators, other groups of their own species, or other hominid species. There was a fantastic variety of carnivores roaming about,

Cracking a limb bone with a stone chopper and anvil to expose the edible marrow. Modern hunter-gatherers prize marrow as a source of fat and energy. During the early Stone Age, marrow processing of carcasses obtained by scavenging or hunting would have been greatly facilitated by the use of simple stone tools.

some of them extremely large and presumably formidable. These included various cats (leopards, some large lions, saber-tooths—who killed by driving their upper canines powerfully into their prey—and false saber-tooths), hyenas (including now extinct groups of hunting hyenas well adapted to running, and ancestors and relatives of the modern spotted, striped, and brown hyenas, some of these very large as well), hunting dogs, jackals, foxes, and other smaller carnivores.

As a means of offense or intimidation against other animals, there may have been some ancient, ancestral forms of display and threat, such as opening the mouth wide and displaying their sharp anterior teeth, rushing toward the enemy, hooting and calling, or waving arms about, perhaps accentuating these gestures with a stick or branch in hand. Brute strength could have been applied, with the larger, stronger individuals attempting to thrash and beat an opponent. Of course, a few well-thrown cobbles hurled at any opponent would have been a definite asset. Any aggressive displays against carnivores (presumably not the largest ones), perhaps in competition for carcasses, might have employed much the same assortment of displays and assaults, most effectively by a large group of hominids, but would probably have required much more consistent application of real force, such as throwing rocks, wielding clubs, thrusting spears, and so on.

For defense, similar weapons and threats could have been useful, at least for a time, but these may have been supplemented with the use of sharp horns, biting, and the sheer strength of these small but powerfully built hominids.

During the earliest Stone Age, however, there is little evidence for actual weapons used for either offense or defense. It is not until much later in time, approximately four hundred thousand years ago, at the British site of Clacton, that we are finally able to find an artifact that is almost certainly a weapon: a shaped wooden artifact with a pointed tip that appears to be a spear.

CONTAINERS, CARRYING DEVICES, AND SUNDRY TOOLS

There is a great variety of natural materials and objects available on the African landscape that might have been used as containers or means of carrying things such as food, water, or stone tools. In the case of consumable materials, such containers and carrying devices would in essence have served as artificial stomachs for early hominids. Although such devices are universal among modern human societies, they are not seen in the nonhuman, animal world. Many of these could have been used as they are found, though some would require

Fracturing a skull with a stone chopper to obtain the animal's edible brains. With a simple stone technology, early Stone Age hominids—even without the powerful bone-crushing jaws of large carnivores—could have had easy access to such foods as brains and marrow.

fashioning to be effective and useful. Ostrich eggshells are excellent for the storage or transport of water, fats and oils, honey, and other materials; tortoise shells could also have served as containers and bowls of sorts. Pieces of wood could have been hollowed out with stone tools, and unfired clay could have been used as well. Large pieces of bark and hardened pads from the base of elephant feet could also have served as trays or bowls. The possible use of skins for carrying bags has already been mentioned. A shaped branch can also serve as an effective skewer for carrying large quantities of animal meat, either on or off the bone.

THE "SMOKING BARRELS" OF THE EARLY STONE AGE

These ideas and experiments put us squarely and graphically within the realm of possibilities: in the early Stone Age what tasks might have required doing, and what materials were on hand to use? It is all very well and good to conduct experiments that test the relative efficiency of different sorts of Stone Age implements for a range of hypothetical tasks that our ancestors may have accomplished, and observations of tool use among modern nonindustrial humans and nonhuman primates can also give us invaluable ideas of the range of activities for which early hominids may have used stone tools. But what is the hard evidence for such stone tool use in the prehistoric past?

An unlikely container: Glynn Isaac and a Tanzanian colleague hold the footpads from an elephant carcass. Such items, as well as ostrich eggshells, gourds, tortoise shells, bark trays, and animal hides, could have served as carrying devices.

MICROWEAR ANALYSIS

Did choppers actually chop, did scrapers really scrape? How can we archaeologists, hundreds or thousands or millions of years later, figure out what a prehistoric tool had actually been used for? During the 1970s a major methodological breakthrough was made in Paleolithic archaeology. Archaeologist Lawrence Keeley began to examine microscopically the edges of stone tools that had been used experimentally

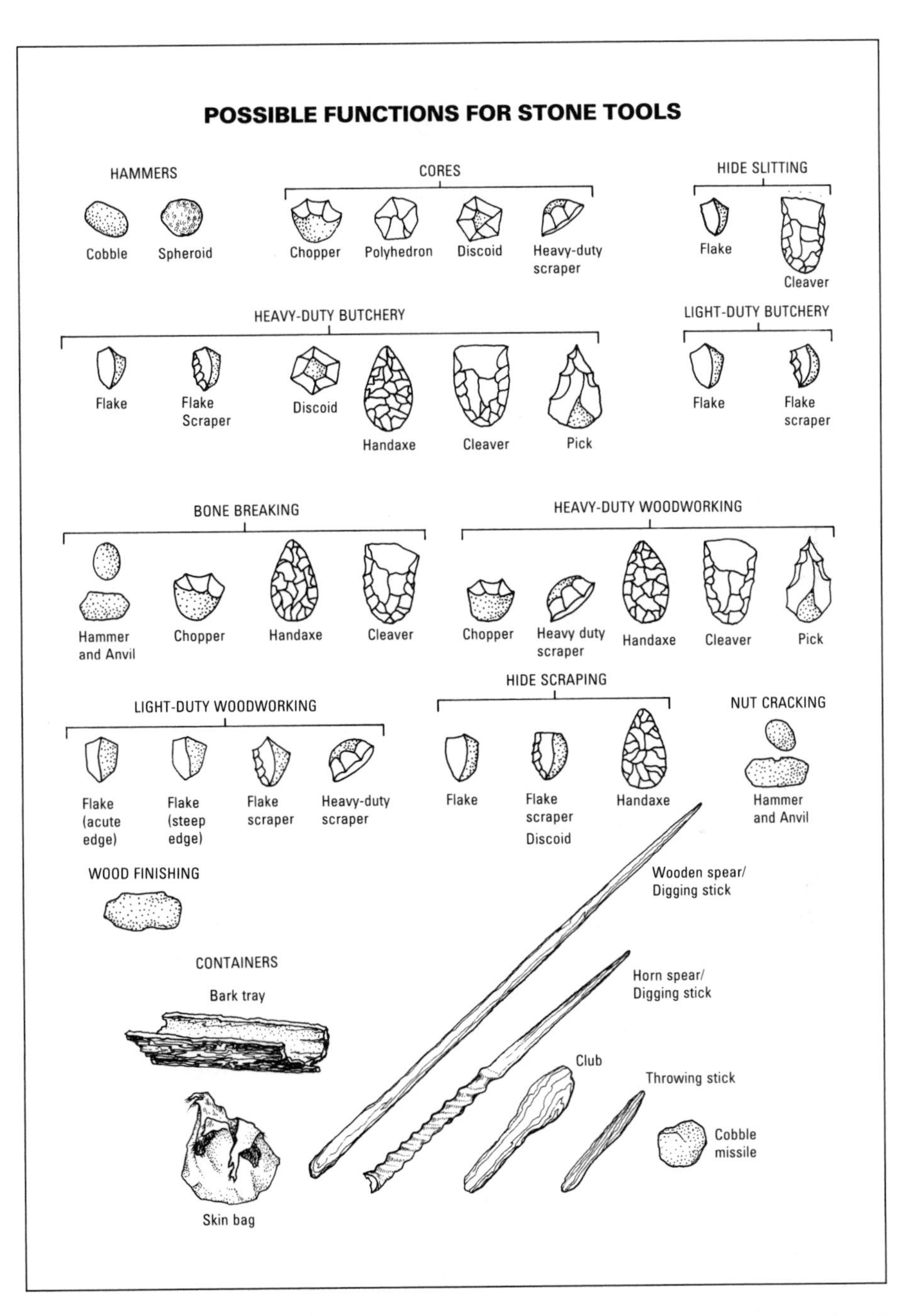

Tools of their trade. The potential functions of Oldowan and early Acheulean stone tools, based upon feasibility experiments. Such tools could have expanded the range of foods that could potentially be exploited, decreased the amount of time required to process them, and increased the amount of food that could be carried by early Stone Age hominids, all important adaptive strategies.

for various tasks. This approach, after pioneering work by Russian archaeologist Sergei Semenov in the 1950s, was finally systematically developed and refined by Keeley, who found that after a stone tool has been used for a significant amount of time, it retains microscopic traces on its edge. Such evidence includes the following:

1. Edge damage: microscopic chipping and snapping of a stone tool's edges (a higher degree of edge damage normally indicating use on a harder material);
2. Striations: scratches on the stone implement, often a sign of the direction of use (striations parallel to the edge of a tool show cutting or sawing activities, those perpendicular to the edge show scraping, planing, or chopping);
3. Polish: actually microscopic alteration of the stone surface of an implement, which varies in brightness, luster, roughness, and pitting according to the material worked (hide, wood, meat, bone, antler, soft plants, and so on).

Since many of the diagnostic properties of use polishes were *optical* in nature, it was found that a high-powered optical light microscope was generally much more informative, at the present state of knowledge, than a higher-powered scanning electron microscope, which uses a beam of electrons to scan surface details.

After we became involved in the Koobi Fora project, a real opportunity to test the limits of this technique *and* test our early stone tools presented itself. Although Keeley had detected use traces on British stone artifacts going back to perhaps three hundred thousand years ago, no attempt had been made on artifacts going back to the Oldowan. So we decided to see if we could find use-wear traces on some of the early Stone Age material from this important site area.

Unfortunately, most of the tools in the early Stone Age are made from materials too rough and coarse to show this microwear: a smooth, fine-grained, silica-rich rock is best. A small number of the artifacts at Koobi Fora, however, are made from small cobbles of glassy materials such as chert. We targeted about fifty of these freshest specimens for study, and Keeley managed to find microwear traces on nine of them. What functions did we see? Quite a variety, especially for such a small sample and so long ago: four tools seemed to have been used for cutting meat, two for cutting soft plants, and three for working wood, all from sites about 1.5 million years old. This sample is small but nevertheless intriguing and suggests that a variety of tool-using activities was taking place in the vicinity of Oldowan sites.

Examining microwear traces on stone artifacts, using the microscopic methods developed by Lawrence Keeley of the University of Illinois, Chicago. A high-powered metallurgical microscope can help identify polishes and striations produced from different activities such as hide working, woodworking, and animal butchery.

STUDIES OF BONE MODIFICATION

Bones can yield important telltale marks and patterns of the agents that modified them, whether hominids with a stone tool or carnivores with powerful teeth. During the past two decades an appreciable amount of actualistic and prehistoric research has been devoted to examining various traces of bone modification.

Cut Marks

The scanning electron microscope has become a valuable tool for analyzing marks made by tools during the butchery of an animal. With an optical microscope, the depth of field (the amount of the image in focus) becomes very shallow when magnifying something over one hundred times, so that much of the image observed on a cut-marked bone is blurred. With a scanning electron microscope, though, using

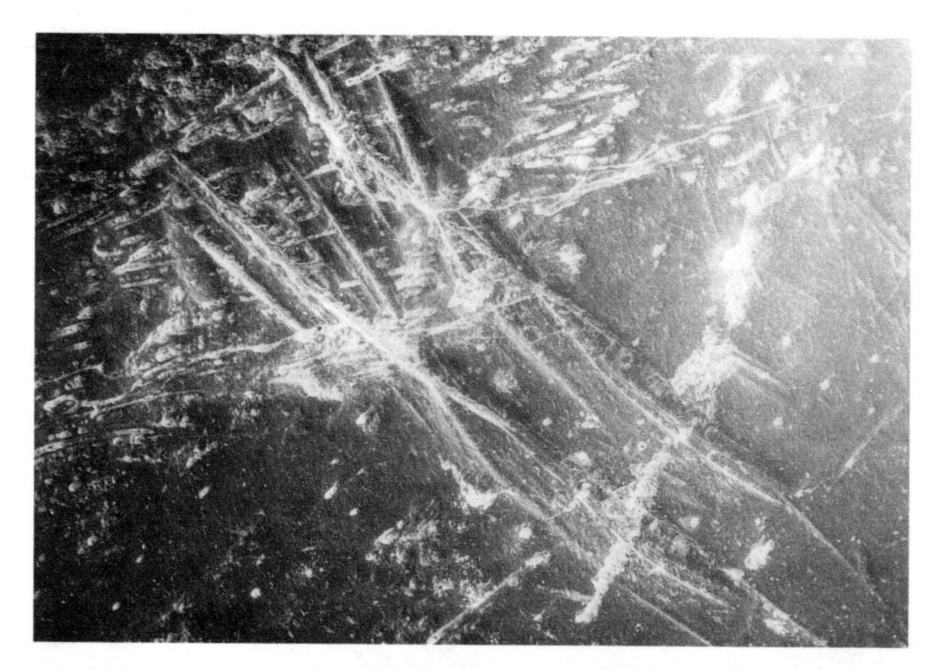

Hominid calling card. Scanning electron microscope picture of prehistoric cut marks made by a stone tool and found on a mammal bone from a million-year-old Stone Age site. Cut marks tend to be narrower than tooth marks, with multiple striations in the main groove. Although most cut marks appear as single grooves, irregularities on the edge of the stone knife that made these marks (probably a retouched flake) have produced a characteristic X-shaped "signature" from this tool when it cut through the bone, repeated twice on this fossil specimen. Magnification approximately 14x.

A carnivore in action. Kodiak, the authors' 110-pound Alaskan Malamute, demonstrates how tooth marks are produced on a limb bone as "guest lecturer" in an introductory human origins class.

electrons rather than light to create an image, there is an almost unlimited depth of field, so the entire image is held in focus. This greatly enhances our ability to detect and decipher evidence of what ultimately happened to an animal. Studies have shown that cutmarks from stone tools tend to be narrow, often V-shaped in cross section, with multiple, parallel scratches within the main groove. Cut marks also often occur in sets at areas along the bone where skinning, jointing, or defleshing would have been concentrated.

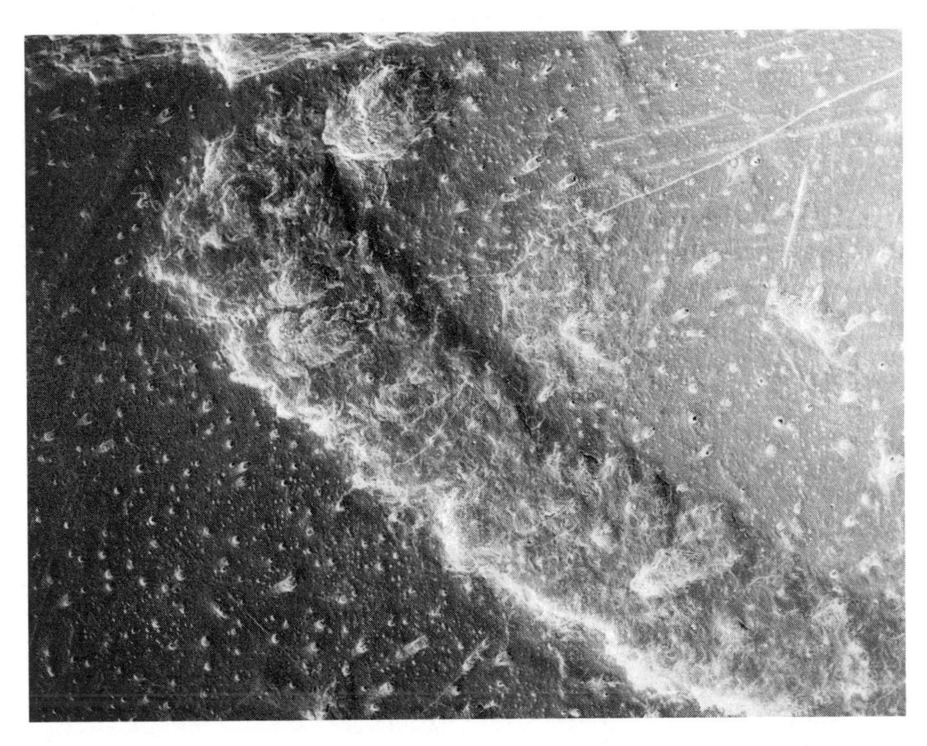

Carnivore calling card. Scanning electron microscope picture of a carnivore tooth mark on another bone from an early Stone Age site dating to approximately one million years ago. Tooth marks tend to be smoother and broader relative to depth. Magnification approximately 14x.

Marks of Carnivores and Other Natural Agents

Do the scratches and striations on bones found at early archaeological sites represent signs of animal butchery by early hominids? Not always. There are other animals with designs on these bones and other agencies that intrude on them before they come to rest in the prehistoric record. Not all carnivores are obsessed with bones—most cats, in particular, have specialized deboning teeth that are not well suited to the really heavy-duty bone crunching we see, for instance, among many hyenas. But whether they are after mainly the meat or the whole flesh and bones of an animal, all carnivores can leave a variety of signs on the bones. It is fairly easy to identify grooves made by the teeth of carnivores as they gnawed the flesh and chewed on the bones of these animals: they produce much broader, shallower striations than the slashes made by stone tools, along with pits made by their canines. The small rectangular sets of parallel grooves left by rodent gnawing is also distinctive, easily distinguished from both a tool's cut marks and a carnivore's toothmarks. Scratches can also be produced on bones as they lay on the ground, trampled by animals or worked by churning sands as floods swept over the site. As animals may carry sharp bits of rocks adhering to their hooves (angular sand grains and such), trampling can produce marks on bones that mimic cut marks from stone tools. Paleontologists Kay Behrensmeyer from the Smithsonian Institution and Peter Andrews from the British Museum of Natural History have each shown that such marks can occur on modern bones in natural, nonhuman settings and may be microscopically indistinguishable from shallow cut marks from stone tools.

But such marks should not be confused with clear tool cuts made in removing meat from a bone. First of all, we should be naturally suspicious of bones contained within sandy sediments: the geological forces that buried the bones could also have abraded them. Furthermore, scratches by trampling and abrading sediments may individually imitate marks made by stone tools, but in random ways that do not show the functional patterning produced in defleshing or disarticulating an animal. Tool cut marks tend to concentrate in more or less parallel sets at places where muscles attach to the bones, where bones connect to one another, or where skin is cut away.

BONE FRACTURE PATTERNS

The archaeological visibility of bone fracture patterns is fairly clear and unmistakable to anyone who has bashed open many bones. Studies by archaeologists Henry Bunn of the University of Wisconsin, Richard Potts of the Smithsonian, and Rob Blumenschine of Rutgers University have itemized these clearly: pits and percussion marks on the upper- and under-surfaces of the bone (from the force of the hammer blow and the underlying anvil stone), sometimes repeated in a line along the shaft and sometimes with bone flakes knocked off (the flake scars can be seen in the inner walls of the bone). Scratches from a hammer stone can sometimes be seen microscopically as well. Bone that is well fossilized and in good condition at archaeological sites presents a very distinctive, telltale pattern for archaeologists to read.

STUDIES OF ORGANIC RESIDUES ON STONE TOOLS

Most modern archaeologists are part social scientist, concerned with human culture and its roles in our adaptation, and part physical scientist, employing myriad new analytical techniques to examine prehistoric materials. One of the promising new approaches to prehistoric materials is the study of organic residues preserved in the archaeological record. Archaeologist Thomas Loy ventured into these largely uncharted waters several years ago with his study of possible hemoglobin (a blood protein) residues on the edges of prehistoric tools at North American sites dated to several thousand years ago, and more recently with a study of prehistoric Australian artifacts.

The essential molecules of life—genetic molecules (DNA)—could finally prove to be one of the richest sources of evidence of the functions of stone tools. Incredible as it sounds, during the past few years scientists have been able to identify very ancient traces of DNA under special conditions of preservation. Thus far, the oldest evidence is about 15 million years old, from the chloroplast structures of ancient magnolia leaves so well preserved in mudstone that they still retained their green color. Other identifiable ancient DNA has included a thirteen-thousand-year-old ground sloth and eight-thousand-year-old brain tissue from Paleo-Indian skulls in Florida. Thus far, it would appear that the ideal conditions for the preservation of DNA are cold, anaerobic environments (with little or no oxygen) at the time of burial.

This archaeological approach is still in its infancy, but it has tremendous potential for allowing prehistorians to identify organic residues that have adhered to the surfaces of prehistoric stone tools. This would give us a very direct, dramatic link between stone artifacts and their actual uses in the prehistoric past. One of our present graduate students, Bruce Hardy, has embarked upon a dissertation project to identify prehistoric organic residues on stone tools.

THE POSSIBLE SIGNIFICANCE OF MEAT EATING

How might these early tools show a new beginning, a new type of adaptation for hominids? Might they show a dietary shift in the new *Homo* species that included more animal resources, through small-scale hunting or the scavenging of fresh carcasses? Wouldn't such an adaptation involve a premium on tools for skinning, dismembering, defleshing and processing the marrow of these carcasses? Anatomically these hominids were almost as ill-equipped in their essential biology as are modern humans for hunting and meat eating. Hominids lack the large canine teeth for stabbing and pinning down prey, and the shearing carnassial teeth for ripping through meat, hide, and gristle. Our jaws and teeth are better-suited to grinding foodstuffs than to slicing through skin or meat and crushing bones. But adding the use of even simple stone tools to our behavioral repertoire would have greatly enhanced our ability to process animal carcasses—to cut them open and carve them up and to break open marrow-rich bones.

Among mammals, primates, and especially apes, are known for their relatively high levels of intelligence. High intelligence has also been selected for in the carnivores. It may have been that, as an already fairly intelligent hominid began to move into niches occupied by carnivores through the use of stone tools, this brought about selection for even higher thresholds of intellect in that primate species. As natural selection favored those bipedal primates that were competent in gaining access to animal food resources either through predation or competitive scavenging—strong in predicting, stalking, pursuing, and even cooperating among themselves and able to deflesh or dismember a carcass quickly and efficiently with their tools—this would have effectively added an intelligent predator-scavenger adaptation on top of the already intelligent primate substrate.

In the next chapter we will explore some of the competing ideas about what these early sites mean, what they signify about hominid behaviors. For the moment we will highlight the pressing issue we face: stone tools and fossil animal bones are so commonly found together at early archaeological sites that we must deal with the question *why*. Most archaeologists agree that this cannot be mere coincidence, that our ancestors were somehow involved doing *something* with animal carcasses or parts. We believe there is good evidence that stone tools became important for our success largely because of the ability they gave us to add significant animal foods to our diet.

TECHNOLOGY AND HOMINID NICHE EXPANSION: TECHNO-ORGANIC EVOLUTION

Between three and two million years ago some early hominids were beginning to develop an evolutionary trajectory that had never before occurred in the history of life: the production and use of a flaked stone technology. This would have increased their diet breadth, especially during times of environmental stress, and increased their reproductive fitness and, ultimately, their evolutionary success relative to other hominid groups. As modest as these early stone technologies were, they nevertheless herald a new direction in evolution, one that stressed the development of technology to supplement the biological capabilities of these early hominids. Through experiments we have been able to explore the possibilities these tools may have opened up for the early hominids, and through minute examination of specific scraps of prehistoric evidence we can see positive, tangible signs of what they were doing.

Although it is quite likely that the earliest of these stone tool makers were, at first, not fully *dependent* upon lithic technologies for their survival, they were slowly able to exploit certain resources *more efficiently* than hominids that did not possess a flaked stone technology. Although other animals have, through instinct or learning, also incorporated some use of tools in their behavior, these early hominids were at an evolutionary turning point in which the open-ended, nonstereotypic use of technology became a recurrent facet of many of their activities and ultimately became *essential* to their adaptation.

It is not difficult to appreciate how tools could have broadened the hominid niche appreciably and strengthened their hold on long-standing food resources. Consider how our experiments reveal possible advantages that tools might have conferred on the hominids, *expanding their niche multidimensionally*, giving or enhancing capabilities of many different, even more specialized animals:

1. Stones thrown as missiles, or stones, bones, or wood used as clubs or simple spears, would have had a devastating effect on smaller mammals (overlapping the niche of smaller carnivores such as jackals, hunting dogs, and some cats).
2. The use of stone-cutting tools in the processing of animal carcasses would have enabled hominids to cut open larger, thick-skinned animals, to rapidly dismember a large animal carcass into transportable units, and also to cut meat away from the bone (giving us capabilities of the carnassial or meat-shearing teeth of some larger carnivores such as lions, hyenas, and leopards). In allowing them to process even large pachyderms, these tools would have helped hominids to exceed the normal bounds of most present-day carnivores, which generally must wait until a carcass is sufficiently putrefied before they can successfully gain

entrance (although hyenas have sometimes been able to enter through the anus of a putrefying rhino or elephant to gain early access to the carcass).

3. A stone hammer and anvil are all that is needed to simulate powerful bone-crushing carnivore jaws and to disclose the valuable marrow contained within limb bones (giving hominids some of the capabilities of the robust-toothed, bone-crunching hyenas such as *Crocuta*, the spotted hyena).
4. A simple digging tool made from an antelope horn, fractured long bone, or sharpened branch would have enhanced the hominids' ability to dig for underground roots and tubers (as bush pigs, warthogs, and some baboons do), to dig for water (as elephants do with their feet), or to break open termite mounds for a rich bounty of white ants within (moving into the specialized aardvark's domain).
5. The use of stone hammers and anvils likewise would give the hominids access to hard-skinned fruits and perhaps hard-shelled

Techno-organic evolution: synthetic artifacts allowed hominids to exploit new resources and to move into new niches in competition with other animals.

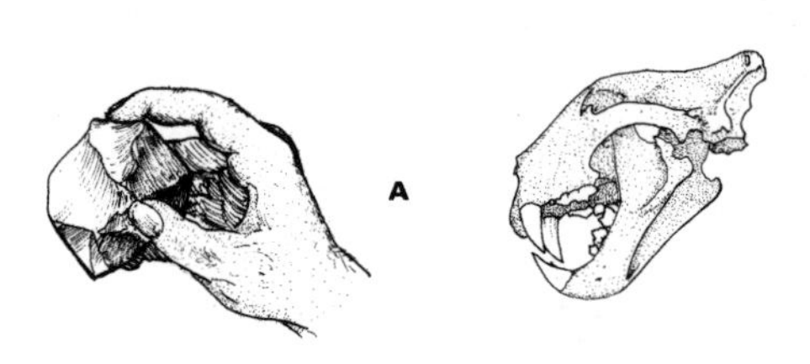

a. Stone flake. Analog: Carnivore flesh-cutting carnassial teeth.

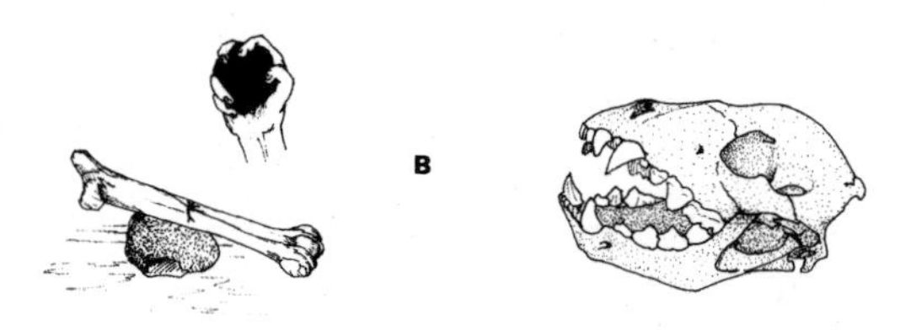

b. Hammer and anvil. Analog: Hyena bone-crushing teeth and jaws.

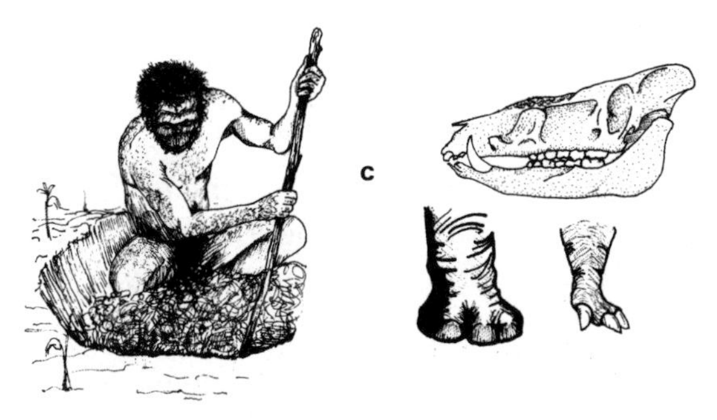

c. Digging stick. Analog: Bushpig snout and tusks; elephant and aardvark feet.

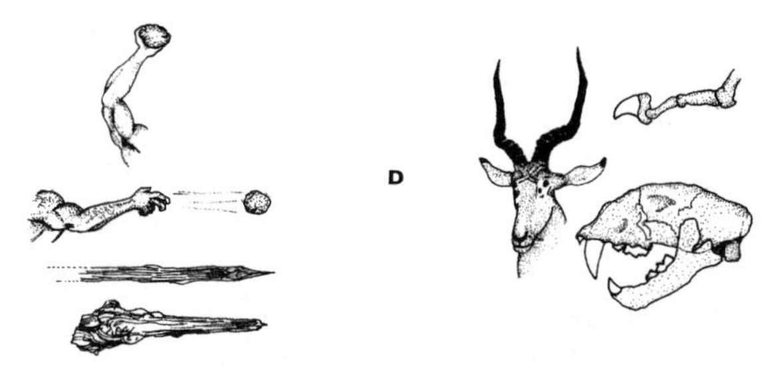

d. Missiles, clubs, spear. Analogs: Carnivore canines and claws, antelope horns.

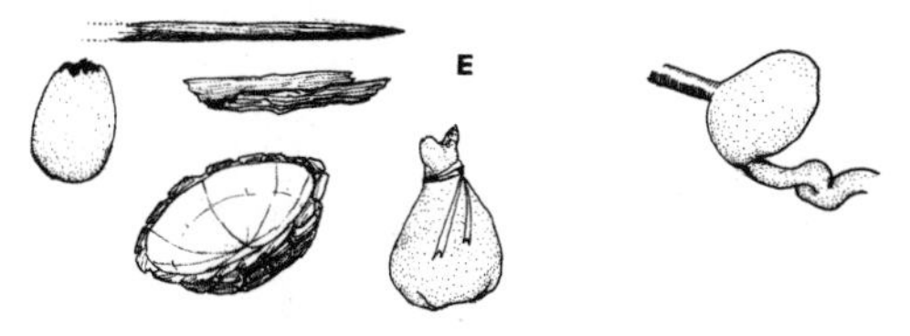

e. Carrying devices and containers: wooden skewer, bark tray, ostrich and tortoise shell, skin bag. Analog: animal stomach.

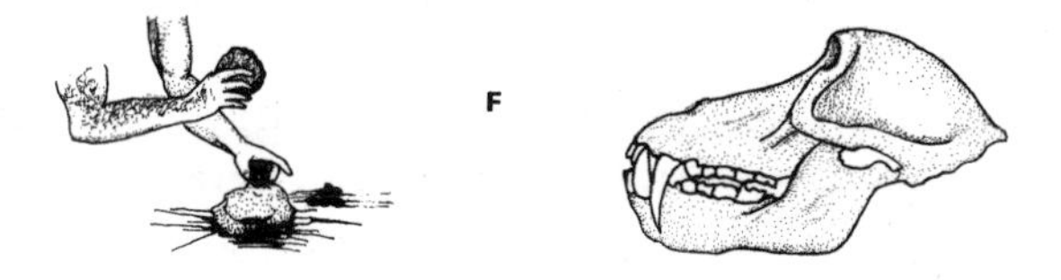

f. Hammer and anvil. Analog: Baboon nut-cracking cheek teeth.

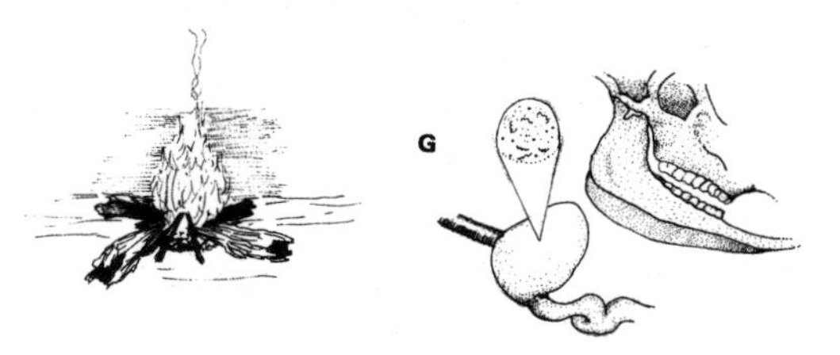

g. Fire. Analog: Herbivore jaws and teeth, stomach with microorganisms to aid in consumption and digestion.

nuts eaten by various monkeys and apes. Although apes normally tear their way into hard-rinded or tough-podded fruits with their teeth and hands, use of tools would have spared the hominids this physical strain and perhaps improved their efficiency.

6. The possible use of tools as prying levers or wedges to strip bark off trees could have provided access to edible gums or invertebrates (a niche of birds, reptiles, and small mammals).

We have called this replacement of biological organs with synthetic, artificial tools "*Techno-organic*" Evolution, a phenomenon that has only become crucial in one major lineage in the history of life: tool-using hominids. It is probably as profound an evolutionary step as the first self-replicating life, or the first eukaryote cells, amphibians, reptiles, or mammals. The human lineage is now adapting and evolving not only through the biological transformations it is able to make over time to adjust to its environment, but also through the technological developments and changes we are able to bring about. And a large part of the environment we are adjusting to is fashioned or shaped by our technology. We cannot escape the biological, organic nature of our existence, but it is now bound up inextricably in the technological realm we have created.

The advantages given by tools were almost certainly focused on improving our feeding abilities, but their ramifications went far beyond a better meal. The circular feedback loop had been set in motion, with advantageous tool use favoring a larger brain to plan, imagine, and oversee even more tool use; which favored intelligent manipulation and more complex communication, planning, and social interaction; whose advantages would select for a larger, better-connected brain (with more complex neural connections); and so on. As we will see, this process seems to have occurred in steps, perhaps as new thresholds of brain size and neural organization stabilized and natural selection worked out remaining kinks.

We must look beyond the immediate, direct aspects of tool functions to consider the larger picture: What might these early archaeological sites indicate about hominid behaviors during this phase of our evolutionary adaptation?

CHAPTER 6

THE NATURE AND SIGNIFICANCE OF EARLY STONE AGE SITES

At the first light of dawn, the hominids stir from their sleeping places in the boughs of several large fig trees, yawning and stretching as the eastern horizon begins to glow with a golden hue. A young mother nurses her infant, only a few weeks old. Slowly they all begin to climb down from the trees or drop from some of the lower branches onto the ground. They walk back to the area of the wildebeest butchery the day before, about a half kilometer away.

Each of the adults and adolescents picks up a few artifacts to carry on their feeding forays during the day: a stone hammer, a core or two, and a few large, sharp flakes. The great majority of the stone artifacts, however, are left on the ground, along with the shattered bones of animals from previous meals. The distinctive footprints of hyenas, which showed up the night before to chew on the ends of the broken bones littered about, are still visible on the sandy, silty ground.

The site is not revisited by any hominids for weeks. With the onset of the long rains, during a major cloudburst the river begins to swell and rise, its color becoming a muddy brown from the countless silt particles suspended in its surging flow.

The water crests the river's natural levee and begins sweeping onto the floodplain, a number of watery tongues that swiftly join together until the flat ground is covered by a sheet of running water. The fine debris on the land surface—leaves, twigs, insects, tiny flakes, and bone fragments—are quickly swept downstream in the direction of the lake. The larger, heavier pieces of stone and bone are engulfed by water but remain behind, the floodwaters flowing over and around them, nudging them until they lie broadside of the flow, some caught up in low vegetation, or spinning them until they submit, conforming to the direction of the current, and are jostled no farther.

For centuries, thousands of years, and hundreds of millennia, the region continues slowly to subside, annual floods adding increments of silt, sand, gravel, and ash to the meandering river deposits. The river gradually builds a long sequence of layers, many meters thick.

Later, new patterns of earth movements uplift the region, for the first time exposing the ancient sediments as erosion begins to cut cliffs and gullies through the banded layers. In one locality a dry, gully-ridden landscape studded with short commiphora trees and straggly brush has replaced the earlier, verdant floodplain of the past. Here a recent gully has cut through this ancient site, creating a scatter of stones and fossil bones littering the slope below the sediment layer these hominids had once occupied.

An archaeological team, males and females, darker- and lighter-skinned creatures with very inconspicuous hair on their naked-looking limbs, moves up the gully. They have covered much of their bodies with a variety of clothing: bush hats, sunglasses, cotton shirts, backpacks, shorts, sneakers, desert boots, or sandals. They come across a profuse scatter of stone artifacts and fossil bones on the surface, toward the bottom of the gully. They trace it up the slope, a stone here, a fossil there, to the level of the archaeological horizon, where several pieces of bone and stone can still be seen sticking out of the light brown, silty layer.

The archaeologists talk excitedly to one another, pointing to certain stone artifacts and fossils, some with cut marks and evidence of hammer stone fracture. They jot down observations in small field notebooks, make sightings with their compasses, and draw sketches of the major stratigraphic features, using a tape measure to determine the thicknesses of different layers. They collect some sediment samples in cloth bags, using the pick end of a geological hammer to clean an exposure and obtain fresh specimens.

They take photographs with a thirty-five-millimeter camera as well as a Polaroid and jot additional notes directly on the instant prints ejected by the camera. Using a colored felt pen, they note the locality with a dot and an identification number on a plastic sheet overlaying an aerial photograph of the area. They discuss plans for a small test excavation before a major, large-scale dig is mounted.

At midday, under the sparse shade of a thorn tree, they sit down for a lunch of some canned foods opened with Swiss Army knives: beans in tomato sauce, corned beef, and plums. A half-melted chocolate bar is portioned on its aluminum wrapper with a knife and passed around. The paler members of the group take out plastic containers and spread a white fluid, warmed by the heat of the morning, on their exposed skin, seeking the protection from the sun's rays that was lost by their recent ancestors in the northern realms from which they have come. A canvas water bag had been hung from a tree and has slowly become covered with bees sucking the moisture emanating from the outside of the bag as it cools its contents through evaporation. Water is poured into aluminum cups. Finally they rise, their legs groaning from a morning of hard hiking, and begin the long trek back to their gray Land Rover parked at the top of a distant ridge.

> Tongues in trees, books in the running brooks,
> Sermons in stones, and good in every thing.
>
> **William Shakespeare (1564–1616),**
> ***As You Like it***

> Nature teaches more than she preaches. There are no sermons in stones. It is easier to get a spark out of a stone than a moral.
>
> **John Burroughs (1837–1921)**
> ***Time and Change: The Gospel of Nature***

There has probably been more disagreement and debate over the nature of early Stone Age sites in the past decade than in the entire century preceding it. A healthy (if sometimes acerbic) skepticism has emerged regarding previously held notions of how our early ancestors lived. New schools of thought challenging the traditional interpretations and dogmas are appearing, along with means of testing new ideas and hypotheses concerning the prehistoric record.

We have, in effect, taken a giant step backward and have posed the basic epistemological question: How do we know what we know? Which of our interpretations regarding the earliest stone tool makers seem valid and which are erroneous? Were they hunters or scavengers? Did they have home bases? Did they build huts? Did they use fire? How intelligent were they?

In this chapter we will explore through experiments how Stone Age sites enter the archaeological record and examine conflicting interpretations of the significance of these early Stone Age localities in terms of the behavior and adaptation of the early hominid tool makers.

THE MAKING OF A STONE AGE SITE

THE EXPERIMENTAL VISION: SEEING SITES FORM

A basic problem in trying to understand the past is that at the very outset of our mission we are irrevocably distanced from our subject: we cannot go back in time. There is no way to turn back the clock and see directly and vividly what long-extinct ancestors were doing, how they were managing to eke out their existence on an ancient landscape, and what they were doing to create the litter of stone artifacts they left in their wake. But these sites give us an invaluable key to unlock some of these mysteries of the past: if we can understand what goes into their making, explore what sorts of activities and events can produce these enigmatic arrays of artifacts, then we are in a better position to decipher behaviors that formed these sites two million years ago.

This approach, once again grounded in experiments, gives us some guidelines when we look at our sites, wonder about the artifacts we find, and ask where, how, and why they were made. For instance, can we say all the artifacts found at a site were made there, or did the hominids bring them in from elsewhere? How did all the flakes and fragments get there—from making the cores found at the site? If not, then we must consider *why* not: were hominids bringing in and taking out lots of artifacts, or did later floods and other natural agencies alter the look of what we see? Experiments making stone tools can give us a valuable baseline for developing our expectations: if tools were made on the spot, and there has been very little geological disturbance since the hominids were there, what would we expect?

BURYING THE PAST

People often ask, "How do archaeological sites become buried?" The television news often presents current events that graphically show us some of the ways: a river floods, turning streets into torrents of running water, and when the floodwaters recede, people are shown shoveling thick deposits of mud out of their streets, driveways, and living rooms; lake and sea levels rise, inundating real estate for some period of time underwater, where muds or sands begin to accumulate; a volcano erupts, blanketing the landscape with deep deposits of gray ash, which chokes lakes and rivers. In fact, modern humans are constantly battling nature to prevent their domestic spaces from being buried or washed away: in the United States, without flood control measures such as building levees along rivers, or constructing dams, dikes, and breakwaters (or, in emergencies, sandbagging), the annual toll of flooding would be much worse.

As a general rule, areas of low topography, such as river valleys, deltas, and lake basins, are where sediments normally collect (called deposition), with the oldest layers laid first, at the bottom, and the youngest last, at the top. (Imagine throwing a year's worth of newspapers, day by day, into a deep, narrow pit: after a year, the most recent issue, December 31, will be at the top of the pile, while the oldest, the previous New Year's Day edition, will be at the bottom.) In areas of high topography, such as hills and mountains, erosion reigns supreme, and objects on the surface have a low chance of being buried and retained in geological deposits; they are usually eroded or washed away. So naturally we tend to look for sites in the ancient low-lying areas where sediments have gradually been accumulating over time.

Certain parts of the world are especially prone to deposition. Earth movements (earthquakes and faulting) are making some portions of the earth's crust sink to form large-scale basins, natural traps for sediments to accumulate, layer by layer, over time. A prime place for this is the Great Rift valley system in Africa, where continental movements over the past several million years have created a gigantic trough over two thousand miles long with numerous lakes and rivers contained within.

This is the type of place we go to find the evidence we want of early human prehistory. Doing field archaeology can be a labor of love, with the accent on labor. When the archaeologist discovers a Paleolithic site buried hundreds of thousands, or millions, of years ago, it has usually been sealed under massive quantities of later sediments and subsequently exposed by more recent geological activity, usually cuts by a river or erosion along a cliff or escarpment. Nature has started uncovering the site, etching away at its edges, but we must finish the job.

We excavate these early Stone Age materials by painstakingly stripping away thin veneers of overlying sediment, eventually exposing stone artifacts, fragments of fossil animal bones, and other associated materials. In modern excavations we plot these objects in three dimensions using surveying equipment, impregnate fossils with a chemical preservative if they are fragile, label all finds in pen and ink with a catalog number, and carefully remove and wrap them for shipment back to a museum or university. The idea is to preserve every scrap of information that has lasted through the ages and may show us something about what these hominids were doing at these sites. But a critical question has to be asked: When we find these prehistoric materials and uncover them in their ancient deposits, are we seeing them as they were left by the hominids? Are they situated in the same place and with the same associations—this bone next to that artifact, these cores sitting in a tight group—as they were when the early tool-making hominids were living? Do we find them as they were left by the hominids a couple of million years ago, or have they been disturbed and rearranged, by floodwaters or by other geological or biological forces, since the site was occupied?

Historically there has often been an assumption that many early Stone Age sites are undisturbed or minimally rearranged, so archaeologists have tended to treat most of them as if their patterns have guaranteed behavioral significance. Sometimes a dichotomy has been drawn between "primary context" sites, usually found in finer-grained sediments, and "secondary context" sites, normally in coarse river gravels containing many large cobbles or boulders. Rarely did anyone attempt a critical assessment of the *degree* to which a site might be disturbed or undisturbed.

In the 1960s Glynn Isaac, while working at the Acheulean site of Olorgesailie in southern Kenya (to be discussed in the next chapter), began a pioneering study of how sites can be affected by rivers. He set out concrete casts of artifacts in seasonal river beds and monitored what happened to them during the next bout of seasonal rain. This was one of the first attempts to examine directly the question of Paleolithic archaeological site formation. Subsequently, one of us (KS) launched a long-term experimental investigation (lasting several years), taking a close, in-depth look at how Paleolithic sites might have formed.

HOW DO SITES FORM?

The first step in trying to understand how Stone Age sites formed is to examine closely the entire system of stone artifact manufacture. What do we expect when an Oldowan core is flaked? What sorts of stone debris are produced? To make the typical Oldowan chopper, for example, would we expect five flakes and fragments? Ten? Fifty? Two hundred? One thousand? Clearly this will depend upon how extensively a core was flaked, but after numerous experiments in producing typical Oldowan core forms, a simple rule of thumb was found: the average ratio is approximately one core to thirty or forty flakes and fragments (above the size of the typical five-millimeter mesh screen used in excavating these sites). Making an average Oldowan core form thus should generate a few dozen of these smaller bits of flaking debris at a site. More thoroughly flaked cores will produce more, others less, but this gives us a ballpark figure with which to work.

Building a Site, Stone by Stone

What sizes of stone debris would we expect from flaking these Oldowan cores? Tool-making experiments can also show us this very clearly and directly. To our surprise, it turned out that there was a very consistent and predictable pattern here as well. The approximate size distribution of these materials broke down as follows:

Over 8 centimeters: 1%
4 to 8 centimeters: 8%
2 to 4 centimeters: 23%
1 to 2 centimeters: 41%
Less than 1 centimeter: 27%

(The smallest size group again goes down to the size of the excavation screen—without this cutoff, the number of smaller fragments would keep increasing in an ever-accelerating curve down to microscopic dust.)

This shows dramatically that flaking just a few cores at a site should generate quite a lot of stone debris—most of it so small that it would not be of much use to an early hominid. But it can be of great use to the archaeologist. This observation was an important one: it meant that besides the geological evidence, the stone artifacts themselves could potentially give an independent means of assessing whether a site had been significantly disturbed by water action, which would have preferentially winnowed away smaller, lighter artifacts from the site. This assessment, combined with geological analysis of the sediments from the site, yields important clues about the degree of disturbance of early Stone Age sites. Most important, it helps show us which sites best preserve evidence of ancient hominid behaviors.

Once we started looking at sites this closely, we wanted to see if we might be able to identify specific places at our sites where the hominids actually stood, stooped, or sat and made their tools. This can also be explored through experiments. We flaked experimental Oldowan cores and examined the spatial scatters of flakes and fragments created, plotting them on graph paper. Again, typical patterns emerged: the chips knocked off our cores flew out in a scatter usually less than five feet in diameter, with the larger flakes falling closer to the tool maker. This gave us another expectation to use when examining patterns of pieces of stone from Paleolithic sites.

Plotting the spatial scatter of experimental flaking debris to be compared to prehistoric scatters. A one-meter square strung at ten-centimeter intervals is laid on the ground to assist in accurately drawing the scatter. (A BBC television crew films the process.)

The next major question: What would have happened to a site after the hominids had moved on, leaving it exposed to the myriad forces of nature—rain, wind, floods, trampling animals, scavengers, and so on? Again, experiments provided a means of exploring this, of seeing how nature transforms a site as it is being absorbed into the archaeological record. A major part of this study involved creating dozens of simulated early Stone Age sites, seeding them with hundreds of Oldowan artifact replicas as well as modern animal bones, and carefully placing them in some area in the modern environment (usually covering several square meters, though the largest one had five thousand artifacts in a one-hundred-square-meter area). Some of these were painted yellow for easy visibility, while others were coated with aluminum foil so they could be found with a metal detector after geological burial.

The experimental workshop (and the whole campsite) was unexpectedly inundated by a flash flood. Investigations of what happened after the flood were the beginning of the experimental program examining site formation.

The experimental sites were set out in a wide range of geographic locations in northern Kenya analogous to the types of environments in which Oldowan sites are found: river floodplains, river gravel bars, channels, delta floodplains, delta channels, low-energy lake margins, and high-energy lake margins (with strong wave action). Artifacts and bones were left out to see what nature would do to these sites, to see what happened to them over the course of weeks, months, and years, and to see how the lucky ones became buried in geological deposits. We were ready to witness their fate.

We monitored these sites for a period of several years. During this time the majority were acted upon by a number of forces of nature: trodding hippos, scavenging birds, rising and falling lake levels, winds blowing sands and silts onto them, and flooding rivers sweeping over them. A large number of sites became buried during this time, and we subsequently went and excavated them. We found fascinating differences in their end results, ranging from sites buried with little or no change in their contents to others that had virtually disappeared, their artifacts and bones dispersed and scattered downstream. These observations gave us important clues to help interpret prehistoric Stone Age sites.

Before the flood: an assemblage of stones and bones set out on a river floodplain. The stones were painted for easy identification, and a sample of the stones and bones was coated with aluminum foil to assist with recovery after burial.

The largest experimental site. Comprising one hundred square meters, this site contained about five thousand artifacts and animal bones, including most of the skeleton of a giraffe.

An experimental site about to be inundated in a high-energy lake margin environment.

An idyllic setting: a stream meandering toward a delta on the lake. This photograph clearly shows a stream eroding the bank on its outer curve and depositing sediments on its inner curve, or "point bar." An experimental site is set up on the inner curve.

Using a metal detector to locate an experimental site after burial. A sample of the stones and bones was coated with aluminum foil so it could be found after burial and possible transport by water.

Left: A stream at Koobi Fora during a flood that followed a cloudburst. The rising water had just crested the floodplain. *From left:* Kathy Schick, Glynn Isaac, and Yusuf Juwayeyi, currently the director of antiquities in Malawi.

An experimental site set up on a broad river delta.

The same delta site a year later during excavation. Note the change in landscape and vegetation; several inches of sediment had built up after the rainy seasons. Almost all of the bones had been washed away, along with most of the smaller stone artifacts. The clustering of some of the larger pieces is one clue to disturbance by flowing water.

The excavation of an experimental site in progress. Note the antelope mandible and numbered stone artifacts.

An extreme cautionary tale: in northern Kenya, the fractured skull and cracked marrow bones of a goat that was processed with stone tools were laid out in rough anatomical position with stone artifacts. Within one day scavengers (that may have included crows, jackals, hyenas, dogs, or lions) had removed the great majority of the bones from the site. In prehistoric times many of the animal bones that had accumulated from hominid feeding behavior may have been removed by scavengers.

A Closer, More Controlled Look: Laboratory Experiments

In these field experiments nature held the upper hand, as we had to wait and see what happened to our sites. They gave us valuable insight into what really can and does happen in natural settings, but it was also useful to get a closer look at what happens to these artifacts when flowing water moves across a site in more controlled situations. To do this we conducted some experiments in a laboratory channel (called a flume) at the University of California, Berkeley, School of Engineering, where we could directly watch the whole process as it actually happened.

So we loaded a bed of sand into the flume (a plastic trough about forty feet long with clear sides for observation), got out our modern stone replicas of Oldowan-type cores and flakes, and turned on the water pump. We were able to see how artifacts become engulfed in moving sand ripples when water moves slowly past them and how they begin to be buried. We also watched artifact "derbies" at higher water velocities and saw how different sizes and shapes of artifacts moved at different rates, so that certain types would tend to move away from a site while others would tend to be left behind. These studies helped us understand more clearly what happened in the experiments left out in the natural settings.

SITE FORMATION: COMPLEX AND DYNAMIC, BUT UNDERSTANDABLE

From all these experiments, plus what we know about what happens in various natural environments, we are able to say something about how archaeological sites form and judge what information a particular site can give us about early hominid behaviors. We have been able to identify many different lines of evidence, some from geography, others from geology, and many from the archaeological sites themselves, which together can give us a good idea of what went into the making of the site.

For each line of evidence there are certain patterns to watch for. Some tell us that the site is pretty much as the hominids left it and thus can reveal more to us about *hominid behavior*, and others flash warning signals that the site has been drastically disturbed and altered after it was occupied and tell us more about *geology* than behavior. The important ones to consider include:

1. Geographical context.

   a. Less disturbance: Sites in low-energy river floodplains, delta floodplains, low-energy beach deposits, or buried by wind-blown sands or silts tended to have little or no significant geological disturbance and to be buried rather gently;
   b. Greater disturbance: Sites located in river or delta channels, high-energy river floodplains, high-energy lake margin environments, and hill slopes were much more likely to have been significantly disturbed and sometimes totally washed downstream to be dispersed or reconcentrated.

2. Sedimentological context.

   a. Less disturbance: Sites buried with fine-grained sediments, such as silts or clays, tended to suffer less disturbance, since the exclusive presence of these fine particle sizes would suggest gentle water or wind action;
   b. Greater disturbance: Sites disturbed by high-energy geological forces tended to be found in coarse-grained sediments such as sands or gravels with heavier pebbles or cobbles incorporated in them. (It is possible, however, for a powerful river *only* to have access to fine-grained sediments in its catchment area, so in these cases the nature of the stone tool assemblage may be an important additional line of evidence.)

3. Artifact composition.

   a. Less disturbance: Stone Age sites where lots of flaking had taken place tended to have a low ratio of cores relative to the numbers of flakes and fragments, as well as a lot more fragments than whole flakes;
   b. Greater disturbance: Sites with high proportions of cores to flakes or fragments, and high proportions of whole flakes to fragments, were a warning of substantial disturbance (or else we have to accept that most artifacts were imported ready-made and few were flaked at the site: making just one core should produce a few dozen flakes, most of them very small).

4. Artifact size.

   a. Less disturbance: Sites that retained many small flakes and fragments, particularly lots of useless, very small chips, had not been worked by very strong flood waters;
   b. Greater disturbance: Running water winnowed away most of the smaller flaking debris, leaving behind mainly larger flakes along with some cores.

5. Orientation of stones and bones.

   a. Less disturbance: Sites did not show strong alignment of elongate artifacts or bones, or only the smaller pieces seemed to have been swept into alignment by slow-moving waters;
   b. Greater disturbance: Many of the elongate pieces showed similar alignments, some of them in one direction (in line with the flowing water) and others at right angles to this.

6. Clustering of stones and bones.

   a. Less disturbance: Stones and bones tended to be scattered about but with some specific concentrations, especially in areas of stone artifact manufacture or use;
   b. Greater disturbance: In some instances the materials were swept downstream and dispersed. In others traffic jams of stones and/or bones would occur around obstacles (such as a large stone, a fallen tree trunk, vegetation, and so on). Sometimes artifacts would overlap like shingles on a house, with the stone artifacts dipping (sloping down in the upstream direction). Often a heavy stone or bone would have smaller pieces underneath it as well.

7. Conjoining.

   a. Less disturbance: Sites with a substantial portion (say, 15 percent or more) of artifacts fitting onto other pieces indicate minimal geological disturbance. The number of pieces that can be refit would be influenced by how much flaking occurred at the site and how large an area was excavated. Some of these refitting pieces would be fairly close to one another;
   b. Greater disturbance: Sites that had few or no refitting pieces suggest that they might have been geologically disturbed.

So we now have ways of deciphering many of the patterns we see at early Stone Age sites and identifying those that show convincing signs of hominid behavior: tool making, hominid transport of materials to and from the site, butchery and bone processing, and use of different parts of the site for different activities. Those early Stone Age sites that appear to have been minimally affected by geological disturbance offer the best potential for inferring hominid behavioral patterns. At some of these sites at Olduvai Gorge and Koobi Fora, the dense concentrations of stone artifacts and animal bones have generated rival hypotheses about the possible behavioral significance of early Stone Age sites.

BIG GAME HUNTERS OR JUST OUT ON SCAVENGER HUNTS?

We assume that our early stone tool–using ancestors were opportunistic omnivores, primarily vegetarians, with some animal food resources (meat and marrow) included in their diet as well. Were they, like modern hunter-gatherers (and, in a much more limited way, like predatory chimpanzees), getting animal foods mainly by hunting? Or did these hominids scavenge the leftovers of large animal carcasses that had been killed and eaten by other predators?

The animal bones at Oldowan sites usually show a wide range of species (different sizes and from different habitats) and body parts (shafts of limb bones, skulls and teeth, ribs and vertebrae, feet and lower legs, and so on). Some of the bones have cut marks from stone tools and fracture patterns consistent with hammer stone fracture, making it clear that stone tool–using hominids directly exploited at least some parts of animal carcasses, which ended up at the concentrations of stone artifacts and bones that we call early Stone Age sites. How did they acquire these animal remains, and how important were they in their diet and overall adaptation?

Traditionally there had been an assumption in archaeology that when animal bones were associated with stone tools, this implied that the tool-using hominids were hunters who had killed prey and then butchered them with their stone tools. This further implied that the roots of human hunting went deep into our prehistory, essentially to the beginning of the archaeological record. Since nonhuman primates tend not to scavenge (a rare instance among chimpanzees has only recently been reported), and modern human foragers do so very infrequently, many thought that an evolutionary stage of scavenging before hunting was unlikely. Recently, however, this perspective has been changing.

Archaeologist Lewis Binford (now at Southern Methodist University), one of the most outspoken and controversial figures in archaeological studies, has explored a series of major archaeological questions over the past three decades. During the 1980s he began to consider the possible meaning of early Stone Age sites in Africa, in particular the significance of the bones they contain.

Binford examined the lists of animal bones and body parts published from Mary Leakey's excavations of early Stone Age sites at Olduvai Gorge. He concluded from his analysis that most of these sites contained high proportions of what he called low-quality body parts—that is, those having little meat and marrow (such as lower limb bones and skulls). Based on this pattern, he concluded that these Oldowan hominids were primarily scavengers, processing carcasses that had been ravaged and then abandoned by carnivores. As these carcasses would have been cleaned of most of their meat, bone marrow would have been the primary resource available to the hominids, obtainable by cracking open the thick limbs and other bones that had not been destroyed by the carnivores.

Some of Binford's suggestions have been examined by Robert Blumenshine of Rutgers University, who studied the abundance of relatively fresh animal carcasses as well as animal remains that could still be scavenged during the wet and dry seasons on the Serengeti Plain and in the Ngorongoro Crater. His results suggest that the best scavenging opportunities would have been during the dry season along the margins of river courses, where edible parts of carcasses, especially lower limb bones containing fatty marrow, would often have been available.

A different interpretation was drawn by Henry Bunn and Ellen Kroll of the University of Wisconsin after detailed examination of animal bones from sites at Olduvai Gorge and Koobi Fora. One of the Olduvai sites that has received the most attention is called FLK Zinj (named after the A. *boisei* cranium found there). Animals found at this site are mostly antelope, such as ancient forms of waterbuck, hartebeest, and springbok, as well as pigs, zebra, and some carnivores. Many of these fossils are from impressively large mammals, weighing more than one hominid could possibly carry as an entire unit. The most abundant body parts here are lower jaws, the upper and lower forelimbs, and the upper and lower hind limbs. The patterns of cut marks found on the bones, according to Bunn and Kroll, indicate that early hominids used stone-cutting tools to process the more nutritious parts of the animal carcasses, especially the upper limbs, which yield the most meat and marrow. They conclude that Oldowan hominids were either adept hunters, killing and butchering whole large animal carcasses, or else formidable scavengers, able to get to animal carcasses before other animals did or to chase predators or scavengers away from carcasses before the carnivores' consumption had gone very far.

Other researchers, particularly Pat Shipman of Johns Hopkins University and Richard Potts of the Smithsonian, using scanning electron microscope analysis, contend that the number of striations that can be identified as cut marks are fewer than Bunn and Kroll have argued. Shipman prefers a scavenging model for early hominid behavior, while Potts concludes that it is not yet possible to ascertain exactly how these animal bones were acquired and concentrated at Oldowan sites.

Richard Klein of the University of Chicago has pointed out that it is unlikely that Oldowan hominids could have been very competent big game hunters, as the archaeological evidence suggests that even much later hominids were not yet very adept at large-scale predation. Relatively efficient human hunting may be a fairly recent evolutionary phenomenon, developing only within the last few tens of thousands of years. Even recent African hunter-gatherers, equipped with efficient hunting technologies including bows and poison arrows, metal-tipped spears, an incredibly complex knowledge of their environment, and the ability to plan and execute communal hunts with modern language abilities have only about a 30 percent success rate in hunting large mammals and only about one-third of their calories are obtained through animal food sources.

In sum, we have not yet achieved a consensus as to how so many bones from different animals ended up at these early Stone Age sites. Characterizations of our ancestors, often from the same data base, have ranged from impoverished scavengers to formidable hunters. It may be that this dichotomy between hunting versus scavenging models obscures a flexibility that may have existed in hominid social groups in the early Stone Age. A more moderate and reasonable view may be that these Oldowan hominids hunted smaller game (say, the size of a gazelle) on a regular basis and scavenged from larger carcasses, sometimes managing to find or claim the entire prize but coping with leftovers when the opportunity presented itself. Large mammals that were sick, wounded, or isolated from their social groups may have been hunted and dispatched, especially if their defenses posed no serious threat to early hominids.

In recent years we have made a great deal of headway in recognizing and understanding how predators and scavengers obtain carcasses and what they do with them. We still need to refine our ability to recognize patterns of hominid hunting or scavenging—if the prehistoric evidence makes that possible—before we can make more definitive statements on this issue.

HOME BASES OR OUT IN LEFT FIELD?

Why were hominids bringing so many artifacts and bones to these sites and then leaving them behind with all these animal bones? What aspects of early hominid behavior do they represent?

Where preservation of bone is good, and geological disturbance seems to be minimal, there is almost certainly more potential for deciphering the behavioral patterns of early hominids. Some of the sites at Olduvai Gorge (such as DK and FLK Zinj) and at Koobi Fora (such as sites 50, 20, and 20 East) have received the most attention in recent years. Excavation and analysis of these well-preserved Oldowan sites normally exhibit the following features:

1. Numerous stone artifacts, often numbering in the hundreds or thousands of pieces; some of these pieces fit back together to reconstruct partial cobbles or chunks of raw material, showing episodes of flaking stone at the site;
2. These stone artifacts, when geological sourcing is possible, indicate that these rocks have been transported by hominids up to several kilometers. Often these rocks have come from more than one source on the landscape and concentrated at these sites;
3. Numerous animal bones, usually representing diverse species and sizes; some of these bones exhibit clear signs of having been processed by hominids: cut marks, especially on limb bones, which suggest dismembering and at least some deboning, and percussion marks from stone hammers, which suggest marrow processing.

But how can we know what these sites tell us about the behavioral patterns of early hominids?

THE HOME BASE OR CAMP MODEL

The most conventional interpretation of these early Stone Age sites is modeled after modern hunter-gatherer campsites, which are essentially home bases or central foraging places, where the members of a band congregate at the end of the day to distribute food accumulated from gathering plants and hunting. These camps may be occupied for days, weeks, or, in very lush seasons, even months at a time. More than feeding goes on at these camps: a lot of social interaction, food preparation, and tool manufacture or maintenance normally takes place. The group is usually composed of several family units, individual families have their own area in the camp, and group members sleep there at night.

The litter of day-to-day activities of life is normally scattered about such campsites, including the bones or bone fragments of animals they have consumed, inedible parts of plant foods, and the waste debris of stone manufacture and the discarded stone tools themselves.

Even in the nineteenth century, antiquarians investigating Lower Paleolithic occurrences in Europe tended to view these sites as evidence of early human camps. This perspective continued well into this century. To Mary Leakey, in her monograph of the excavations in Beds I and II of Olduvai Gorge, Oldowan sites represented campsites, with the presence of animal bones there bearing testimony that *Homo habilis* was an adept hunter.

During the 1970s and early 1980s, Berkeley archaeologist Glynn Isaac also suggested that these early Stone Age sites were such camps, or home bases. (He later changed the term to "central foraging places," as modern humans sometimes react emotionally to the word *home*, which conjures up visions of a cozy, sedentary life.) He added or made explicit important new elements to this model: it was likely that Oldowan hominids were beginning to show a pronounced sexual division of labor, with females gathering plant foods and males doing the hunting, and these early hominids were actively using these central places to share food with other members of the social group, a feeding strategy not seen in most other animal species.

Beginning in the late 1970s several researchers began to question the home base or camp interpretation.

THE SCAVENGING STATION MODEL

Binford's study of bone accumulations at Olduvai Gorge site concluded that 1) early hominids at Olduvai were not hunting most of the large animals whose bones were found at the sites but were primarily scavenging the marrow from carcasses left by other predators; and 2) the stone artifacts and bones had become concentrated at Oldowan sites not because the sites were "camps," but because the hominids periodically visited places where carnivores had abandoned their leftovers. In subsequent publications, both Binford and Pat Shipman have suggested that these early sites represented scavenging stations where hominids brought parts of large animal carcasses for processing and consumption—breaking bones for marrow or cutting dried, relict meat off bones. These scavenging stations would have been situated in safe places on the landscape where hominids could monitor the movements of potential competitors, such as lions or hyenas, when they were feeding. Binford has raised the possibility that these sites were resting places for scavenging tool makers during the peak of heat in the middle of the day. In this perspective and in contrast with the home base model, such behaviors as hunting, sexual division of labor, and food sharing were not necessary components of these early Stone Age sites, but might have developed later in human evolution.

A hyena's dining room. A modern striped hyena den investigated by the authors with the collaboration of Dr. Mujahed Muheissen in the Jordanian desert. This dense scatter of bones, at the entrance to a small cave, was discovered by Donald Johanson and Tim White's 1984 expedition to Jordan in search of fossil hominids. Hyenas (as well as leopards, porcupines, and owls) are among the few animals besides hominids that collect and concentrate bones in substantial quantities as a result of feeding activities. In this example, the bones included camel, goat, sheep, dog, horse, donkey, gazelle, badger, and birds.

THE STONE CACHE MODEL

Potts has proposed that early Stone Age sites were actually strategic places in a hominid territory where significant amounts of stone were intentionally cached by early hominids, in places where usable stone did not occur naturally and where significant stone tool–using behavior, especially the processing of animal carcasses, occurred. In this model, sites would have become islands of usable rock on an ancient African landscape.

In Potts' scenario, an average of seven such caches of stone would have been an optimal number in a hominid territory at Olduvai Gorge. It is an energy-efficient model, considering the costs and benefits of such behavior. If an animal carcass was obtained by scavenging or hunting, it would be transported to the nearest stone collection, perhaps after initial disarticulation, for final processing and consumption. Potts called these cache areas "proto-home bases" and argued that it was still ambiguous whether the animal bones found here were primarily obtained through hunting or scavenging. Potts also points out that the areas of animal carcass processing and consumption would probably not be the sleeping places of early hominids, as the fresh bone refuse would have attracted numerous carnivores.

THE FAVORED PLACE HYPOTHESIS

One of us (KS) has suggested that many of these sites are actually the residual products of hominids repeatedly occupying the same locations, where they consumed foods, rested, and carried out other individual and group activities. This is similar, in some ways, to the camp, home base, or central foraging place model in that hominids repeatedly occupied a site, foraging in its vicinity, sometimes for a period of time. But no elaborate sexual division of labor would be needed, nor systematic sharing of food.

Rather than eating on the run, as other primates do, food would have often been taken to the site for consumption, not because members of the group had to meet up to distribute the food, but because these sites would have been centrally located in the foraging areas of the hominids—for both plants and animals—and they offered some degree of protection, shade, safety, or a vantage point of game (both potential prey and potential threats). Also, the group may not have been continually splitting up and rejoining as hunter-gatherers do but may sometimes have been foraging as a group. Acting as a group would have been particularly important for them in invading fresh carnivore kills or outcompeting scavengers for available meat resources and defending their bounty. And they most probably slept at sites that provided enough protection (with large-boughed trees inaccessible to carnivores).

In this model not all sites are the same. Some are probably special butchery locations where a large animal carcass was found and cut up. Sites that are particularly dense with artifacts and bones probably indicate places in the midst of very good food supplies, which were used over and over again. Others probably represent areas where food resources were not such a draw and were visited much less frequently. At some of the sites the hominids probably slept in overhanging or nearby trees, from others they moved on to spend the night elsewhere.

If early hominids habitually carried small quantities of raw material with them, prepared to use them as opportunities arose, then we would expect to see a gradual, cumulative buildup of artifacts over time, especially at sites visited repeatedly: artifacts would be brought in, some would be flaked at the site, some would be left behind, and others would be carried on to other places. And this is indeed what we see: at the well-preserved sites, we see lots of stone transport by hominids. An important question here is Why did hominids import more artifacts than they removed? Some were likely refuse: tiny flakes, worn-out tools or cores. Carrying items such as surplus food away could also be a factor. Also, as conspicuous artifact concentrations built up at some sites, these may have relaxed the hominids' "Don't leave home without it" credo, as a bank of potential tools would be available in the general area. The larger sites are the busy intersections of this artifact and bone traffic—some busier and more congested than others. This scenario does not necessitate a complex home base model, nor intentional caching of stone, but it does suggest a variety of functional activities at the favored locations.

There is at present no consensus as to which of these models may be the most valid, and it is entirely possible that each model has some validity, part of the total story of the meaning of Oldowan sites. Future research, it is hoped, will resolve these questions.

A ROOF OVER ONE'S HEAD? THE QUESTION OF ARCHITECTURE

Since nest building in trees or on the ground is a common trait among apes, it would not seem unlikely that early hominids possessed at least that level of architectural sophistication, creating sleeping places by weaving branches together in trees or building simple leaf or grass nests on the ground or even inside caves. Such structures would almost certainly not survive in the prehistoric record, however. Is there any other evidence in the early Stone Age archaeological record to suggest more complex building activities?

One Oldowan site in Bed I of Olduvai Gorge has been cited as a possible example of such construction behavior. At site DK, the earliest archaeological site known in the gorge, a circular feature of lava cobbles and blocks some fifteen feet in diameter was discovered by Mary Leakey. At this prehistoric level many stone artifacts and fragmented animal bones were also discovered, the majority of which were outside the circle. Leakey argued that this feature was an early hominid structure, perhaps a large hut or windbreak structure of branches held down by the rocks. (Modern African pastoralists still keep their livestock in circles of thorn branches held down by rocks at night to protect them from lions and other predators.)

But is this really a hominid-made structure? The problem here is that the site, about 1.8 million years old, immediately overlies a hardened lava flow that was the source of the circle as well as many of the stone artifacts at the site. Several researchers, including Isaac and Potts, have pointed out that if a large tree had been present at the site (perhaps providing shade and shelter for an early hominid group), it might have naturally pulled up lava rocks from below; as the tree grew, these rocks would have been pushed outward by the ever-expanding trunk diameter. For this reason the case is still ambiguous.

As we will see in the next chapter, persuasive evidence for hominid architecture begins to emerge only during the last several hundred thousand years. If earlier hominids habitually built crude structures, it appears that they were normally made out of perishable materials that left little if any trace behind. But the absence of evidence is not necessarily the evidence of absence. It is worth noting that many of the hut structures of modern hunter-gatherers in Africa and Australia would probably leave little or no signs thousands of years in the future. We cannot verify early huts or shelters, but we cannot rule them out, either.

PROMETHEUS UNBOUND? THE QUEST FOR FIRE

The advantages of having a knowledge of fire are many: it not only provides heat and light, but makes many plant foods more digestible by neutralizing their toxins or altering their chemistry; it can kill potential parasites or other harmful organisms in food, especially in meat; and it can drive away insects and keep predators at bay. The cultural importance of fire in human history and prehistory has been appreciated for a long time.

In the intriguing film *Quest for Fire*, dubbed a "science fantasy" set in the prehistoric past, a group of somewhat dim-witted hominids (with, incidentally, no children, suggesting little reproductive success) are portrayed with the ability to conserve, but not to produce, fire. Ultimately, toward the end of the movie, they learn the secrets of fire production from observing another group of more technologically advanced hominids (via their contact, played by Rae Dawn Chong). In the real world, just how far back in the prehistoric record can we detect fire-using hominids?

Raymond Dart, as previously mentioned, erroneously thought that blackened bones from the South African cave at Makapansgat were due to the use of fire by australopithecines, prompting him to give the fossils found there the romantic name *Australopithecus prometheus*. Subsequent research demonstrated that the discoloration on these bones was the result of mineral staining and not fire. Are there any valid candidates for fire use in the early Stone Age record? Is there actually good evidence of fire present in prehistoric deposits? If so, is there some definite association between the fire and early hominid activity that suggests it was used by hominids? And if so, is there any evidence of the *production* of fire by early hominids as well as its use?

At one site in the Koobi Fora area (called *20 East*), dating to about 1.5 million years ago, there are reddened and apparently baked patches of sediment and discolored artifacts that seem to have been altered by heat. At Chesowanja in Kenya, dating to about 1.5 million years ago, chunks of reddened and apparently baked clay sediment were found in proximity to an archaeological site. And at the South African cave of Swartkrans, perhaps dating to about 1.5 million years ago, a few specimens of darkened bone were chemically and microscopically analyzed by C. K. Brain and Andrew Sillen and found to have been burned. In each case there is reasonable evidence that fire had thermally altered materials associated with early Stone Age horizons.

We do not yet have conclusive evidence that early hominids were directly involved with either the production of fire or its use. Without a clear hearth structure (that is, a fireplace), or without patterned burning on animal bones to suggest intentional roasting, it is difficult to exclude the possibility that these instances may simply be natural occurrences such as brushfires or lightning strikes that may have occurred before, during, or after hominid occupation (when artifacts are also baked it would indicate the fire was either during or after). The question of hominid manufacture and use of fire will be addressed again in the next chapter.

NEW THRESHOLDS IN SOCIAL BEHAVIOR?

One of the most difficult aspects of reconstructing the past is dealing with the question of social organization. Contrary to the home base hypothesis, we just do not have evidence for the complex patterns of interdependent social behaviors as can be seen in modern hunter-gatherers: long-term bonds between males and females, sexual division of labor with males out getting meat and females sticking nearer the camp and gathering plants, and organized distribution of food resources when all members of the group returned to the camp.

But what sorts of groups did these hominids live in? What sort of mating structure did they have? What sorts of relationships did different members of the group have with one another? How did different groups link together to form some sort of larger social community? What was the overall population size of different communities, and how large was the territory they exploited? Though we have little direct evidence, it is possible to offer some general hypotheses based upon what we know of other primates and what can be gleaned from the fossil and archaeological record.

Although modern human social organization varies a great deal around the world, some things are more or less constant: formal, long-term bonds between males and females that constitute the core of the family and regulate responsibilities toward offspring, with both sexes contributing substantially to the maintenance and well-being of the family (though these bonds are not always monogamous or permanent). And there are generally larger responsibilities, relationships, and rules of behavior within the group, including rules of distribution of food and other goods between individuals or groups.

We know that nothing like human forms of family organization exist among the primates, even among the apes. Chimpanzees live in groups of related males and unrelated females, and mating is promiscuous. They feed more or less on their own, although an incipient nuclear family might be seen among the females with their young, who generally move along and feed together and have lots of social interaction among themselves for years, until their offspring are adolescents. At this time females move off to other groups and males go off more on their own, generally linking up more with other males. Males often act together to defend the community's territory against other chimpanzees, and sometimes groups of several males carry out a hunt against some smaller animal, such as a baboon or small antelope. Gorillas, on the other hand, live in a social group with one dominant male, subordinate males, and several females and their offspring. The group moves together, foraging across the landscape, eating as they go.

Paleoanthropologist Owen Lovejoy has proposed that the earliest bipeds had developed a formula for success that depended upon a new form of social organization: the development of strong, monogamous bonds between the male and female, in which the male ventured to the outer fringes of the couple's area and brought back food provisions to a more centrally located female, who foraged more locally with her offspring. By this reproductive strategy, their offspring would have a better chance of survival. In this model, bipedal locomotion would thus be a selective advantage as it enables the male to carry out this provisioning behavior.

Some other researchers, such as primatologist Richard Wrangham of Harvard and archaeologist Robert Foley of Cambridge University, have stressed that both their ecology and their ancestral condition must have played strong roles in determining what forms of social organization would have been effective for early hominids. Wrangham, considering patterns seen in all the African apes as a common denominator for our ancestors before we and the apes went our different directions, has proposed a model of early hominids living in groups with polygynous sexual relationships between males and multiple females.

Ecological studies show that savanna-adapted animals tend to live in larger groups of related individuals, largely for protection against the ever-present threat of predation from the many carnivores roaming around and also for defense of food resources against other competitors, hominid or carnivore. Groups of individuals acting together can be enormously beneficial for the entire group—not only does the individual have a better chance of surviving, but one's relatives, who are carrying the family genes, are also helped out. Called kin selection, this is a central tenet of sociobiological theory.

Thus, it is likely that early hominids lived in social groups, though their type of organization is not known. (If we can use the African apes as a model, females may have moved off to other groups when they became old enough to conceive—called "exogamy" by anthropologists —and many male members of the local group may have been related to one another.) The older females of the group would have been related to their offspring and may have formed regular mating relationships with one or more males of the group. Regulation of social relationships in such a group would have been an important factor: our communication skills may have received an evolutionary boost to help manage the interactions, movements, and activities of the group. Group interaction and cooperation would certainly have facilitated hominids gaining access to large animal carcasses, particularly when competing with other carnivores. Likewise, efficient processing of large carcasses would not only permit but even require the concerted action of a group of individuals. Cooperation was probably as vital to the functioning of the group and the evolutionary success of the species as were hostile confrontations with other groups and other species.

GETTING SMART? THE EVOLUTION OF INTELLIGENCE

If we use modern ape intelligence as a rough guide to what the mental capabilities of the earliest australopithecines might have been, it is certain that the mental faculties and communication skills of the first bipedal hominids must have been considerably less than those of modern human beings. Throughout our evolutionary development, there must have been substantial growth in our intellectual capabilities and our ability to communicate even if this development was punctuated by a number of jumps, or rapid changes. Can we recognize in the prehistoric record evidence of these increased mental abilities through time?

During a critical phase of human evolution, between three and two million years ago, two major and very different trajectories in hominid adaptation were being established, the robust australopithecines and the larger-brained forms put in the genus *Homo*. Both of the emerging lineages were upright, bipedal creatures with free hands that could easily manipulate their environments. Both lived in the same general habitats, tropical and subtropical grasslands and woodlands, and apparently coexisted for over a million years. Both may have had similar degrees of sexual dimorphism.

Yet only one lineage ultimately survived the radical climatic and environmental changes that occurred about one million years ago: the larger-brained, more generalized genus, *Homo*. The lineage of robust australopithecines, characterized by a smaller brain, huge cheek teeth, and a massively buttressed skull adapted toward powerful chewing and crushing forces, disappears in the prehistoric record by about one million years ago. At about the same time, the genus *Homo* expands its range and ultimately spreads out of Africa and into much of Eurasia (as we will see in the next chapter).

What was the reason for this success? We believe that the answer rests with several intimately interrelated factors.

1. Increased *intelligence:* an expansion in brain size, an increase in the total number of neural cells, and a restructuring of the hominid brain, as suggested by anatomist Philip Tobias of the University of Witwatersrand and Dean Falk of SUNY, Albany, led to better problem-solving behavior and long-term memory retrieval. Endocranial casts (of the interior of the cranium) of early hominid skulls are usually made by applying latex to the inside of a fossil cranium, then using this latex impression to cast the shape of the brain that the skull had once housed. Such casts give us gross facsimiles of the size, shape, and circulatory patterns of hominid brains, the personal computers that every hominid had to rely upon to learn, think, and problem solve during its life. Studies of such casts have suggested that new morphological features (which *Australopithecus* did not have) appear in the brains of early *Homo* and *Homo erectus:* these include larger frontal and parietal lobes and prominently enlarged Broca's and Wernicke's areas, associated with speech.

   Stone artifacts, out of all lines of prehistoric evidence, probably have the greatest potential for giving us information about the evolution of early hominid intelligence. Once made, they are almost indestructible; they have been produced over some 2.5 million years; they require enhanced skill and foresight to produce, which can be assessed partially through experimental archaeology with modern humans and nonhuman primates; they are beginning to yield important clues about their functions and thus early hominid adaptation; and their distribution on ancient land surfaces can help us understand early Stone Age land-use patterns.

   Let us recap what we see as the major implications of the stone technologies of early Oldowan hominids.

   a. Although these technologies are quite simple, the actual production of these Oldowan stone artifacts appears to take more skill and ability to work with complex three-dimensional shapes than we have seen among modern apes, even in experimental situations.

b. The artifacts found at early Stone Age sites suggest longer habitual transport distances, as well as longer attention spans for keeping potential tools for future use, than are normally seen among nonhuman primates. Connected with this, they seem to be more organized in their use of the environment and to have developed a special spatial focus for some of their activities, habitually carrying and building up concentrations of stone and bone from diverse sources.

c. Later stages of cobble flaking are normally represented at these archaeological concentrations, suggesting that the patterns of stone artifact use are more complex than a spontaneous technological response to an immediate need. In sum, we feel that Oldowan technology represented a new cognitive plateau in primate evolution, one that emphasized planning and foresight, and was a precursor to the thought processes and behavior of humans today.

d. These early stone tool makers appear to have been preferentially right-handed, suggesting that by two million years ago Oldowan hominid brains were becoming more lateralized and specialized for different activities, with the left hemisphere devoted more to time sequencing, language, and controlling the dominant right hand for manual activities like hammering and tool use; the right hemisphere devoted more to spatial perception and mental mapping of the environment. This was a pattern that would continue in the genus *Homo* to the present.

2. Increased *communication* skills: an increasingly sophisticated set of vocalizations allowed hominids to convey more specific information about the world around them. It is possible that these vocalizations began to take on a primeval symbolic character, with some sounds representing specific individuals, food items, locations, sources of danger, and so forth. Besides an increase in overall brain size, the part of the brain called Broca's area, associated with motor control of speech in modern humans, seems to be significantly more developed in early *Homo* than in the earlier australopithecines or the ones contemporary with *Homo*. The possibility has been suggested from this that the vocal communication of *Homo* may have been more complex than that of *Aus-*

tralopithecus. This may not mean talking in the modern sense of language with syntax and complex rules of grammar, but merely a more highly developed set of vocal symbols to communicate more elaborate meanings from one individual to another.

Exactly what types of communicative abilities Oldowan hominids had is not clear; these simple stone technologies could have probably been learned primarily through example rather than vocalized communication. If, as we argue, these early stone tool makers were preferentially right-handed, we may be seeing a more profound lateralization and specialization of the hemispheres of the hominid brain for different tasks and more complex forms of information exchange through vocalization may have been important among these tasks. Physical anthropologists have noted that early genus *Homo* skulls have asymmetries associated with right-handedness, while australopithecines do not. This may be an indication that *Homo* was the major tool maker at Oldowan sites. But, as we will discuss in the next chapter, understanding the evolution of language, even when we are armed with the sophisticated hammers and anvils of modern scientific inquiry, can be a very tough nut to crack.

3. More emphasis on *technology*: tools became more formalized and habitual, with hominids beginning to transport stone and other materials longer distances and with more premeditation and foresight. Flaked stone technology was especially critical in producing cutting implements for the rapid and efficient processing of large mammal carcasses. If, as we believe, early representatives of the genus *Homo* were the principal makers of tools at the dawn of human technology, then the correlation between a mode of adaptation that required stone tools, especially directed toward the processing of animal carcasses, and the departure in hominid evolution toward creatures with substantially larger brains, is a strong one. Unlike chimpanzees, who primarily use tools to assist in the acquisition of stationary and more predictable food resources, such as groves of nut trees or termite mounds and anthills, early stone tool–using hominids were processing animal carcasses through scavenging or hunting. Animals are a mobile and less predictable resource, "food on the hoof," which may be in many parts of the landscape. For this reason the necessity of transporting tools or raw material becomes much more critical here, and hominids who had the intelligence to plan ahead would have had an important edge.

Tools are also important from a human evolutionary perspective because they herald a new, and perhaps not inherently obvious, concept in hominid social groups: the concept of *personal possessions*. There would almost certainly be occasions when raw materials for tools or the tools themselves would be in short supply, and more than one individual would want possession of a tool for a particular purpose. Unlike valued food resources, which are normally immediately consumed in the nonhuman primate world (and only rarely shared) most tools have the potential of being shared, borrowed, stolen, or reused. They are not, like food, just for immediate gratification. The development of new types of social politics and means of communicating information may have become very important during the early dawn of human technology.

4. Enhanced *omnivory*, including more incorporation of meat in the diet, and a generalized adaptation: some early hominid groups expanded their dietary breadth by more *efficiently exploiting* a wider range of foods with the help of technology, and they also extended their niche breadth, for example by using stone knives and hammers for large animal butchery and wooden digging sticks for excavating underground vegetable resources.

   This adaptation would be especially important and critical during times of rapid and profound climatic change, as occurred about a million years ago with the onslaught of a cooling and drying phase in Africa. Many animals that had previously been adapted to a warmer, wetter environment were now overspecialized and faced extinction, while other forms, including representatives of the genus *Homo*, appear to have been more flexible and dependent upon behavioral change rather than biological specialization to adjust to changing conditions. They adapted rapidly to these environmental changes, in fact radiating out at this time into Eurasia and adjusting to environments ranging from the tropical grasslands of Africa to the temperate woodlands and plains of Europe to the tropical forests of southeast Asia.
5. The development of an incipient *home base, camp or central foraging place* type of adaptation, in which social interaction may have taken place, especially in the early morning, midday, and late afternoon. During much of the day hominids would move about and forage, individually, in family groups, and sometimes in larger social units of related families. But some specific sites could have become the focus of many sorts of activities—feeding,

tool making, grooming, and sometimes sleeping—and permitted enhanced social interaction and communication, which proved to have high adaptive value.

This is not, however, to say that this adaptive system was identical, or even that similar, to the elaborate system that has been developed by modern hunter-gatherers, who use complex symbolic communication (modern human language and culture) to categorize, verify, and institutionalize their forms of social behavior.

6. As discussed previously, a unique and intense interplay was set under way between cultural behavior and biological change, setting forth this system of *biocultural feedback*. For the first time in the history of life on the earth a vital feedback loop was created between the biological evolution of a species and its adaptation through learned, cultural behavior, with technology an essential aspect of this culture.

   Why had this feedback loop not occurred earlier in the evolutionary history of the earth? At our state of knowledge, the possible reasons or causes remain speculative. It seems that about two to three million years ago, a vital combination of increased intelligence, manipulation, technology, and social organization achieved a "critical mass," unique in hominid evolution, which precipitated our journey down this techno-organic pathway at an ever-accelerating speed.

FORGING AHEAD INTO OUR PAST

It is clear that all Paleolithic archaeological sites are products of complex hominid and nonhominid processes. And it is essential to identify clues to the forces that formed and transformed the sites we find. During the next few decades new archaeological sites will undoubtedly be uncovered, new techniques of analysis developed, and new theoretical approaches advanced. Although we will never know the entire story about early Stone Age hominids and their life-styles, this accumulated information will nevertheless give us a much more realistic perspective about what had actually happened, what was probable, what was possible, and what was unlikely (or untestable). That challenge, to understand and explain the past to the best of our abilities, is what archaeology is all about and the main reason that most of us have thrown our field hats into the archaeological ring in the first place.

The next major threshold in human technology is associated with the emergence of new hominid forms in Africa, usually called *Homo erectus*, about 1.7 million years ago. With these hominids we see some new and different developments in anatomy, technology, and behavior, with some extremely interesting implications for human evolution.

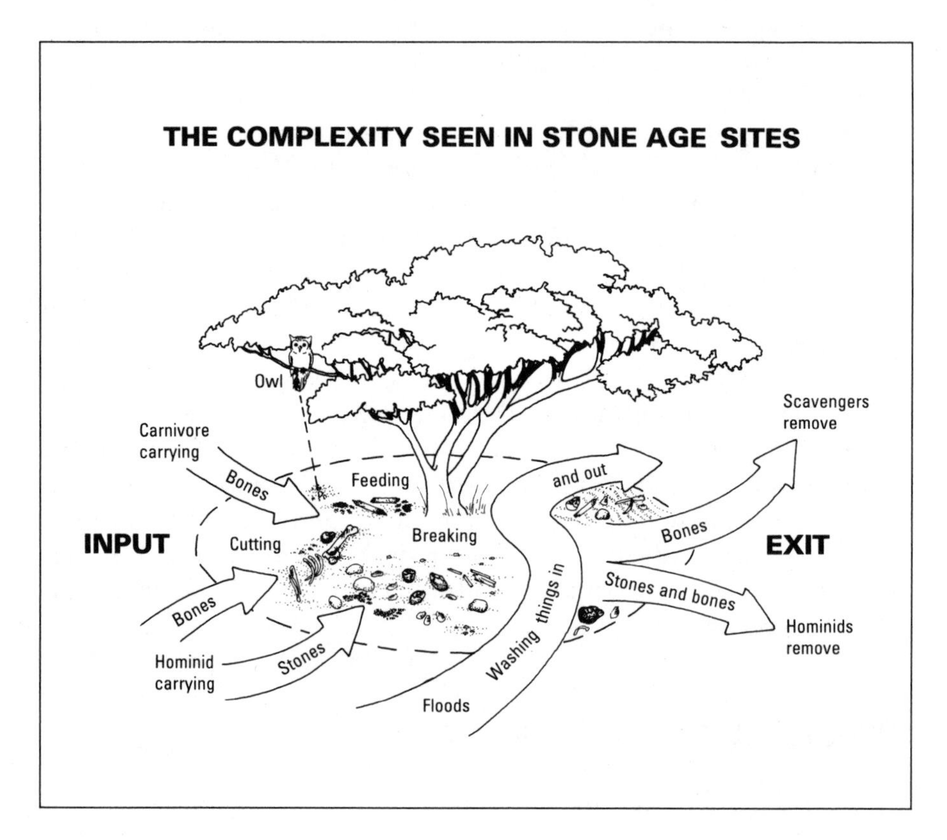

The complex types of hominid and nonhominid processes that can modify an early Stone Age archaeological site. After burial, the site can also be modified by root action, burrowing animals, and other agencies.

C H A P T E R 7

THE HANDAXE MAKERS AND THEIR CONTEMPORARIES

In the volcanic hills overlooking the plains and distant lake there is little vegetation: shrubs and tufts of grass work their way through the rocks here and there, and an occasional thorn tree stands defiant in the coarse rubble. Small lizards dart for safety as the bipedal figure approaches; the uneven black terrain is hot underfoot. Amid the giant rocks of the lava highlands, a figure searches for the ideal size and shape of boulder to use as a core for making his handaxe blanks. He picks up a somewhat smaller stone, raises it above his head, and hurls it violently against the boulder on the ground. The sharp, glassy ring of basalt is heard as a large flake is detached from the boulder core. He hears nearby, but out of sight, the sounds of other members of his group flaking boulders, the distinctive cracks echoing through the rugged hills.

Once this first flake is removed, his core is now ready for serious work. He selects a large cobble of lava as his hammer stone. After positioning the boulder core on top of smaller rocks, he begins striking many large flakes from the core, holding the hammer in his right hand and helping to steady the core with his left. He strikes down hard with the hammer, and he emits a deep grunt as the hammer hits the core a few inches from its edge. After two blows a large flake is removed from the underside of the core and falls on the ground with a dull thud. Three more great flakes are quarried before the hammer stone accidentally splits in half. It is replaced by another large cobble found nearby.

The boulder core is slowly worked down by further flaking blows, repositioned repeatedly as flaking proceeds, until there is very little weathered cortex left on its surface. It is abandoned as a large discoid core, dark bluish gray in color, with bold scars on both sides from flake removals. Spread around on the ground are about a dozen large flakes, as well as hundreds of smaller flakes and fragments, the inevitable waste from such an operation. The hominid's dark, muscular body is glistening in the late morning sun from the exertion of constantly reorienting and bracing the massive core against his powerful blows. His right shin and one of his fingers are bleeding from cuts, a common hazard of this quarrying.

He then selects the best flake blanks from which to make handaxes. Some of the large flakes are too thick, too thin, or too narrow or exhibit the telltale hairline fracture of a fatal, internal flaw. The flakes he chooses are about a foot long, at least six inches wide, and about two to four inches thick. One roughly rectangular flake has a sharp, straight edge along one end. He selects it as a blank for a cleaver.

Carrying five of these blanks, and picking up another, smaller lava cobble for a hammer, he moves into the cooler shade of a lone acacia tree, sitting on his haunches and leaning back against the trunk, where he is joined by several of his fellow tool makers. He now concentrates more on his craft, removing flakes from first one face of the blank and then the other, paying special attention to thinning the large striking platform and bulb of percussion, until slowly an irregular oval form is created, with a sharp edge zigzagging around the circumference of the piece.

He reaches into his large tortoiseshell container and removes his favorite stoneworking tool, a dense giraffe ankle bone bearing the telltale wear marks of having been used as a soft hammer. He deftly strikes it against the edge of the handaxe, detaching longer, flatter flakes, which begin to thin the piece and give it a more even edge and a symmetrical teardrop shape. The final task is carefully thinning and shaping the curved tip and making sure its edge is less sinuous and very sharp. The entire process has taken about twenty minutes. After holding it up for closer scrutiny, he shows the finished tool to his compatriots, sets it down with a contented expression of self-approval, and reaches for another blank.

In a little more than an hour he completes one more handaxe, almost identical to the first, as well as a cleaver. The latter he shapes by removing flakes around most of its circumference to make a U-shaped profile, which will serve as the handle, leaving the natural flake edge at the other end as a straight, sharp, cutting bit. Two other attempts at fashioning handaxes lead to failures: one specimen breaks early because of an invisible flaw in the stone that splits the piece in two, and a forceful, misplaced blow with the lava hammer takes off the entire tip of the other. On both occasions he snorts with displeasure.

He places the bone hammer and the three finished forms in the tortoiseshell container, wedging a wad of coarse grass between each tool to cushion their sharp edges during transport. A natural water hole in the highlands, formed from massive column-shaped walls of basalt, provides refreshment and cooler shade. Within the hour the rest of his group has joined him, each adult with his or her own large cutting tools of surprisingly similar sizes and shapes, while those of most juveniles are much cruder and more irregular. Then they begin their slow descent from the lava highlands to the plains.

A kene ax him sulf he huld.

Robert of Gloucester
***Metrical Chronicle,* 1297**

INTRODUCTION

Two dramatic changes occur in the African prehistoric record between approximately 1.7 and 1.5 million years ago. One is biological: the evolutionary emergence of a new and larger-brained form of the genus *Homo.* The other, just as striking and conspicuous, is technological: the first real departures from the simple Oldowan tools of the previous several hundred thousand years. In this new technology we see larger forms of tools, created with a more standardized appearance or design, called handaxes, picks, and cleavers. These forms are the hallmarks of what archaeologists call the Acheulean industrial tradition, named for the Paleolithic site of St. Acheul in France, which was discovered in the nineteenth century.

The Acheulean spread across vast distances, through Africa, much of Europe, western Asia, and India. During this time we see our first evidence of hominids ranging outside the African continent. All of this occurred when the earth was experiencing a number of pronounced cold phases, which began about 2.5 million years ago but became so severe by about 1.6 million years ago that they became associated with major glaciations, or "ice ages." Recent studies of the oxygen isotope composition of deep-sea geological cores debunked the old view of only four major cold phases alternating with four warmer phases during this time, and they suggest that at least seventeen major cold and warm phases occurred during the past 1.6 million years. At their peak, the planet's ice sheets grew to about three times their present size and covered almost a third of the earth's surface.

This period was important not only in our evolution, but in that of other animals as well. Geologists and paleontologists call this period of oscillating cold and warm phases the Pleistocene, characterized by this overall global cooling and the emergence of new species of animals, including direct ancestors of some modern forms such as horses (*Equus*). They divide the Pleistocene epoch into three major phases:

- The Early Pleistocene, 1.6 million years ago to the paleomagnetic reversal, about 780,000 years ago, when magnetic north began its latest major stint at the North Pole;
- The Middle Pleistocene, 780,000 years ago to 125,000 years ago, when the next-to-last major cold pulse ended;
- The Late Pleistocene, 125,000 years ago to the end of the last Ice Age, about 10,000 years ago.

Strategies for success: a series of quartzite Acheulean artifacts from Stellenbosch, South Africa, illustrating the major stages in biface manufacture. *From left:* two boulder cores, four large flakes removed from such cores, and three hand axes and two cleavers fashioned from such large flakes.

These Pleistocene glacial periods would have been a tough time for many tropical primates: mean annual temperatures dropped significantly, especially at higher latitudes; world sea levels sometimes lowered by as much as five hundred feet (as more of the earth's water was locked up in massive ice sheets a mile or more thick), exposing vast continental shelves and land bridges in places; and more arid conditions existed on a worldwide scale, even at more tropical latitudes. This had profound effects on the plants and animals and on the geographic distribution of hominids, especially in the higher latitudes of the northern hemisphere. Continental drift had dealt uneven hands to the different hemispheres: there is a much greater concentration of landmasses in the north, which allowed more cooling and greater buildup of glacial ice, making these environments much colder than in Africa and other tropical regions during glacial periods.

The Acheulean is remarkable for how long it persisted and how far it spread, which distinguish it from any other technological tradition in the history of humankind. This new technology lasted for well *over a million years*, during which time a new hominid form spread across much of the Old World.

THE EMERGENCE OF *HOMO ERECTUS*

Toward the beginning of this period, a new fossil character enters the scene. Beginning approximately 1.7 million years ago, at the beginning of the Pleistocene epoch, fossil skulls in East Africa exhibit new features that appear to have evolved out of some form of early *Homo*. These fossils bear many similarities to Asian fossils that show up later, between 1,000,000 and 250,000 years ago. Traditionally these all have been grouped by many paleoanthropologists into the same species: *Homo erectus*. (We should note that there is growing dissent whether all these fossils—from Africa to Asia, spanning over a million years and showing some important morphological changes—should be put in the same or different species. But for the purposes of our discussion here, we will consider them as one species.)

These new hominid fossils are characterized by much larger skulls than *Homo habilis*, with a cranial capacity ranging from about 850 to 1,100 cubic centimeters (about two-thirds to three-quarters the size of modern humans). The skulls are thick-vaulted, with a sloping forehead, a massive and projecting brow ridge with a pronounced furrow

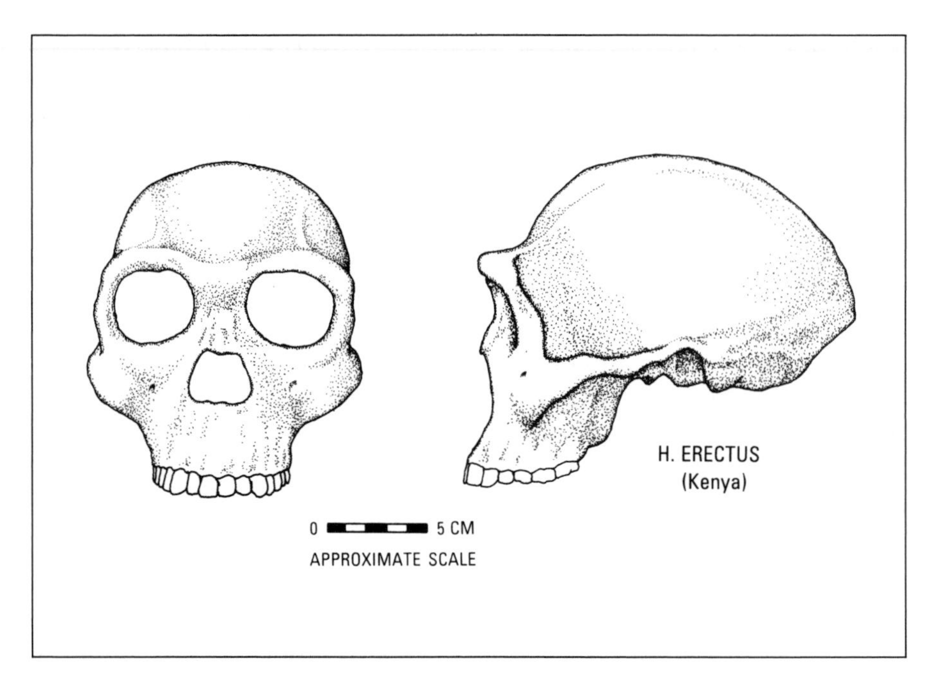

An early example of *Homo erectus* dating to approximately 1.7 million years ago, from Koobi Fora (East Turkana) in northern Kenya. Soon after the emergence of this hominid in the evolutionary record the earliest Acheulean handaxes and cleavers can be found.

behind it, a pentagonal shape when viewed from the rear, and a strong angle to the bone at the back of the skull (the occipital bone) when viewed from the side. Its face and jaws jut out markedly, and the lower jaw is robust. The teeth are smaller and in different proportions than in *Homo habilis*. The limb bones are also robust, with strong markings for the attachment of muscles, suggesting very active and powerful animals. Based on the range of variation known from the fossil record, there appears to be less difference in the overall size of males and females than in earlier fossil hominid forms.

The earliest known fossils of *Homo erectus*, about 1.7 million years old, come from localities on the eastern and western shores of Lake Turkana in northern Kenya. This includes one nearly complete skull with a face and another skull cap from East Turkana. More recently, a famous, almost complete skeleton was found at a locality called Nariokotome on the western side of Lake Turkana. This skeleton, found by a research team led by Richard Leakey, Kamoya Kimeu, and Alan Walker in 1985, is of an exceptionally large youth (estimated to have died at about twelve to fourteen years of age) who would have been an estimated six feet tall when fully grown.

Other important African *Homo erectus* fossil sites include Olduvai Gorge in Tanzania (Beds II and III), the Casablanca area of Morocco (the Sidi Abderrahman and Thomas stone quarries), Ternifine in Algeria, and the cave of Swartkrans in South Africa. During this time period, by about one million years ago, the smaller-brained australopithecine lineages that had persisted in eastern and southern Africa (*Australopithecus boisei* and *Australopithecus robustus*, respectively) both go extinct, leaving the larger-brained *Homo* populations as the sole inheritors of the hominid world. The possible implications of these extinctions will be discussed at the end of this chapter.

THE EARLIEST ACHEULEAN

WHAT DISTINGUISHES THE ACHEULEAN?

By around 1.7 million years ago, the crude and simple cobble-core and retouched-flake technologies of the Oldowan tradition were starting to be supplemented by dramatically new technological elements never before seen in the archaeological record: large tools with pointed or sharp-edged tips, usually made on massive flakes that had been struck from giant cores. Unlike the Oldowan artifacts, which show no definite preconceived shape that their makers were trying to produce, these Acheulean tools show unequivocally that these hominids had specific mental templates of the forms they wanted.

What are these new tools, these products of *Homo erectus*'s ideas and designs? The distinctive Acheulean tool forms are usually divided into three major types:

1. *Handaxes* are bifacially flaked tools, often teardrop or pear-shaped in outline. They are usual flaked around all or most of their circumferences, and the edges are normally carefully flaked and fairly sharp near the tips, an effort evidently made to thin the tools at this end. In past Paleolithic literature they have been called *coups de poing* (French for "blow of the fist"), *bouchers* (after the French prehistorian Boucher de Perthes, previously mentioned), and bifaces (for being flaked on both sides), but the term *handaxe* is most commonly used.
2. *Picks* are usually made on large flakes or on long, flat cobbles and are flaked on one side (unifacially) or both sides (bifacially). They

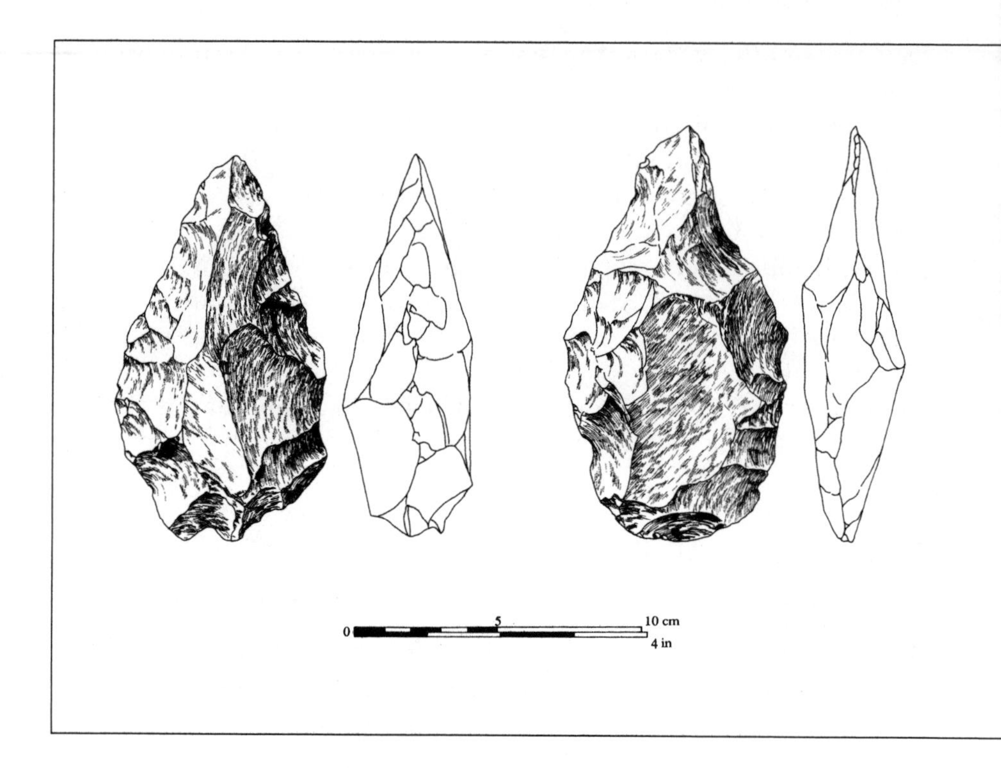

tend to be fairly thick, with fat, angular cross sections (shaped like a triangle or lozenge). There tends to be less emphasis than with handaxes on creating sharp cutting edges at the tip ends, which instead usually form thick, triangular points. (In the very early Acheulean it is sometimes difficult to discriminate between handaxes and picks because both tend to be crude and thick.)

3. *Cleavers* are large tools with sharp bits at one end, usually formed by an acute, natural flake edge. Flakes have normally been removed around the rest of the tools' circumferences, resulting in long, U-shaped outlines. When these tools are made on large flakes, their striking platforms are often removed and the prominent bulbs of percussion are also pared down by flaking, as the hominid tried to thin and shape the pieces.

The terms *biface* or *large cutting tool* are sometimes applied to all three of these forms—handaxes, picks, and cleavers (Mary Leakey used the term *biface* at Olduvai Gorge.) These tools, especially the handaxe, are considered to be emblematic of what prehistorians have called the Acheulean industrial tradition. We can see the very early

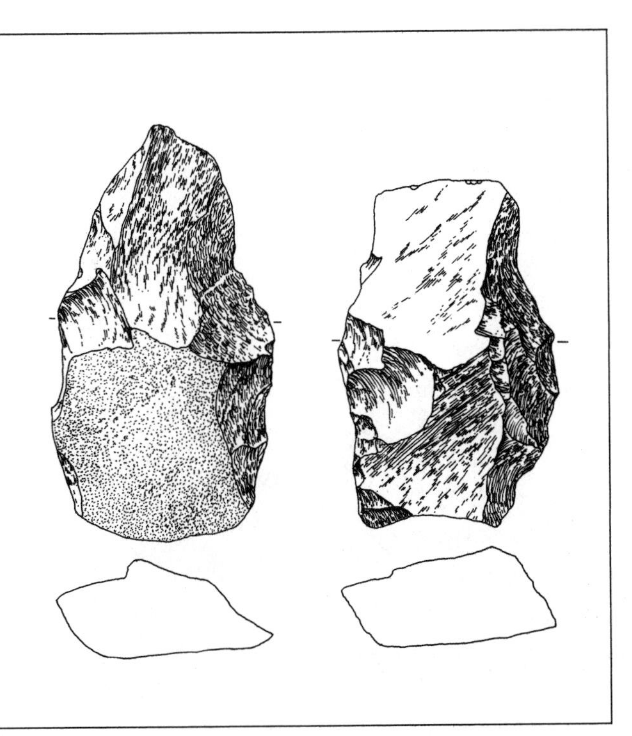

A new sense of style: early Acheulean artifacts from Bed II of Olduvai Gorge, dating to approximately 1.5 million years ago. Three crude handaxes and one cleaver are shown. Such artifacts appear to show more determination in form than most Oldowan artifacts.

phases of this new technology at several prehistoric localities in East Africa, including Olduvai Gorge (mid-Bed II) and Peninj in northern Tanzania, the western side of Lake Turkana in northern Kenya, and the newly discovered site of Konso-Gardula in southern Ethiopia.

EARLY ACHEULEAN EXAMPLES

This early Acheulean locality is located on the western side of Lake Natron in northern Tanzania, not far from the Kenyan border, about fifty miles northeast of Olduvai Gorge. In the area of the Peninj River here, an expedition led by Richard Leakey and Glynn Isaac in 1967 discovered two sites with concentrations of crude handaxes, picks, and cleavers, dated to about 1.5 million years ago. This expedition also uncovered the first known and well-preserved mandible of *Australopithecus boisei* at another locality. Subsequent archaeological research by Glynn and Barbara Isaac, Amini Mturi of the Tanzanian Antiquities Department, and ourselves in the Natron study area in 1982 located several more of these early Acheulean occurrences, believed to be the same approximate age.

A view looking across the Great Rift valley from the western side of Lake Natron in northern Tanzania. The sediments in the foreground have yielded several early Acheulean sites dated to approximately 1.5 million years ago.

At Peninj, as at Olduvai, we can see once more a clear technological shift about 1.5 million years ago: hominids start becoming involved, even preoccupied, with the manufacture of these large, crude Acheulean forms, while also continuing to make the Oldowan-like artifact forms hominids had been making for hundreds of thousands of years (that is, simple cobble cores and casually retouched flakes).

These earliest Acheulean industries in Africa share features that set them apart from earlier technologies. The large, generally bifacial forms of handaxes, picks, and cleavers are generally made on large flakes struck from lava boulders; the Acheulean tool makers were becoming much more ambitious and proficient in their exploitation of raw materials, working with massive pieces of lava rather than just small cobbles. In addition, these tools show much more regularity in form or style than one sees in Oldowan industries.

Other notable earlier Acheulean localities are known in Africa from deposits believed to date from 1.5 to 0.5 million years ago. Most have been discovered in river deposits, suggesting a correlation between the use of handaxes and cleavers and hominid use of river valleys rather than lake margin habitats. These sites include Melka Kunturé and Gadeb in Ethiopia; Koobi Fora, Kariandusi, Chesowanja, Kilombe, and Olorgesailie in Kenya; and Beds II, III, and IV at Olduvai Gorge, Tanzania. Other earlier Acheulean sites include river and lake deposits at Ain Hanech and spring deposits at Ternifine in Algeria; several marine beach and dune localities in the Casablanca area of Morocco; and the later cave deposits at Sterkfontein in South Africa.

One lingering question is why quite different-looking tool industries sometimes occur at sites very near one another in the same time period —at Olduvai Gorge and Olorgesailie, for example. Mary Leakey has documented the coexistence of early Acheulean sites, with relatively high proportions of large bifaces, and what she called "Developed Oldowan" sites, in which bifaces are few and normally show inferior craftsmanship. Acheulean sites tend to be found at this time at Olduvai along river systems in prehistoric highlands, while Developed Oldowan sites are often located closer to the lake, at the center of the sedimentary basin. She has suggested that these differences could be the result of different cultures or traditions of different hominid groups (or perhaps even different hominid species). Other researchers, such as Lewis Binford and the late Glynn Isaac, have suggested that the differences may be due primarily to different activities taking place at different places, with only some of these activities requiring the large handaxes and cleavers that characterize the Acheulean tradition.

A collection of early Acheulean lava handaxes and cleavers found at Peninj, Lake Natron, Tanzania. Most are made on large flakes.

Another possibility has been suggested by two different archaeological studies done at Olduvai Gorge: an analysis done by Dan Stiles and experimental replications done by Peter Jones. These studies suggest that Developed Oldowan bifaces appear to be cruder primarily because hominids at these sites were using inferior or more intractable types of stone. This may suggest that the distinction between Acheulean and Developed Oldowan may be due, *at least in part*, to differences in raw materials, which in turn could be due to different functional requirements or land-use patterns. At sites closer to the lake, the most desirable raw materials for these bifaces may have been too far away, and the lower-quality bifaces of the Developed Oldowan were the result.

HOW WERE HANDAXES AND CLEAVERS MADE?

EXPERIMENTAL STUDIES

Our experiments have shown that the making of handaxes and cleavers is much more complicated and patterned than the production of Oldowan technologies: from start to finish, the technology is directed toward a clear end product, and the amount of skill and often the amount of strength required to make such Acheulean forms is much more formidable.

The technique of striking off large flake blanks from lava, quartzite, obsidian, or flint boulder cores is especially common in the Acheulean of parts of Africa, the Near East, Iberia, and India. It requires a large hammer stone and a lot of force. Not often stressed by archaeologists is the inherent danger. We know from experience that the injuries produced in quarrying massive flakes from boulder cores can be formidable, especially if one is scantily clad. We suspect that death due to loss of blood from a severed artery was probably not unknown in Acheulean times. Accidental injuries from flaking stone may have been one of the most common "occupational hazards" during these times.

Finding suitable raw material does not mean simply locating a source of stone that can be flaked, but also finding stone significantly larger than the dimensions of the handaxes or cleavers to be made. It will be flaked and worked down considerably before the biface is ultimately produced. The most common rock types used are lavas and quartzites, but in some places quartzes, obsidians, flints, or cherts were used as well. The quarry area might be at the primary, bedrock location of the raw material, such as an outcrop or cliff face, or it might be a secondary source, such as a high-energy river gravel containing lots of boulders.

In antiquity, handaxes were produced in two main ways. One, used especially in working lava or quartzite, is to take boulder cores and strike off large flakes to be used as starting blanks for the bifaces. The other, normally used when smaller nodules of flint or chert are available, is to select a large discoid cobble as a starting point for the biface.

Experimental Acheulean quarrying. A lava boulder core is struck with a large hammer stone to detach great flake blanks. The technique of striking flakes from boulder cores to use as biface blanks was used by Acheulean hominids especially where lava and quartzite were the primary raw materials.

Quarrying large flakes from lava or quartzite boulder cores is strenuous as well as potentially dangerous. Normally a large hammer stone is used, between five and ten inches in diameter (the larger hammers are held in both hands). When boulders are massive and not very angular, often the biggest hurdle is removing the very *first* flake from the rock; this can sometimes be facilitated by throwing another boulder against the potential core to initiate fracture. After this, further flakes are then removed from the boulder with a hand-held hammer stone, the core supported with smaller rocks to position it for the flaking blows.

A large flake that was struck from the boulder core, and subsequently fashioned into a handaxe. The ability to produce such large flakes was one of the hallmarks of the Acheulean, beginning about 1.7 million years ago in Africa.

A range of Acheulean forms produced in experimental replicative studies. *From top left:* an ovate handaxe, a pointed handaxe, a cleaver, a pick. *From bottom left:* a spheroid, a flake scraper, three handaxe trimming flakes.

The next step is roughing out an irregular bifacial form. Using a hard hammer, flakes are alternately removed from both faces of the large flake, with special attention paid to removing the bulges in the flake—the thick striking platform and the swollen bulb of percussion. The tip end is generally flaked to make it narrower and sharper than the butt end (which might even retain a lot of the shape of the original flake or cobble used). In the early Acheulean, flaking usually stopped at this stage of refinement, while in later Acheulean times, after one million years ago or so, some handaxe industries begin to show more refinement.

This refinement in the later Acheulean is evident in the great symmetry of some of the bifaces and in how the hominids were able to make them very thin relative to their width. The skill of thinning bifaces, one of the most fundamental abilities that later hominids mastered, is not an intuitively obvious procedure for beginners. In fact, the handaxe edge is usually steepened first by removing small chips so that more force can be directed to the edge when the thinning flake is struck off. This chipping at the edge is called "platform preparation" and is a technological breakthrough that becomes a characteristic of this time. Either a stone hammer, or more preferably a soft hammer of bone, wood, antler, or softer stone, can be used for striking off the thinning flakes.

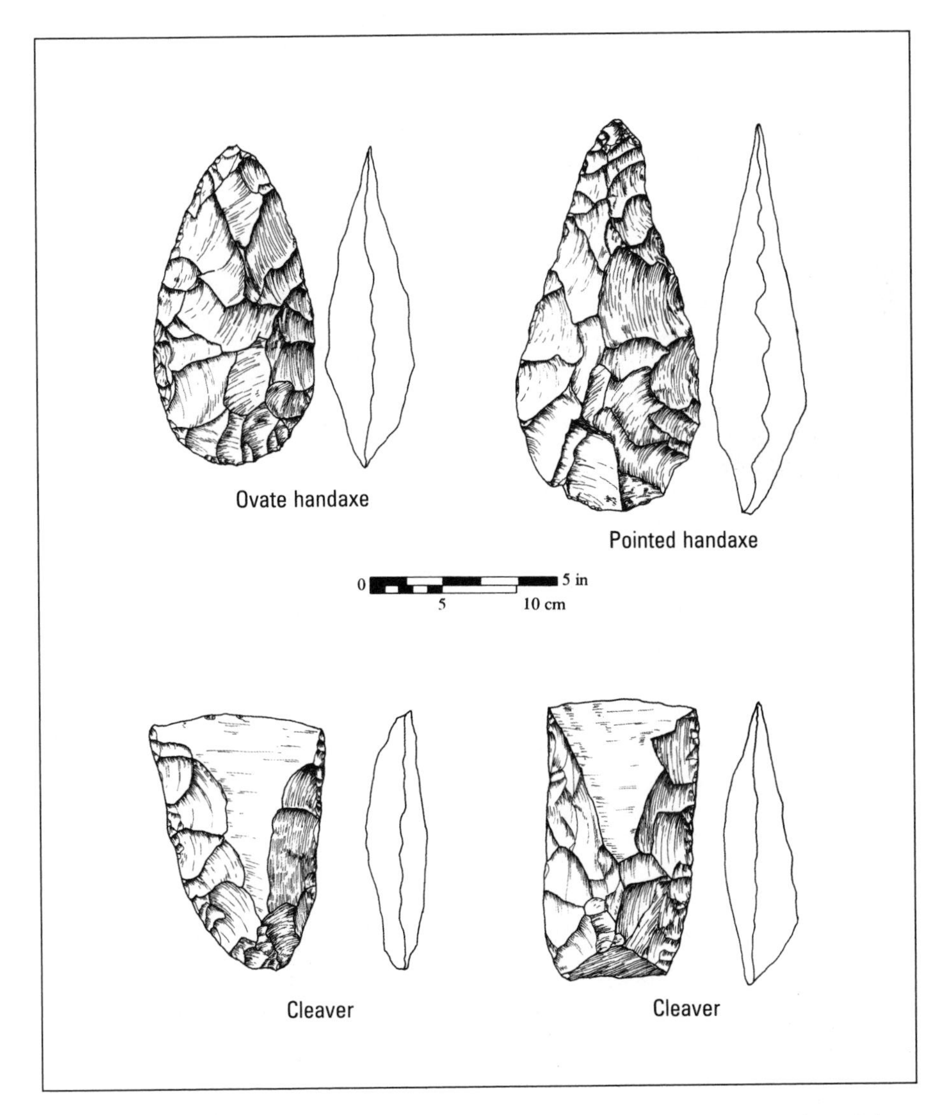

A stronger sense of symmetry, technological finesse, and perhaps even aesthetics: later Acheulean quartzite handaxes and cleavers from Kalambo Falls, Zambia, probably dating to between 400,000 and 200,000 years ago.

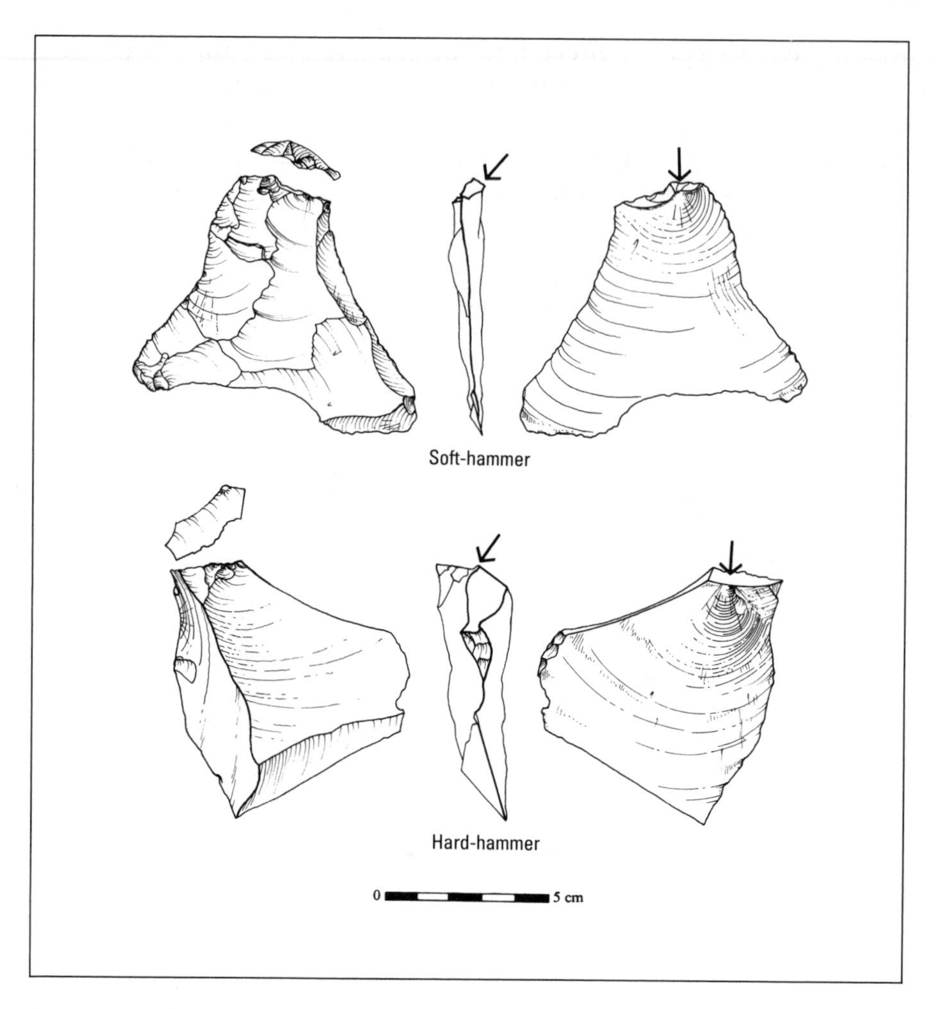

Top, characteristic soft-hammer trimming flake versus *bottom,* a hard-hammer flake. Soft-hammer flakes tend to be thinner with a less pronounced bulb of percussion.

These flakes must be thin and not remove too much of the edge of the biface. Such final shaping of the handaxe normally involves removing smaller thinning flakes from the edges to achieve a more symmetrical shape and straighten the edges. In most cases special care is used when fashioning the tip to make it thin, straight-edged, and sharp. The final shape varies from one Acheulean site to another: sometimes they tend to be more oval or pear-shaped, sometimes more elongated and pointed.

When flaking flint or chert, the ideal starting form is usually a flat, disc-shaped or tabular cobble (rarely are very large boulders of flint to be found). By alternately removing flakes from both faces of the cobble and working around its circumference, an irregular, disc-shaped biface with a flaked edge around its entire perimeter is produced. Thinning and final shaping can again be done using a hard or soft hammer, with platform preparation common for more refined examples. Again, there tends to be great variety in the sizes and shapes of the final forms at different sites. In practice, handaxes of these finer-grained raw materials are easier to produce, requiring less force to fracture the stone.

Using an antler soft hammer to resharpen an ovate flint handaxe during an experimental elephant butchery.

A large piece of flint suitable for handaxe manufacture. One flake has been removed from a corner of the cobble to inspect the quality of the inner stone.

Cleavers of lava or quartzite are usually produced by selecting large flake blanks with one sharp, regular edge, which is kept unmodified as the guillotinelike cleaver bit. The piece is then flaked around the rest of the circumference by hard-hammer or soft-hammer percussion until the characteristic U-shaped outline is achieved. Usually handaxes are made when working flint and chert; cleavers are relatively rare, as nodules do not offer the sharp flake edge required for their bit end. Some hominids, especially in western Europe, did develop a clever

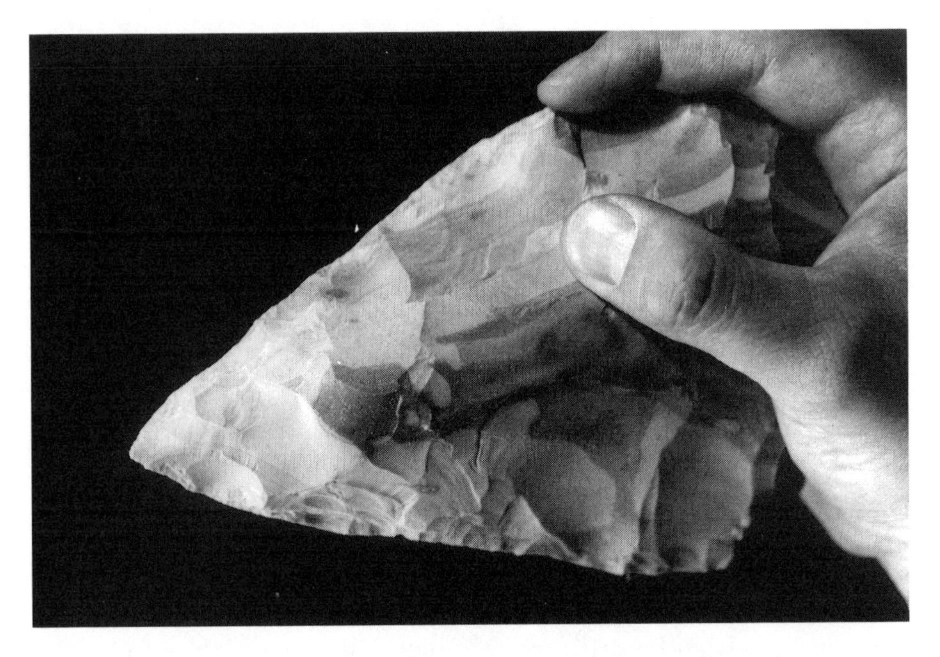

A pointed flint handaxe experimentally manufactured from the flint cobble at left. Final thinning and shaping were executed with an antler "soft hammer," which bites into the edge of the biface, producing long, thin trimming flakes characteristic of many later Acheulean industries. When Acheulean hominids had flint for handaxe manufacture, it was usually in the form of cobbles, nodules, or tabular chunks rather than large flakes as with lava and quartzite.

method to produce this, though: a sharp bit was sometimes made on an oval handaxe by removing an oblique flake that took off the entire tip of the biface (this is called a "tranchet" flake, from the French for "to cut off"). The intersection between the flake scar and the opposite face of the tool forms a sharp cutting edge.

Experiments in manufacturing Acheulean forms are useful for understanding various ways hominids could have produced their handaxes and cleavers and the types of technological problems they would have encountered. In addition, a group of modern stoneworkers from New Guinea provides clues that may shed more light on Acheulean technologies and their hominid makers.

THE ADZ MAKERS OF HIGHLAND NEW GUINEA

From a technological perspective, it would be extremely interesting if any modern human groups made and used stone tools resembling Acheulean handaxes. Unfortunately no such groups have ever been described. There are, in fact, few human groups today that rely on flaked stone tools for a major part of their livelihood. Only in certain remote corners of the world, where metal tools from the industrialized world have not yet made a significant impact, can one find people with a high degree of traditional stone technological skill. The skills of our ancestors are truly dying arts, but there are still a few vestiges.

In 1985 we were startled to find, in an Italian travel magazine, an article written by Italian businessman and anthropologist Giancarlo Ligabue, who had visited a group of people in a remote part of New Guinea who still manufactured and used flaked stone adzes. The photographs showed these people striking large flakes off boulder cores, working them down with hammer stones into narrow bifaces, then grinding them into finished adz heads and hafting them with wooden handles. The early stages of this stone technology, especially the quarrying process and the bifacial roughing out, seemed striking similar to Acheulean technology.

One of us (NT) had the privilege of joining a second international expedition to Irian Jaya (the western half of the island of New Guinea) in 1990. Organized once again by Giancarlo Ligabue and sponsored by the Centro Studi Ricerche Ligabue, our expedition, including J. Desmond Clark of U.C. Berkeley, traveled high into the central mountains by helicopter to visit and study a group of people called the Kim-Yal, who live in Langda village at an altitude of approximately six thousand feet above sea level.

These adz makers, all men, are members of a horticultural community of about two hundred people whose principal domesticated food resources include sweet potato, chicken, and pig. Their diet is supplemented by hunting wild animals, usually with bows and arrows, and gathering wild plants that grow at that altitude. Their hafted stone adzes are used for felling trees and clearing fields, as well as other woodworking tasks. They are also used in ceremonial exchange and traded to other groups that do not have suitable stone for adzes.

The principal raw material for the adz heads is a blue-gray lava found as boulders in a river bed deep in a valley one half of a mile below Langda. A precipitous path winds from the village through terraced fields and mountain forest to the river in the valley below.

An especially fascinating technique is typically used by the Kim-Yal to quarry the large flakes for adz manufacture. They stand with the boulder core to be flaked behind them, take a large cobble as a hammer stone, and with both hands swing the hammer through their legs (rather like an American football center hiking the ball) and strike the

Clues from the present: a New Guinea stoneworker quarrying large flakes from a lava boulder core, as did Acheulean hominids. This unusual technique of swinging the large hammer stone through the legs probably reduces the chance of serious lacerations to the legs (or other body parts) and could have been used by prehistoric hominids as well.

core behind them. The flakes that are detached are sometimes more than a foot long. A great deal of debris is created by the quarrying process at the river, from which the most suitable flake blanks are chosen for further work. Sometimes one boulder is thrown against another to initiate fracture.

With a smaller stone hammer, the flake blank is then shaped into a crude, elongated oval bifacial form, reminiscent of a narrow handaxe. The adz makers strike the blank with force, removing large, thick flakes from both faces as they work around the periphery of the piece. These rough-outs are then usually transported back up to the village to be finished, a number of them wrapped in leaves and carried up the mountainside in woven net bags slung from the backs of the villagers.

A skilled New Guinea adz maker has selected a large flake blank and, with a hammer stone, roughs out a large biface that resembles a handaxe during the early stages of manufacture. Note the waste flakes on the ground, similar to those produced by Acheulean hominids at prehistoric sites.

Up at the village, the final shaping is done with a yet smaller stone hammer and more delicate and precise flaking. Stone knappers like to sit in a line as they flake, talking to each other and uttering the exclamation *ha-dok* when an ideal flake has been detached. This refined flaking is a very involved task: edges are carefully prepared by chipping and grinding until a suitably steep edge is achieved, then a flake is detached from the opposite face. When flaking is finished, the final product is a long, narrow adz with a curved bit and a triangular cross section. In their habitation area, the only signs of this whole technological process are the waste flakes from the last stages of stone reduction along with the broken discards from adz manufacture and use.

Master craftsmen at work: a line of New Guinea stone workers at various stages of stone adz manufacture. Note the large handaxelike form in the *left* foreground and the finished elongated forms at the bottom *right,* ready for grinding and hafting. All of the flaking is done with stone hammers and produces large quantities of waste flakes and fragments in the mountain village.

A farmer's technology: the later stages of adz manufacture by the New Guinea stoneworkers are usually associated with the rise of farming communities in the prehistoric record. A flaked adz has been ground and polished and is being tightly hafted to a wooden handle using split liana vine as a binding material. These tools are very efficient for chopping down trees, woodworking, or clearing fields or for using as expedient weapons. Ground and polished axes and adzes became common only in the last ten thousand years of the archaeological record.

The final phases of adz manufacture involve techniques one normally finds only in the last ten thousand years or so of the prehistoric record, with the emergence of farming communities in the Old World and the New World: grinding and polishing the adz against another rock with water as a lubricant to create a regular and durable working edge. The adz is then lashed securely with strands of split cane to a wooden handle and is ready to use.

Even though these skilled stoneworkers are anatomically modern humans like ourselves, with most of their food coming from a farming economy and *not* by hunting and gathering, their stone technology nevertheless gives us clues to the possible ways that Acheulean hominids could have approached technological challenges. Especially interesting are their means of quarrying large flakes and flaking these into large bifacial forms. We felt privileged to document and study these New Guinea crafstmen, and they, as we were told through our translator, were amazed and honored that anthropologists were willing to travel halfway around the world just to meet them and study their technology.

OUT OF AFRICA (AND INTO EURASIA)

Several hundred thousand years after *Homo erectus*'s first appearance in Africa, this new species was on the move. It began a dramatic expansion in its geographic range, spreading not only throughout Africa, but far beyond. Hominid fossils and Stone Age sites finally begin to appear outside of Africa, showing up sporadically at sites in Eurasia, documenting the first hominid migrations out of the African continent.

WESTERN ASIA

This hominid expansion out of Africa seems to have begun sometime before one million years ago. This may well have been during a time when sea levels were low, exposing a substantial land bridge in the area north of the Red Sea between the Nile and the Sinai peninsula (across Egypt, through where the present-day Suez canal runs). From the Sinai, during this Pleistocene exodus, hominids could have entered into southwest Asia through the Jordan Valley, and it is precisely here that we find what is probably the earliest evidence of hominids out of Africa: the prehistoric site of el 'Ubeidiya.

Ubeidiya is a site on the western flanks of the Jordan valley, not far from the Sea of Galilee. It is composed of ancient lake and beach deposits, alternating between cobble beaches and finer-grained lake sands deposited sometime between 1.4 and 1.0 million years ago, based on paleomagnetic studies and faunal correlations. More recent earth movements have inclined these beds from their original horizontal position to a near vertical one, giving an impression of looking at a wall rather than a floor.

Here Oldowan-like as well as early Acheulean handaxe levels were uncovered, with stone artifacts made from lavas, flints, and silicified limestones and numerous fossil animal bones. Although only a few hominid teeth have been found, presumably from these levels, they are consistent with a *Homo erectus* morphology.

EASTERN ASIA

As hominids moved from the Near East into eastern Asia, they would have encountered formidable barriers: the cold Eurasian steppes as they moved to the north, the Himalayas as they advanced eastward into the Indian subcontinent, and, farther to the east, the dense tropical forests of Southeast Asia. Perhaps the most likely route of hominid migration would have been from the Near East through present-day Saudi Arabia, Iraq, Iran, Pakistan, India, Bangladesh, and Burma; from there, hominids could have migrated north into China and Korea, east into Thailand, Laos, and Vietnam, and south into the Malay Peninsula, and into Sumatra and Java during times of low sea levels when these islands and others formed a subcontinent, Sundaland.

Recently, claims have been made for a remarkably early hominid occupation in Pakistan. Archaeologist Robin Dennell of the University of Southampton has reported that possible quartzite artifacts (simple cores and flakes) have been discovered from river gravel deposits from Riwat in northeast Pakistan. Based upon paleomagnetic studies and a fission-track date, it has been reported that the deposits in which these stone materials were found are approximately two million years old. If these claims are substantiated, they would drastically alter conventional wisdom about this period of human evolution, suggesting that *Homo habilis* or some other Oldowan tool-making hominid existed outside Africa during the late Pliocene and thus their range was much more extensive than is currently believed by most authorities.

The status of this find is still unclear. Some prehistorians have questioned whether the stone materials are truly artifactual or whether they might be the product of natural fracture of quartzite cobbles in a high-energy geological setting such as cliffs or waterfalls upstream from the site. There have also been questions as to the actual date of these gravel deposits. If the age was actually later in time, the site might fall into the known time period of hominid occupation in this part of the world; if the site ends up to be substantially earlier than two million years, then it would be even more of an anomaly, predating any known occurrence of stone artifacts anywhere in the world. Further fieldwork should ultimately resolve this question.

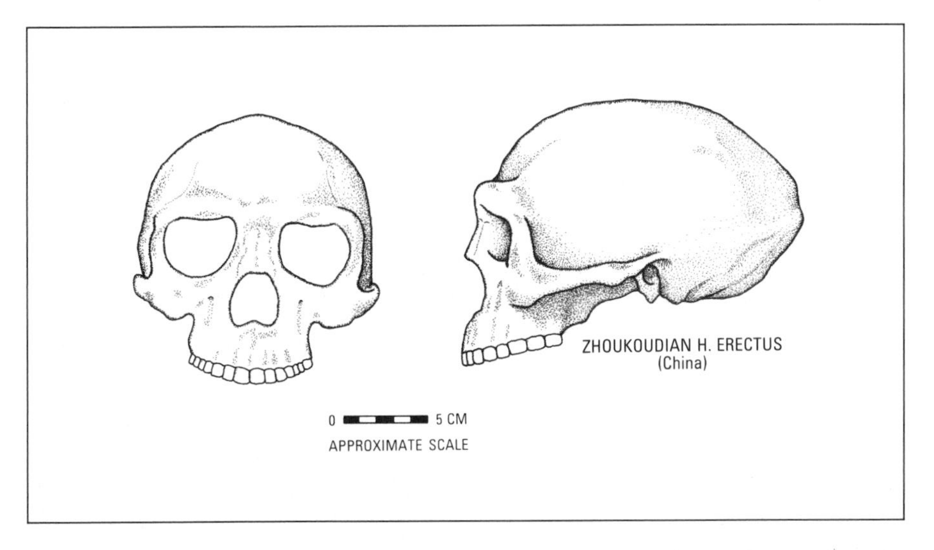

An example of an east Asian *Homo erectus* from Zhoukoudian cave (Peking Man site), north-eastern China.

Elsewhere in eastern Asia, the evidence is less jarring: early hominid occupation appears to go back to about one million years ago. In China, Paleolithic sites in the Nihewan basin (to be discussed in more detail later in this section) and *Homo erectus* fossils from Lantian have been estimated to be approximately one million years old based upon paleomagnetic studies. From somewhat later in time, the famous Peking man locality of the Zhoukoudian cave to the west of Beijing yielded numerous fossils of *Homo erectus* as well as many tens of thousands of stone tools, mostly made of quartz, in a series of deposits that began an estimated 500,000 years ago and ended about 250,000 years ago. These early Chinese sites are essentially Oldowan-like industries with simple cores, casually retouched flakes, and the associated flakes and fragments.

And on the island of Java, connected to the Southeast Asian mainland during Pleistocene lower sea levels, have come *Homo erectus* fossils from several localities believed to be slightly younger than the earliest Chinese sites, perhaps 700,000 years old. Curiously, the early hominid localities in Java do not appear to have stone artifacts definitely associated with the fossil-bearing deposits, although simple, Oldowan-like artifacts have been discovered at other localities which are much more recent or of uncertain age.

A spectacular and almost surreal view of the prehistoric sediment outcrops in the Nihewan basin of northern China. Nonhandaxe stone industries with fossil animal bones—dating back approximately one million years ago and representing the earliest secure evidence of early Stone Age hominids in Eastern Asia—have been excavated by Chinese researchers.

An Early Stone Age Example from Asia: The Nihewan Basin, China

For the past four years we have been participating in archaeological investigations that bear upon the problem of the earliest inhabitants of eastern Asia and the question of nonhandaxe industries in that part of the world. Diligent research by Chinese prehistorians over the past several decades has dramatically increased our knowledge of human evolution and adaptation in that part of the world.

Among the archaeological sites considered to be the oldest known in eastern Asia are several in the Nihewan basin region, approximately one hundred miles to the west of Beijing. Since 1991 we have been participating in an international collaborative project here by the invitation of Professors Jia Lanpo and Wei Qi of the Institute of Vertebrate Paleontology and Paleoanthropology (IVPP) in Beijing and Dr. Xie Fei of the Institute of Cultural Relics in Hebei Province. Wei and Xie, as well as a number of other Chinese scientists, have spent many years exploring, discovering, and excavating prehistoric paleontological and archaeological localities here. Dr. J. Desmond Clark (the U.S. team leader), and graduate students Dennis Etler (U.C. Berkeley) and Dong Zhuan (Indiana U.) have also been our colleagues on this project.

The Nihewan area is a large sedimentary basin presently being exposed by the Sangaan River, which has carved out a valley about fifty

A winter visit to the Donggutuo site in the Nihewan basin, China, dated to approximately one million years ago. This well-preserved site contained layers with simple Oldowan-like stone artifacts and many fossil animal bones, some with cut marks. *Far left,* the authors, Nicholas Toth and Kathy Schick; *third from left,* Dennis Etler (University of California); *fourth from right,* J. Desmond Clark (University of California); *third from right,* Wei Qi (the excavator, from the Institute of Vertebrate Paleontology and Paleoanthropology in Beijing).

miles long by five miles across and three hundred feet deep, a sort of giant-size Olduvai Gorge. The view from the rim of the valley is breathtaking: badland exposures and terraced fields, with horizontal beds clearly marked in different-colored bands down to the fertile green valley floor. These deposits range in age from the late Pliocene (perhaps two million years ago) up to recent times. The earliest archaeological occurrences in the sequence are estimated, based on the paleomagnetic events in the sedimentary record, to between about 1,000,000 and 780,000 years ago.

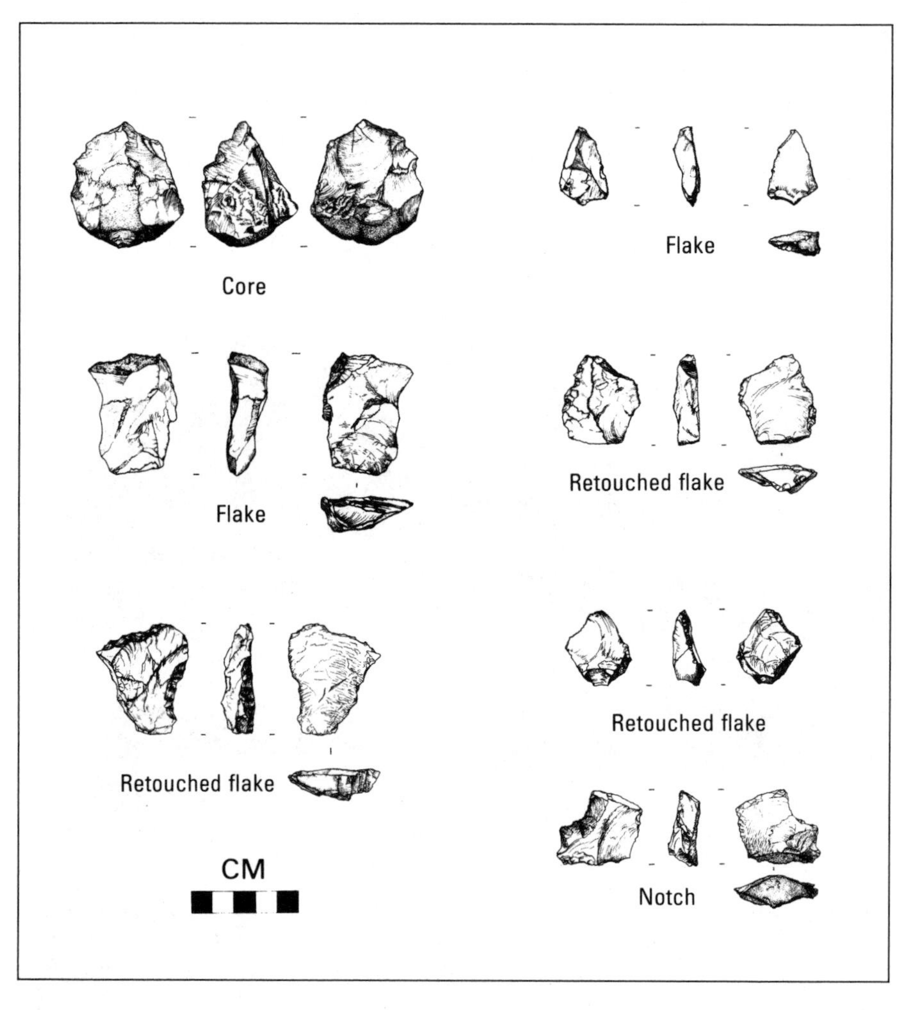

Simple, Oldowan-like artifacts from Donggutuo, which characterize the east Asian early Stone Age.

The stone industries from the oldest excavated sites, including the one we are presently investigating, called Donggutuo, are fascinating. Most artifacts are made of local chert and quartzite available in bedrock exposures near the sites. The technology, typical of many of the early Asian Paleolithic sites, is essentially an Oldowan-like industry composed of casual cores, flakes, and fragments as well as some flakes that have been shaped to have irregular, or denticulated, edges, with an occasional convex protrusion, or nose.

Many animal bones have also been recovered from these excavations, including fossil forms of horse, bison, elephant, rhino, and deer. Analysis is presently under way concerning how they might have ended up at the site and any signs of modification by carnivores or hominid tools. No hominid remains have yet been discovered, but it is probably just a matter of time before such discoveries are made here and would almost certainly represent the Asian variant of *Homo erectus*.

EUROPE

The first peopling of Europe may have occurred during a period of low sea level by hominid groups migrating westward along the Anatolian peninsula and crossing near the area presently inundated by the Straits of Bosporus and onto the Greek mainland. Alternatively, they could have come around the northern coast of the Black Sea and into the area of the Balkans. A recent discovery of a jaw from Dmanissi in Georgia, perhaps well over one million years old and reported to be an early *Homo erectus*, may document such a hominid migration.

Other possible (but perhaps less likely) routes were across the Mediterranean, "island hopping" during periods of low sea levels between North Africa, Sicily, and across to Italy, or crossing from North Africa to the Iberian peninsula at the Straits of Gibraltar. In these scenarios, however, there still would have been significant distances of water to have been traversed.

In any event, the earliest reasonably well-dated and definitely artifactual sites in Europe appear toward the end of the Early Pleistocene. The site of Isernia in central Italy, where potassium-argon dates and paleomagnetic studies indicate reversed magnetism, suggests an age of more than 780,000 years ago. Interestingly, in this, one of the earliest sites representing the spread of hominids into Europe, the tools are once again Oldowan-like, nonhandaxe industries. They are found in ancient lake and river deposits, and along with them are fossil animal bones. No hominid fossils have yet been found at this site; presumably they would represent a *Homo erectus* grade of human evolution, based upon what we know of contemporaneous fossils in Africa and Asia.

Other archaeological sites that may be significantly older than 500,000 years, based on paleomagnetic studies and fauna, include lake deposits at Przeletice and cave deposits at Stranska Skalá in Czechoslovakia, the travertine spring deposits at Vértesszöllös in Hungary, river gravels at Kärlich in Germany, and perhaps two French sites: Vallonet cave on the Riviera, and the open-air site of Soleihac in the Massif Central.

Significantly, most of the earliest Stone Age sites in eastern Asia and in Europe appear *not* to be Acheulean technologies, but rather simpler, Oldowan-like industries. This is a curious pattern, since we know that the Acheulean stretches back at least to 1.5 million years ago (and from recent reports, perhaps even to 1.7 million) in Africa, and it would seem likely that hominid groups that moved out of Africa between 1.5 and 1.0 million years ago (and later groups as well) would have come from a long-standing handaxe tradition. The question of why the earliest eastern Asian and European hominids did not appear to possess Acheulean technologies will be considered later in this chapter.

WHAT WERE ACHEULEAN TOOLS USED FOR?

Few Paleolithic artifact forms have generated as much controversy and speculation as handaxes. In part this is because no recent ethnographic groups had been documented making and using very similar stone tool forms, so that we have no modern analogues for such artifacts. This is especially surprising when one considers that the distribution of handaxes in prehistoric times included much of Africa, Europe, the Near East, and parts of India, from around 1.5 million to 100,000 years ago in Africa. These were made and used for tens of thousands of generations of hominids over much of the inhabited world. But the secrets of their importance have all but died with their makers.

There are many theories of what handaxes were used for, including

1. all-purpose tools (rather like the Swiss Army knives of prehistory) for cutting, scraping, chopping, hacking, digging, and so on;
2. heavy-duty butchery tools for dismembering and cutting meat off medium-to-large-size animals;
3. digging tools for obtaining plants, burrowing animals, or water;
4. highly stylized cores for producing usable flakes;

5. bark-stripping tools for feeding on the cambium layer of a tree (this has been suggested by J. Desmond Clark);
6. projectile weapons to be thrown in a similar manner as a discus ("killer Frisbees") for big-game hunting (this scenario has recently been proposed by Eileen O'Brien and Charles Peters of the University of Georgia and elaborated upon by neuropsychologist William Calvin of the University of Washington).

During the past decade and a half we have conducted a range of functional experiments with handaxes, cleavers, and picks. These experiments have included animal butchery, woodworking, hide scraping, digging, and bark stripping, as well as throwing. Based upon our assessment of these forms, we have come to the conclusion that the primary function of handaxes and cleavers was what we believed sparked flaked stone technology in the first place: the processing of animal carcasses.

Our experiments have suggested that out of the range of possible functions handaxes would have had, their morphology is most consis-

Handaxes make excellent heavy-duty butchery tools. Here, a lava handaxe is used to detach a rear leg of an animal, exposing the ball-and-socket joint of the femur and pelvis.

tent with butchery knives for larger mammals. Handaxes are usually fairly large flaked stone artifacts, averaging about five or six inches long (although examples of less than three inches and more than a foot in size are known), usually somewhat ovoid and symmetrical in shape, and from one to two inches thick. They are normally flaked on both faces around much or all of their circumferences, with the most careful attention paid to the thinner tip ends.

The tip end can cut through meat and joints, while the butt end (sometimes blunted or even left unflaked in the Acheulean) serves as the handle for one or both hands. Before the advent of hafted stone tools, handaxes would have been an ideal form: the sharp, serrated edge can keep cutting longer than an unmodified flake and can be resharpened easily by further flake removals, and its large bulk supplies a handle to grip and wield the tool with force. Special attention was often paid to the tip end of the tool, which was made very sharp and regular, suggesting an emphasis on cutting through some relatively soft material.

The few flint handaxes thus far subjected to microwear analysis by Lawrence Keeley have shown a wear pattern consistent with animal butchery, the first direct evidence of their probable function that we have. It is hoped that other investigations in the future, perhaps with new, improved methodologies, will be directed toward specific questions surrounding the purposes of handaxes.

Acheulean cleavers are an interesting artifact class as well. Experiments show their efficiency for a number of tasks, notably animal butchery and wood chopping. But when cleavers are used on a relatively hard material such as wood, substantial chipping and flaking occurs along the bit edge. This type of damage is rarely found on prehistoric cleavers, which often retain sharp, crisp edges. This suggests that cleavers were used on a *soft* material, and hide slitting, dismembering, and meat cutting would appear to be likely tasks.

The technological shift toward these highly stylized, large cutting tools for heavy-duty butchery is, in our view, probably an adaptive response to the dietary shift among early hominid populations in some parts of the Old World toward more habitual and systematic butchery, especially the dismembering of large animal carcasses obtained by scavenging or hunting. Picks, on the other hand, appear to emphasize the pointed tip rather than sharp edges. Such forms make excellent digging tools in hard earth and woodworking tools for shaping and hollowing wood, and also would have been lethal weapons for dispatching game with a handwielded blow to the cranium.

THE TRANSITION FROM *HOMO ERECTUS* TO EARLY *HOMO SAPIENS:* THE TECHNOLOGICAL PATTERNS

Between approximately 500,000 and 150,000 years ago, Middle Pleistocene fossil hominids from Africa, Europe, and Asia begin to show features that diverge in many ways from the classic *Homo erectus* morphology and seem to be heading toward the more modern human condition. Unfortunately, many of these fossils are poorly dated: younger than the last paleomagnetic reversal, but too old for carbon dating (which can only be used back to about 50,000 years ago) and lacking the volcanic deposits required for potassium-argon dating. Faunal correlations and less well-tested dating techniques such as electron spin resonance, uranium-series dating, and thermoluminescence are some of the principal methods used to establish chronological placement of prehistoric sites in this period.

Despite a lot of variation in hominid fossils from this time period, a gradual overall shift can be seen among them, particularly an increase in the size of the braincase, which often exceeds 1200 cubic centimeters. Fossil skulls exhibiting this brain enlargement (encephalization)

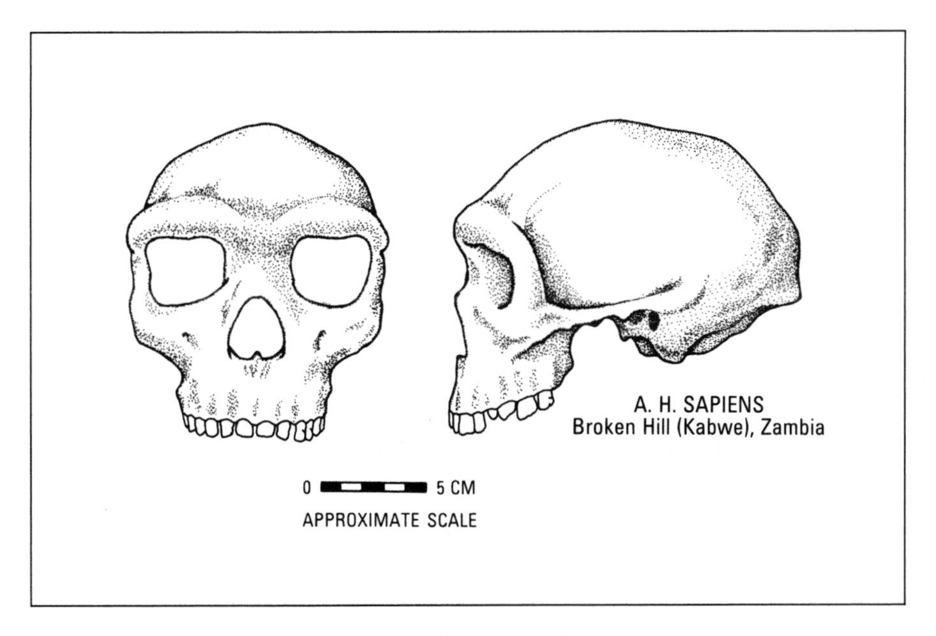

An example of an early form of archaic *Homo sapiens* from Broken Hill, Kabwe, Zambia (formerly called the Rhodesian Man). Studies have suggested a later Middle Pleistocene age, perhaps around three hundred thousand years ago.

tend to be put by human paleontologists into our own species, *Homo sapiens*, but are called archaic because they retain varying degrees of more primitive features that hark back to their *Homo erectus* ancestry.

These early forms of archaic *Homo sapiens* tend to be characterized by skulls with prominent brow ridges, sloping foreheads, and thick cranial walls; robust, jutting jaws, with mandibles lacking prominent chins. But in contrast with the classic *Homo erectus* fossils, these skulls usually have more expanded sides and less angled (more rounded) backs, heralding a more modern human condition. The postcranial bones, as in *Homo erectus*, tend to be robust, with strong markings for the attachment of powerful musculature.

But there is such a high degree of geographical variation among these hominids that paleoanthropologists are currently engaged in heated debate over their evolutionary significance. This debate, in part, concerns the origins of anatomically modern humans, or *Homo sapiens sapiens*. Did modern humans originate in only one geographical area and spread out from there, replacing and/or interbreeding with more archaic-featured hominids? Or was there regional evolution occurring in various parts of the Old World, with anatomically modern humans in several regions evolving simultaneously from archaic populations?

At present there is no consensus concerning the relationship of archaic *Homo sapiens* to modern humans, and the debate continues; this will certainly linger as one of the hot topics of human evolutionary studies during the next decade. We will discuss the emergence of modern humans, and the rival theories to explain this phenomenon, in the next chapter.

Let us look at the evolutionary patterns and technological changes of the archaic forms of humans that occurred during this time.

AFRICA

A number of hominid fossils are considered to represent late *Homo erectus* and early *Homo sapiens* in Africa. These include the Broken Hill skull (the Rhodesian man) from the Kabwe mining operation in central Zambia; the Saldanha skull (also called Elandsfontein or Hopefield) from a wind deflation hollow in South Africa; the Bodo skull from river sand deposits in Ethiopia; and the Ndutu skull discovered near Olduvai Gorge in Tanzania. There is a lot of variation among these fossils, but some paleoanthropologists, such as Chris Stringer of the British Museum and Gunter Bräuer of the University of Hamburg, believe that it is especially the hominids in Africa that seem to be evolving toward anatomically modern humans during this period of time.

Technological advances were made as well. Handaxe and cleaver industries tend to become more refined during this period of time, at times with finely made, thin, symmetrical bifaces probably made with soft hammer technique—some of them virtual works of art. Very finely made Acheulean industries made in lava come from such sites as Bodo in Ethiopia (from the same beds as the Bodo skull), made in quartz from the Masek beds at Olduvai Gorge (above Bed IV), made in quartzite from Kalambo Falls in Zambia, made in silcrete from Elandsfontein in South Africa (the locality of the Saldanha skull), and made in flint from sites in the Nile valley, such as the site of Arkin.

The tool makers seem to have employed new techniques. Experimental studies have suggested that these later Acheulean hominids may have used hammers of some softer materials to finish these forms. Bone, wood, ivory, or a softer stone could have been used to bite into the edge of a handaxe or cleaver, removing flat flakes from both faces and dramatically thinning the tool.

Other tools show more planning and more finesse. Retouched flakes begin to show greater standardization and more careful, regular patterns of removing flakes from the edge of the tool. The most common retouched form is a side scraper with a straight or convex edge, but denticulates (sawtooth-edged retouched flakes) and other forms are often found as well. Flake industries, characterized by few or no handaxes and cleavers but with well-made flake tools, become more frequent at the end of the Acheulean in Africa, perhaps by two hundred thousand years ago in some places.

In parts of East and central Africa new and different developments eventually emerge out of these technologies. Instead of typical handaxe and cleaver industries, we find new tools at some sites, including levels immediately above the Acheulean ones at Kalambo Falls in Zambia. They are still large tools but are now thicker, with pointed or rounded edges, and are called "picks" and "core axes" (these industries are called "Sangoan" after a Ugandan site). J. Desmond Clark has suggested that these large forms represent woodworking tools by late Middle Pleistocene hominids adapted to more wooded environments.

Another new technological trait (which becomes even more important in the succeeding Stone Age period, to be discussed in the next chapter) that is occasionally found in the later Acheulean is the removal of flakes from prepared cores. These are fairly large discoidal cores flaked around their circumferences in preparation for the finale flakes—large circular or oval flakes that take off the top of the core and are sharp virtually all the way around their edges. This is known as the Levallois method, named after a town on the western outskirts of Paris where these artifacts were first described in the nineteenth century. In the later Acheulean this technique is especially common in North Africa but is also occasionally found in other parts of the continent.

An African Later Acheulean Example: Kalambo Falls, Zambia

Near the southern end of Lake Tanganyika and on the eastern side of the Great Rift valley is a spectacular escarpment, a cliff face of quartzite rising up almost a half mile from the valley. Large marabou storks make their nests on these cliffs and can be seen soaring below the escarpment rim. The Kalambo River meanders in the area, forming a winding border between Zambia and Tanzania, then suddenly and spectacularly cascades off the edge of the escarpment and falls some 670 feet, creating Kalambo Falls, the second highest waterfall in Africa. Just upstream from the falls is an ancient sedimentary basin where ancient hominids once lived and whose deposits the present-day Kalambo River is cutting through, exposing archaeological layers from recent prehistoric times back several hundred thousand years.

J. Desmond Clark and M. R. Kleindeinst excavated here in the late 1950s and early 1960s, uncovering one of the longest continuous archaeological sequences known anywhere in the world: Acheulean, Sangoan, middle Stone Age, later Stone Age, and Iron Age occurrences were all documented here. Although bone was not preserved, fossil pollen was found in most layers, and fossil wood and other plant materials had also survived in the lower, waterlogged levels. This has allowed an unparalleled reconstruction of environmental changes from Acheulean times onward. The Acheulean levels, found in prehistoric river sands, are not precisely dated, but estimates of between 200,000 and 400,000 years ago seem reasonable.

In the summer of 1988 we went to Kalambo Falls to conduct some test excavations and experimental archaeological research. With Desmond Clark and his wife, Betty, we were actually able to use the same genial campsite under a gigantic fig tree that Clark and his team had used some twenty-five years earlier, which impressed upon us that in prehistoric times certain localities probably attracted humans for considerable periods of time. Our field crew was composed of some of the local villagers living along the Kalambo River, and included several of the older men who had been excavators with Clark in the original seasons as well.

Kalambo Falls, northern Zambia, the second-highest waterfall on the African continent. Here the Kalambo River (which marks the Zambian/Tanzanian border) spills over a spectacular seven-hundred-foot quartzite cliff that forms an escarpment of the Great Rift valley, then flows to Lake Tanganyika. The Acheulean occurrences (most of these handaxes and cleavers are made out of the same type of quartzite) are just several hundred yards upstream from the falls, on the right side of the river in this view.

Excavation of an Acheulean floor at Kalambo Falls, Zambia, by J. Desmond Clark. This dense concentration of handaxes, cleavers, and other stone artifacts was found almost at the level of the present-day Kalambo River, requiring the use of a pump to keep the site from flooding. Note the dark fossil wood logs at the top right of the photo.

Rarely does one see a higher standard of Acheulean craftsmanship than at Kalambo Falls. Exquisite handaxes and cleavers are made from large flake blanks struck from river-polished boulders of the local rose-colored, tan and gray quartzites (the same types of rock that form the escarpment) as well as from a range of other less plentiful materials. These large Acheulean forms tend to be symmetrical, carefully thinned and shaped, and surprisingly fresh in appearance.

Even more striking is the density of Acheulean material. One site that had been excavated by Desmond Clark in the 1960s and which we

A close-up view of an Acheulean floor at Kalambo Falls, Zambia, showing the dense concentration of handaxes, cleavers, and other materials. (The scale is in inches.) These enigmatic concentrations may be due, in part, to geological forces as well as to early hominid behavior patterns.

subsequently analyzed, called Kalambo B5, had exposed approximately two hundred square feet in one field season. (In fact, this site is actually part of a much larger excavated area of over fifteen hundred square feet.) On one level of the site, about 550 artifacts were found, including almost 100 handaxes and cleavers—an average of one handaxe or cleaver for every two square feet of the site!

This puzzling pattern, seen at some other Acheulean sites as well as at Kalambo, of incredible quantities of artifactual materials (including high proportions of handaxes and cleavers) on prehistoric horizons have sometimes been called "living floors," implying that the prehistoric occupants left behind all these tools at a place that had been fairly intensely occupied. Besides Kalambo Falls, other sites with comparable densities of material include Gadeb and Melka Kunturé in Ethiopia, Olorgesailie in Kenya, and Isimila in Tanzania. For decades archaeologists have speculated on what types of hominid behaviors were responsible for these dense concentrations.

When one examines the sedimentology from such sites, it seems clear that the majority of the dense Acheulean concentrations come from contexts that have been geologically disturbed. To some extent it is possible that these large bifacial forms were reconcentrated from more diffuse scatters along ancient stream courses where early hominids used and discarded them. Glynn Isaac suggested that this may have been going on at the Acheulean site of Olorgesailie, and cited geologist Luna Leopold's observation of the kinematic wave effect: cobbles in sandy bedded rivers tend to congregate in fairly stable bars at intervals along a stream course, with individual cobbles moving in and out just as cars move in and out of traffic jams on city streets.

Why were Acheulean bifaces discarded, often still in a very large, useful form? These large handaxes could have been easily resharpened a number of times until they became too small for practical use, but often this did not seem to be the case. It would seem instead that they were left in places that hominids planned to frequent so that they could be used again (but never did for some reason); or that they had served their primary function: in the case of large animal butchery, the meat-bearing bones might have had higher priority for transport (for instance, to an area where the food could be safely consumed.)

WESTERN ASIA

We have already discussed the site of el 'Ubeidiya in the Jordan valley as the earliest probable evidence of hominids in Eurasia between 1.4 and 1.0 million years ago. A significant number of later Acheulean sites have been discovered in the Middle East, including the site of Latamne in Syria, Jisr Banat Jacub, and Tabun in Israel.

The site of Tabun, a large cave and one of several important caves on Mt. Carmel, was first excavated by British archaeologist Dorothy Garrod in the 1930s (she became Cambridge University's first female professor) and subsequently by Arthur Jelinek in the 1960s. A deep sequence of deposits here spans from the later Acheulean to much later times. The site looks to the west, out over the coastal plain to the Mediterranean Sea. This would have been an excellent vantage point for early hominids in the area.

The Acheulean materials from the lowest levels at Tabun are estimated to be about 200,000 years old. Numerous fossil animal bones are found within these deposits. Typical artifacts include small flint, oval handaxes, and retouched flake scrapers and denticulates, of careful and regular craftsmanship, common in many parts of the Old World after the end of the Acheulean.

EUROPE

In Europe there are, as yet, no hominid fossils classified definitely as typical *Homo erectus* remains; all of the more complete skulls have been classified by most authorities as forms of archaic *Homo sapiens*. Fossils believed to date from between 500,000 and 200,000 years ago include the Mauer mandible (Heidelberg man) from a sand quarry in Germany; the Vértesszöllös skull (rear portion) from a travertine deposit (formed near an ancient spring) near Budapest, Hungary; the nearly complete Petralona skull from a limestone cave deposit in northeastern Greece; the Arago cranium, jaws, and other postcranial remains from a limestone cave deposit in the foothills of the Pyrenees in southwest France; the Swanscombe skull (rear portion) from river gravels in southeast Britain; and the nearly complete Steinheim skull from river deposits from a quarry in southern Germany.

Some authorities believe these European fossils represent their own evolutionary branch, developing in the general direction of a form of archaic *Homo sapiens* still to come: the classic Neanderthals, or *Homo sapiens neanderthalensis*. (The Neanderthals, who thrived in this part of the world during the earlier part of the last Ice Age, will be discussed in the next chapter.)

As mentioned earlier, most or all of the earliest stone technologies in Europe are nonhandaxe, Oldowan-type, with simple cores and casually retouched flakes. These industries continue through this period of archaic *Homo sapiens*, but finally handaxes also emerge on the scene by about half a million years ago and soon become quite common, especially in western and southern Europe and southern Britain.

European sites without handaxes in this period include Clacton-on-Sea in southern England; the lower gravels at Swanscombe in England; the cave deposits at Arago in France (associated with the partial skull and other hominid remains); in river and lake deposits at Bilzingsleben in Germany (also associated with fragmentary hominid remains); and Vertesszöllös in Hungary. Sometimes the term *Tayacian* is applied to some of these European nonhandaxe industries, named after a site near the town of Les Eyzies-de-Tayac in southwest France. Well-known sites with handaxes include Abbeville and St. Acheul in the Somme River gravels of northern France; marine beach and dune deposits at Terra Amata on the Mediterranean coast of France; the middle gravels and middle loams at Swanscombe in southern England (the fossil hominid skull was also found in the middle gravels); lakeside deposits at Hoxne in central England (the site first described by John Frère in 1799); the Boxgrove site in England; Torre en Pietra in Italy; and the lake and marsh deposits at Torralba and Ambrona in Spain.

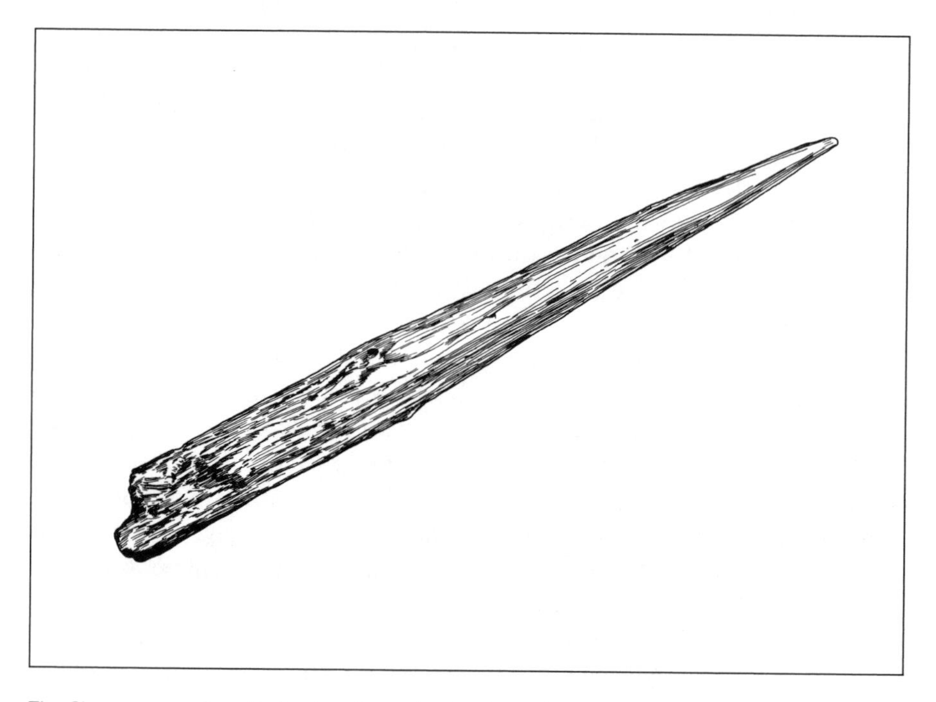

The Clacton spear from England, estimated to be about three hundred thousand years old.

Of special interest was the preservation of prehistoric wood at Clacton-on-Sea on the southern coast of England, in particular the broken shaft and tip of a spear made out of yew. This is the earliest definitive evidence of wood technology in the prehistoric record. Microscopic examination by archaeologists revealed clear striations from the wood having been shaped with a stone tool about 300,000 years ago. At the Acheulean site of Kalambo Falls in Zambia, probably between 200,000 and 400,000 years old, a possible wooden club was discovered among the fossil wood specimens. And microwear analysis by Lawrence Keeley of flint tools from the English sites of Clacton and Hoxne shows clear use-wear patterns from woodworking on some implements, from hide scraping on others. Artifacts made out of wood and hide are inferred from these polishes: as previously discussed, wood could have served as spears, digging sticks, pegs, or containers, while scraped hides could have served as containers, clothing, or as elements of architecture.

This meager but tantalizing evidence suggests that there was probably a range of perishable materials employed as tools, and again suggests a rich invisible technology that rarely survives in the earlier prehistoric record. Among recent Stone Age hunter-gatherers, tools made from organic materials, such as wood and hide, are very common. The stones give us the tip of the iceberg, perhaps, but an invaluable tip it is.

A European Later Acheulean Example: Ambrona, Spain

Nestled in the rolling hills of northern Spain is the Acheulean site of Ambrona, estimated to be approximately four hundred thousand years old. In the 1960s and 1980s, F. Clark Howell, Leslie Freeman of the University of Chicago, and the late Martin Almagro uncovered one of the largest early Stone Age sites yet excavated. We joined the project in 1981 and 1983 and have studied the technological patterns of all the stone artifacts excavated over the years.

The fine limey silts, with some layers of sand stratified within them, suggest that this site was formed in a natural depression containing a small bog or lake. Here archaeologists uncovered thousands of Acheulean stone artifacts, made in quartzite, flints and chalcedonies, and silicified limestones. Most notable are handaxes and cleavers of moderate to high craftsmanship, often made on large flake blanks. A range of smaller and simpler core forms, as well as flake tools and simple flakes, make up the great majority of artifacts from this site.

Fossils here include a range of species, with elephants predominat-

F. Clark Howell, director of the research at the Acheulean site of Ambrona, Spain, stands above a deep excavation section. This site, believed to be between three hundred thousand and four hundred thousand years old, was associated with many fossil elephant bones as well as other prehistoric mammal forms.

ing. Other fossils include prehistoric horse, deer, wild ox, and rhino, as well as a range of carnivores—lion, hyena, wolf, and lynx. No hominid skeletal remains have yet been found.

Especially interesting is the large number of elephant bones, more than from any other known archaeological or paleontological site of the Middle Pleistocene. The interpretation of the site in the 1960s was that this was a big-game hunting and butchering station—that Acheulean hominids drove these elephants into the bog, perhaps with fire, and then finished them off and cut them up. More recently, archaeologist Richard Klein has analyzed the fossils from the site and suggested that many of the elephants there may have died of natural causes. The site may represent a complex combination of hominid as well as animal behaviors near a source of water, the animals living, drinking, and sometimes dying, the hominids perhaps preying upon or at least scavenging from some of the animal carcasses there.

A later Acheulean level was also excavated at Ambrona, found in channel sand deposits. The vast majority of fossils from this level represent horses, although some remains of deer and wild ox are also present. In this later level not only the animals have changed, but also the artifacts: the tools tend to be better made, the handaxes and cleav-

A number of fossil elephant tusks and bones associated with Acheulean artifacts at the Ambrona museum; this building was erected over the excavations here, and the materials are still in situ.

ers smaller and more refined, and the flake scrapers much more carefully made than from the layers below.

Within sight of Ambrona, about one and a half miles away on the other side of the modern river valley, is the site of Torralba, also excavated in the 1960s by Clark Howell and Les Freeman. Here were uncovered sediments, artifacts, and fauna very similar to those found at the lower levels at Ambrona.

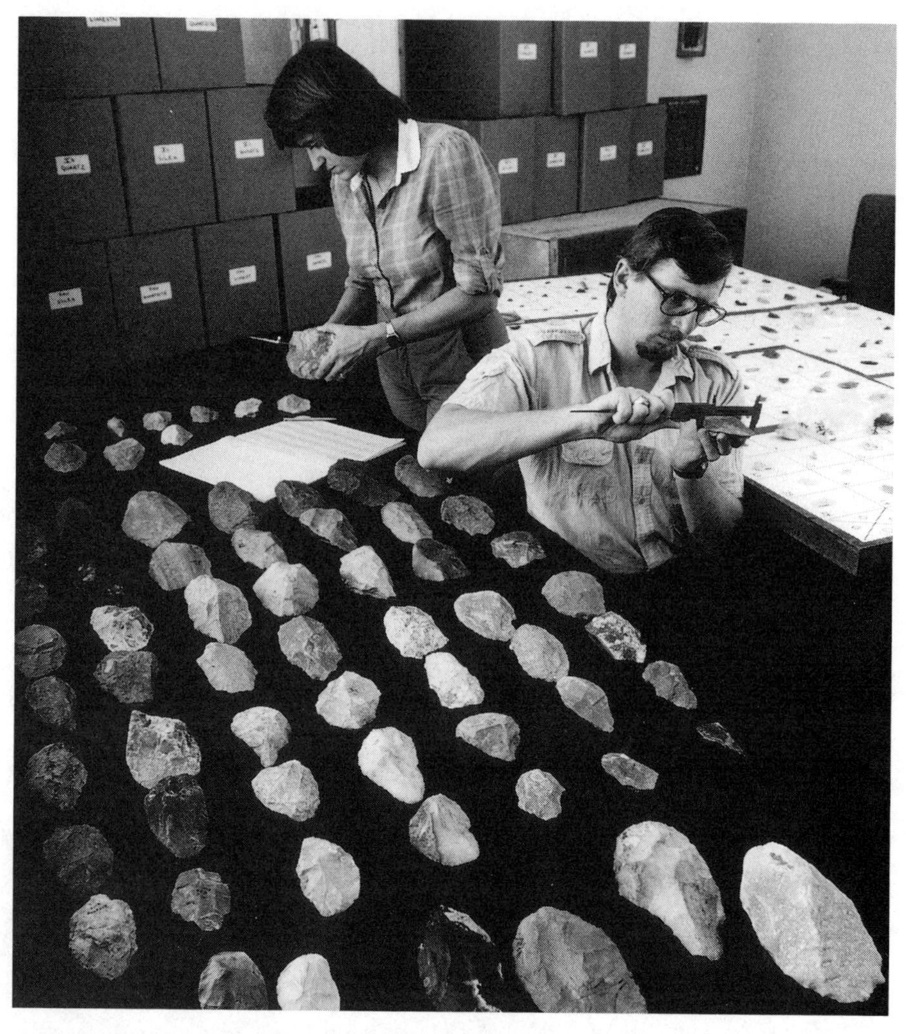

The authors measure and record technological characteristics on handaxes and cleavers from Ambrona, Spain, at the National Archaeological Museum in Madrid. Smaller flakes, fragments, and retouched forms rest on the table behind. (© David L. Brill, 1985)

EAST ASIA

In Asia, fossil evidence suggests that classic *Homo erectus* populations may have survived for a longer period of time without much morphological change. Fossil skulls from sites at Ngandong (Solo man) and Sambungmachan in Java, hominid fossils from cave deposits at the Zhoukoudian cave in China (between 500,000 and 250,000 years ago), and the Hexian skull (perhaps 250,000 years old), retain many or most of the classic *Homo erectus* features, which may have lasted longer there than in other parts of the Old World.

Other eastern Asian fossils, such as the nearly complete Dali skull (which is perhaps 200,000 years old) from river deposits in northern China, the partial Maba skull (which is probably a bit younger) from a cave deposit in southern China, and the Narmada skull cap from river gravels in northern India, suggest more modern features, including an expanded braincase, and are considered by many to be early archaic forms of *Homo sapiens*.

Archaeologically, there is not much change here in this time period. Most of the stone industries in eastern Asia remain remarkably simple and unspecialized throughout the Middle Pleistocene, with Oldowan-like cores, casually retouched flakes, occasional spheroids, and large quantities of unmodified flakes and fragments the most notable technological features.

The most famous Middle Pleistocene archaeological site is Zhoukoudian, a cave forty miles southwest of Beijing, where Peking man was found. Excavations in the 1920s and 1930s were undertaken by W. C. Pei (China's first great Paleolithic archaeologist) and Jia Lanpo in conjunction with Canadian Davidson Black and German Franz Weidenreich (who took over after Black's death in 1936).

Numerous fossil hominid remains were uncovered by excavations in the cave sediments in this time range, all attributed to *Homo erectus*. Some tens of thousands of stone artifacts were also retrieved, made primarily out of vein quartz. Simple Oldowan-like cores and casually retouched pieces are characteristic of the technology, with bipolar technique a common way by which hominids made their tools.

But then a tragedy occurred: all the original *Homo erectus* remains from Zhoukoudian were lost during World War II. In 1941, because of Japanese occupation and the threat of a larger world conflict, it was decided to have United States Marines stationed in Peking remove the fossils for safekeeping in the United States until the war's end. They were packed and taken by a train traveling to the coast, where they were to be put on the US *Benjamin Harrison*. As fate would have it, the train departed from Peking the weekend Pearl Harbor was attacked: Japanese soldiers boarded the train, made the marines prisoners of war, and the Peking man fossils have never been seen again.

In the 1960s and 1970s Chinese scientists resumed excavations at Zhoukoudian and uncovered a smaller sample of *Homo erectus* remains and stone tools. Today tourists can visit the site, stand before a one-hundred-foot vertical excavated face with the stratified cave layers exposed, and contemplate these hominid pioneers of eastern Asia. A fascinating prehistory museum is now situated a short distance from the cave, where we have recently conducted workshops on stone technology with Chinese archaeologists.

Out of Bounds? East Asia and the Problematic Movius Line

In 1948, Hallam Movius, professor of Paleolithic archeology at Harvard University, was the first formally to point out a puzzling discontinuity in the distribution of Lower Paleolithic industries in the Old World. As we have seen, in much of Africa, the Near East, and Europe, handaxe technologies develop over time and then prevail for hundreds of thousands of years. In eastern and Southeast Asia, however, early Stone Age industries are characterized by what the late Professor Movius called "chopper/chopping-tool industries," essentially Oldowan-like technologies, with simple cores and casually retouched flakes the most prominent artifacts in these assemblages.

Movius divided the Old World into two major geographical and technological areas, with the line running through eastern Europe, the Anatolian peninsula, and northern India. To the south and west of this boundary, later to be known as the "Movius line," Acheulean sites were common, whereas to the north and especially to the east, in China, Southeast Asia, and Java (then connected to Asia when lower sea levels exposed the Sunda shelf) they are absent.

We now know the distinction is not quite as clear as Movius had proposed: as we have seen, *Homo erectus* or early archaic *Homo sapiens* on the west of the Movius line did not *always* make handaxes or cleavers. Nonhandaxe industries contemporaneous with Acheulean sites are also known from both Africa and Europe.

There are some unusual sites now reported, such as Dingcun in China and Chon-Gok-Ni in South Korea, believed to be Middle Pleistocene in age (between 700,000 and 125,000 years ago), with artifact forms that could be called picks, cleavers, and occasionally handaxes. They tend to be on the crude end of the technological spectrum, being thick, with bold flake scars characteristic of hard-hammer percussion. These rare, sporadic occurrences in eastern Asia, in the view of J. Desmond Clark, are not really a part of the Acheulean phenomenon found in Africa, the Near East, and Europe but rather are separate, independent technologies probably directed toward heavy-duty woodworking tasks.

A number of explanations have been forwarded to explain this discrepancy in technologies in different parts of the world. These include

1. *chronological barriers*: Hominids may have moved out of Africa and into Eurasia *prior* to the invention of Acheulean technologies in Africa. If so, then the earliest Eurasian occupants would have had only simpler Oldowan technologies. The handaxe and cleaver industries that emerged later in Africa would never have reached eastern Asia (perhaps for some of the other reasons listed).
2. *geographical barriers*: Once non-Acheulean hominids had established themselves in eastern Asia, they were separated from the West by the Himalayas, the cold and bleak Eurasian steppes, and the tropical forests of Southeast Asia: formidable barriers to the spread of later handaxe technologies from the west.

A New Guinea horticulturalist uses a stick knife made out of split bamboo to butcher a pig. Bamboo can produce razor-sharp edges.

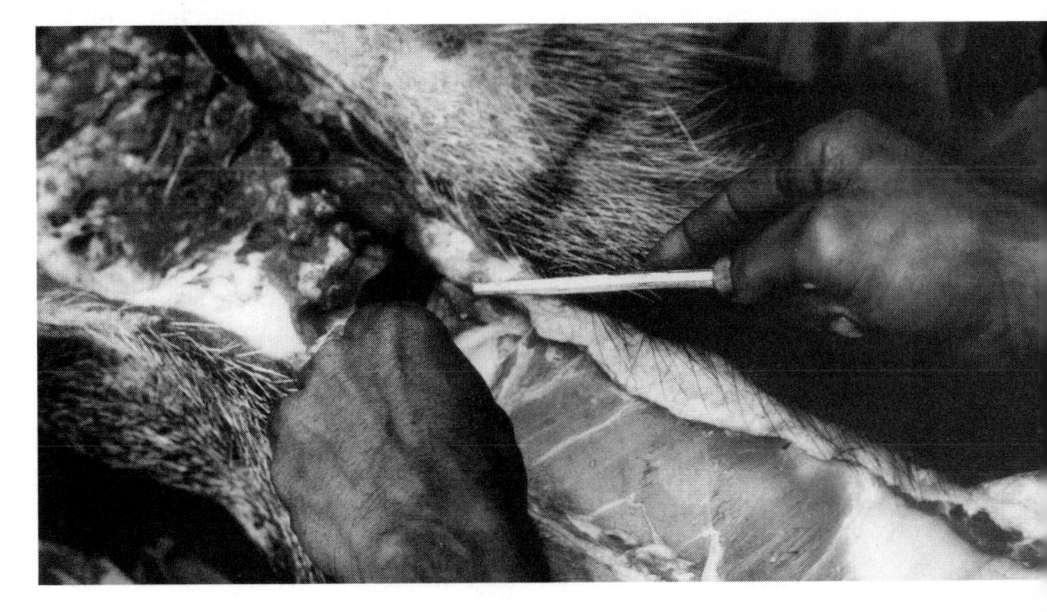

3. *use of bamboo as a raw material:* During prehistoric times, much of eastern Asia had bamboo woodlands and forests. Bamboo is unusual in that when split it produces incredibly sharp edges that can be used for a wide range of cutting, hacking, and scraping activities, as some societies in the world still do today. For this reason, stone technology may have taken on a reduced or even secondary role for these hominid groups, giving them less incentive to develop or maintain elaborate and more sophisticated stone tool types.
4. *unsuitable lithic raw materials:* According to this argument, many areas of eastern Asia—unlike much of Africa, the Near East, and Europe—did not offer suitable raw materials for making handaxes and cleavers, materials such as massive outcrops of lavas or quartzites that could be flaked or exposures of large nodules or cobbles of flint or chert. Smaller, often poorer-quality materials such as vein quartz were much more common, so technologies tended to be simpler in appearance, lacking the standardization seen in the Acheulean.
5. *different functional requirements:* Some prehistorians have argued that handaxes, viewed in this scenario as butchery tools especially for large mammals, were used primarily in open habitats. In much of Southeast Asia, on the other hand, habitats are believed to have been more closed and forested (as they still are today), so that large, gregarious mammals would have been uncommon. Hominids, therefore, would have had much less opportunity to exploit large mammals, and large cutting tools such as handaxes and cleavers would thus not have been necessary components of their technology.
6. *different species:* The *Homo erectus* in eastern Asia may have been, in fact, a different species from the hominids found in Africa, the Near East, and Europe. Some researchers, such as physical anthropologist Peter Andrews of the British Museum of Natural History and archaeologist Robert Foley of Cambridge University, have suggested that the Asian hominids, a group long separated genetically from those in Europe and Africa, may have had different cognitive abilities and produced different types of stone artifact assemblages as well.
7. *low maintenance of technological traditions:* We have recently suggested that due to limited language capabilities of early hominids, transmission of elaborate technological traditions may have required continual practice in order to be passed on to the next

generation. The manufacture of tools such as handaxes and cleavers, learned primarily by example, would have required continued access to suitable raw materials to maintain the tradition. If migrating hominids spread to uninhabited areas and did not have access to suitable raw materials for one or two generations, the technology would die out, since there would be no example from which to learn. This model would also help explain the absence of handaxes in the earlier occupations of Europe: the migratory hominids were new to these environments and did not have enough knowledge of raw material outcrops to maintain their handaxe traditions. It was only later, after substantial hominid settlement of Europe had mapped out rock sources, that migrants were able to bring in the handaxe tradition once again.

It seems quite likely, though, that any one of these explanations would oversimplify what was probably a very complex phenomenon and that a combination of these or other factors could explain the general absence of handaxe industries in eastern Asia. As prehistoric investigations in Asia and elsewhere intensify, a clearer picture should emerge that will shed more light of the technological divergence we see there.

BIG TALKERS? THE QUESTION OF LANGUAGE

> A sharp tongue is the only edged tool that grows keener with constant use.
>
> **Washington Irving**
> ***Rip Van Winkle* (1819–20)**

It would be wonderful if phonemes, morphemes, nouns, verbs, adjectives, prepositions, rules of syntax (or their primeval antecedents), and so on could be found as fossilized items in the prehistoric record. They could then be excavated, cleaned, fitted back together, and analyzed through time and space to see long-term evolutionary patterns. But, alas, as we have previously lamented, language does not fossilize in the prehistoric record.

Imagining the developmental stages of human language—those

transitions in complexity from communication levels seen in apes to the rich, elaborate, and dynamic systems seen in modern humans—is very difficult. Most anthropologists and linguists shrug their shoulders when asked what protolanguage for *Homo erectus* would sound like. Ironically, some of the most interesting (if unsubstantiated) interpretations of early hominid communication come not only from academic books or journals, but from the cinema in such movies as *Quest for Fire* and *Clan of the Cave Bear*.

There may be glimmerings of some increased ability for language, or at least the communication of specific, concrete symbols, to be seen in the stylistic complexity and specific designs among Acheulean tools. However, if the extremely stereotypical, incredibly conservative stone technologies we see persisting throughout the Acheulean are any guide, these communication systems may have had some severe limitations. As Desmond Clark has observed: "I don't know *what* they were saying, but I bet it was the same things over and over and over again!"

HEARTHS AND HOMES?

As in the Oldowan, evidence for clear architectural features and fireplaces is still rather ambiguous during the time period of *Homo erectus* and archaic *Homo sapiens*.

Fire can be documented as concentrations of charcoal or ash at a number of sites, but again the question is whether hominids were involved directly in their use and/or production, since unambiguous hearth structures are normally not present. Evidence of the presence of fire can be found, for instance, at Zhoukoudian in China, Vértesszöllös in Hungary, Bilzingsleben in Germany, Torralba and Ambrona in Spain, and Pech de l'Azé, the Lazaret cave, and Terra Amata in France. At Terra Amata, depressions with burned material were discovered with rocks on one side, suggesting a hearth with a windbreak (facing the direction of the prevailing winds today).

The presence of charcoal and ash, and sometimes burned bones and stone at many sites, does suggest that hominids were more involved with fire use during the last 500,000 years than were earlier hominids. But the lack of any evidence of charcoal, ash, or burned bones or stones at many deeply stratified and well-preserved sites between 500,000 and 200,000 years ago makes us wonder just how universal the use of fire was. As we will see in the next chapter, the constant use and presumed production of fire seems to emerge in the archaeological record only with later forms of *Homo sapiens*.

The question of domestic architectural features such as huts is also somewhat controversial. As we have discussed previously, substantial hut or tent structures might leave little evidence for the archaeologist —the most convincing clues would be features constructed with large stones or circular patterns in the soil showing the holes where posts had been shoved into the ground.

At the site of Bilzingsleben in Germany, estimated to be about 250,000 to 300,000 years old, a nonhandaxe industry was associated with an oval feature composed of large rocks and large animal bones such as elephant, which roughly delineated an area eighteen feet long by nine feet wide. Charcoal was especially dense at one end of the feature, suggesting a fireplace just outside a structure, possibly with poles or hides held in place with the stones and bones.

Again, at the Acheulean site of Terra Amata on the French Riviera, about 300,000 years old, post-hole patterns and stones, along with the inferred hearth structures mentioned above, suggested to the excavator, Henry de Lumley, that there were several hut structures, most with a fireplace inside. And at the cave of Lazaret in France, estimated to be about 200,000 years old and also located on the French Riviera, an Acheulean assemblage was associated with a pattern of stone blocks measuring about thirty feet by twelve feet along one cave wall, with two concentrations of charcoal inside. Several areas with quantities of small marine shells suggested to the excavators that hominids were bringing dried seaweed (and, as hangers-on, the tiny shells) for bedding material. Once again the evidence is sparse and, often, controversial. It is likely that structures from this time period were ephemeral and would not have left behind much archaeological visibility.

YOU ARE WHAT YOU EAT? THE QUESTION OF CANNIBALISM

Some suggestions have been made for cannibalism during the time period of *Homo erectus* and archaic *Homo sapiens*. It was proposed back in the 1930s that the Zhoukoudian (Peking man) fossils from China may indicate these hominids were cannibalizing each other: the faces of their skulls were often gone, the bottom of the skulls sometimes broken away, and some possible cut marks were reported on the remaining bone. In hindsight, it is quite possible that hyenas or other carnivores visiting the cave could have produced some of the damage and eaten away missing parts of the skulls. Unfortunately, the disappearance of these important fossils during World War II prevents any further exploration of whether the scratches on the hominids were made with stone tools.

The second major piece of evidence is on the Bodo skull from the Middle Awash valley of Ethiopia. When Tim White was cleaning this skull, carefully removing the crust of sediment adhering to the fossil surface, he found a number of very curious marks: sharp, narrow grooves that appear to be cut marks made by stone tools. These were located on the forehead of the individual and around and inside the left eye socket. The strokes of the apparent stone knife seem to have been cutting flesh off the skull (scalping) and removing tissue from the face. Why would another hominid have done such a thing? We really don't know. Whether it represents cannibalism, head-hunting, or some other behavior must unfortunately remain ambiguous at present.

SYMBOLISM AND AESTHETICS?

A rudimentary definition of a symbol is something used to stand for or signify something else. Do we have any evidence of the use of symbols during this time period? Various claims have been made for use of different objects as ornaments or the use of ocher (a natural iron oxide pigment, often red but sometimes yellow or brown), which develops an apparent symbolic importance in later prehistory that was maintained among many recent and modern cultures. This evidence is, however, again quite ambiguous during this period of our prehistory.

The handaxes themselves might give some indication, though, that hominids by this time may have developed some sense of an aesthetic, of something that was pleasing and meaningful to them because of its shape and form. This is especially so in the later stages of the Acheulean, from about 500,000 to 200,000 or 100,000 years ago. Bifaces in this time period are sometimes undeniably, strikingly beautiful (whether you're an archaeologist or not). They can be extraordinarily symmetrical (showing bilateral symmetry in both their outline and their cross section) and show very even, regular edges—all the while maintaining a sharp, cutting efficiency.

This is even more remarkable considering all the technological operations and decisions that went into their manufacture and the fact that these hominids were in effect sculpting a rock with its own peculiar properties of fracture. Furthermore, there seem to have been strong rules operating at many sites as to exactly what a handaxe should look like: at some sites we see very similar forms produced over and over again by the hominids that occupied it, while another nearby site may show another ideal type or types for which the hominids were striving. These stylistic patterns seem to go far, far beyond mere need, or beyond the demands of the functions of these tools, and are clear indications of learning by example and observation.

Given these properties of the bifaces they made, it seems likely that these hominids had developed an enhanced ability to visualize and conceptualize specific forms and end products. It is possible that not only intelligence, but, to some degree, low levels of symbolic communication or language skills may have developed to some critical level by this time, as verbal symbols may have enhanced and solidified these tool ideas they seem to have shared.

THE SIGNIFICANCE OF THE ACHEULEAN

For the first time in the Stone Age, in the Acheulean we see very definite *end products*, tools deliberately produced and shaped by their hominid makers, using a new set of techniques.

Might these large, well-made bifacial tools have developed special significance within the hominid groups themselves, standing as symbols of their success in being able to butcher large animals? Might they even have signaled success on a personal level, a hominid brandishing a beautifully made tool that boasted his or her ability to control the medium of stone and use it to access significant resources for the group? One gets the definite sense of style and aesthetics in these technologies.

Another striking feature of this period of our evolution is the geographic and environmental expansion our ancestors were able to accomplish. A tropical animal that could break out so rapidly from the environment in which it had been born and evolutionarily bred is a creature that had made behavioral adaptation and flexibility an important part of its way of life. Interestingly, the migration of *Homo erectus* out of Africa coincides with a similar biogeographic spread of carnivores such as lions, hyenas, and leopards. Anthropologist Alan Walker has suggested the *Homo erectus* may have become a member of the "carnivore guild," and that successful hunting skills may have facilitated such large-scale movements among these species.

Yet a final question to consider is why we see so *little* change in Acheulean artifact types over time and space. As previously mentioned, there is a gradual tendency for bifaces to become more refined over time, but considering that we see the Acheulean spread across thousands of miles, lasting almost one and a half million years, and continuing through considerable biological change among hominids, this conservatism is absolutely astounding. We never see anything else like it in our recent prehistory or history.

How can we understand this uniformity? (We seem better suited to explaining change rather than sameness.) In the modern world we are now used to a rapid pace of change in technology, life-styles, values, jobs, and so on. Although the rate of cultural and technological change was much slower in recent history and prehistory, one *can* see real change over the course of hundreds or thousands of years. The changelessness of the Acheulean has no real prototype in our world, and it probably has something to do with the cognitive and cultural capabilities of the hominids themselves. It seems that they may have depended heavily upon learning in their technology, but this may have been largely through imitation without much emphasis upon innovation. The stylistic constancy one sees in some Acheulean tool assemblages may indicate some fixation to which these hominids were prone, or cultural ruts in which they became entrenched.

THE END OF THE BEGINNING

This period of the Acheulean and its contemporary technologies makes up the bulk of our past as a tool-making hominid. The makers of these technologies, *Homo erectus* and archaic *Homo sapiens*, now leave the scene, passing on a genetic and cultural legacy to their descendants. Why did the Acheulean end, and what took its place?

In the next phase of our prehistory the tempo of human cultural change picks up considerably, changing more rapidly over time and space and leading ultimately to late Ice Age peoples who seem very much like us in many ways. We are now ready to join this account at the point where the earliest fossils of anatomically modern humans *(Homo sapiens sapiens)* appear.

C H A P T E R 8

THE HUMAN THRESHOLD

All progress is based upon a universal innate desire on the part of every organism to live beyond its income.

Samuel Butler (1835–1902),
Note-books, Life

Up to this point the ancestral creatures we have considered have been somewhat strange and removed from our modern experience. To explore their ways of life, their thinking, their feeding behaviors, and their use of technology, we have persistently compared and contrasted early hominids to living apes. In the later periods we have just looked at, approaching 100,000 to 200,000 years ago, they seem somewhat more "like us" but still too distant for us really to identify with them.

It is during the next phase of human prehistory, only within the past 100,000 years or so, that we begin to see extremely strong biological similarities between ourselves and the evolving hominids in various parts of the world. Then, within the past 40,000 years or so, we are faced with a creature that not only looks like us, but also seems to act and think very much as we modern humans do. We see the emergence of fully modern humans, not only in their biology but in their culture, their thinking, their living sites, their symbolism, and their art. Thirty thousand years or so later, when the Ice Age abates and environments permit, these humans set forth on the most recent experiment in human adaptation, in which they start manipulating and shaping a very different world for themselves.

In this new departure, within the past 10,000 years or so, humans began harnessing the energy of various plant and animal species to support large populations. Many of these reached a critical threshold of abundance and opportunity sufficient to develop what we call "complex societies": cities with the urban gluts, class structures, bureaucracies, specialists in crafts, arts, sciences, and religion (and possibly even the traffic jams, ulcers, and high blood pressure) that characterize modern "civilization." The patterns discernible in the prehistoric record during this time period set the stage for the world we now live in.

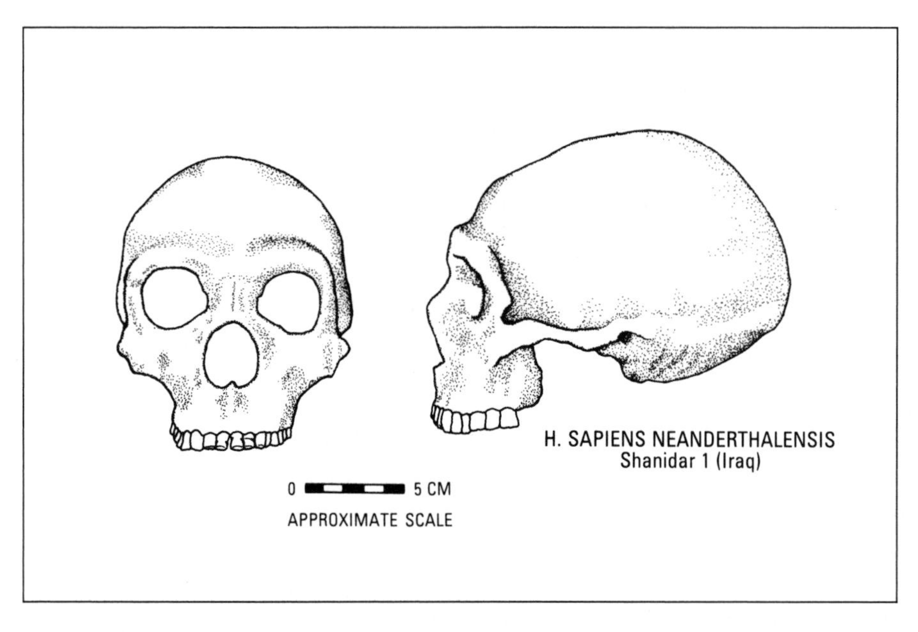

The skull of *Homo sapiens neanderthalensis,* a later form of archaic *Homo sapiens,* from Shanidar, Iraq, believed to be about fifty thousand years old. The Neanderthals from Shanidar were associated with Middle Paleolithic tools.

TRANSITIONS: EVOLVING *HOMO SAPIENS* AND THE MIDDLE REACHES OF THE STONE AGE

In the first stage of this transition, new biological forms of hominids appear along with some new forms of technologies.

THE FOSSIL RECORD

Between 100,000 and 40,000 years ago, a new form of archaic *Homo sapiens* emerges in Europe. Hominid fossils recovered from caves, rock shelters, and open-air sites are characterized by a large brain (basically of modern human size), a thick-vaulted skull, sloping forehead, beetling brow ridges, a broad nose, a jutting lower jaw, and a robust, stocky body. Named *Homo sapiens neanderthalensis*, the so-called Neandertal man has become the archetypical "caveman" or "troglodyte" (from the Greek for "one who creeps in holes or caves") in Western literature and films. The evolutionary status of these creatures has been controversial ever since their discovery: are they part of the ancestry of modern humans, or are they an extinct offshoot from the main branch of human evolution?

The European Neandertal forms seem to disappear fairly rapidly after the appearance of anatomically modern humans, first seen here about 40,000 years ago. Within 5,000 to 10,000 years, the Neandertals basically disappeared from the face of the earth.

In Africa there appears to be more of a gradual and earlier transition from archaic *Homo sapiens* to anatomically modern human forms. Fragmentary evidence suggests that early anatomically modern (or near modern) fossils may date to approximately 100,000 years ago, and the modern human form seems to characterize hominids found subsequent to this.

In the Near East recent studies have suggested that both Neandertals and anatomically modern humans may be present between 90,000 and 40,000 years ago, raising interesting questions about reproductive isolation and possible differences in adaptive strategies. In Eastern Asia the fossil record is less complete, but archaic forms of *Homo sapiens* between approximately 80,000 and 150,000 years ago have been reported from a few sites (the earliest reported anatomical moderns from China are about 25,000 years old).

At present, then, the overall pattern seen in the fossil hominid record suggests a relatively early emergence of *Homo sapiens sapiens* in Africa and possibly the Near East and their subsequent appearance elsewhere in the last 40,000 years. How might this transition have occurred?

At present, three major schools of thought exist regarding the origins of modern humans:

1. a total replacement of archaic forms by the ancestors of modern humans coming out of Africa or the Near East;
2. a spread of modern humans out of Africa or the Near East with some interbreeding with archaic populations in Eurasia;
3. regional continuity from archaic to modern forms of populations in Africa, Europe, and Asia, with some interbreeding as these populations evolved.

Studies of DNA or genetic material in modern human populations have fueled arguments for the first of these theories. Researchers have examined differences in DNA sequences of the mitochondria of humans in different geographical areas. (Mitochondria are cellular structures controlling metabolism and energy production for an organism; these mitochondria have their own DNA independent of the nucleus of the cell.) This mitochondrial DNA (mtDNA, for short) is inherited only through the mother, so that in theory all structural changes over time should be due to random mutation.

Their results, as well as several other independent studies of modern human genetic diversity, suggested that there were two major branches of modern human mitochondrial DNA variation: an exclusively African branch and another branch made up of all other human groups (Amerindians, Asians, Australian aborigines, Europeans, New Guineans, and so on, as well as some other African populations).

Using the first known archaeological appearance of human colonization of Australia (about 40,000 years ago) and the Americas (about 12,000 years ago) to calibrate the probable rate of mutation of the mtDNA, and making the assumption that this rate is fairly constant through time, they extrapolated that the date of the last common ancestor (popularized in the press as the Mitochondrial Eve) of these two main branches of modern human variation was approximately 150,000 years ago. They went on to suggest that the origins of modern diversity occurred on the African continent and that the subsequent spread out of Africa by our direct ancestors genetically replaced other archaic hominid populations that existed in Eurasia (including the Neanderthals of Europe and the Near East).

This controversial and provocative model of human evolution has recently come under fire on methodological grounds, and other anthropologists have argued that there is fossil evidence for regional continuity from archaic forms to modern humans in several parts of the world, such as Africa, Europe, and eastern Asia. It is going to take some more time, more work, more fossils, and more genetic studies to resolve these issues to everyone's satisfaction.

THE ARCHAEOLOGICAL RECORD

Stone Tools

In many parts of the Old World there was a gradual technological shift away from the large handaxes and cleavers that characterized the Acheulean. This transition began between 200,000 and 100,000 years ago, and later in some regions. Handaxes, cleavers, and other large-core tools virtually drop out of many tool kits, replaced by smaller tools made on flakes. These are often very regular, standardized flake tools, some of which had been struck from special types of prepared cores (the Levallois cores introduced in the last chapter).

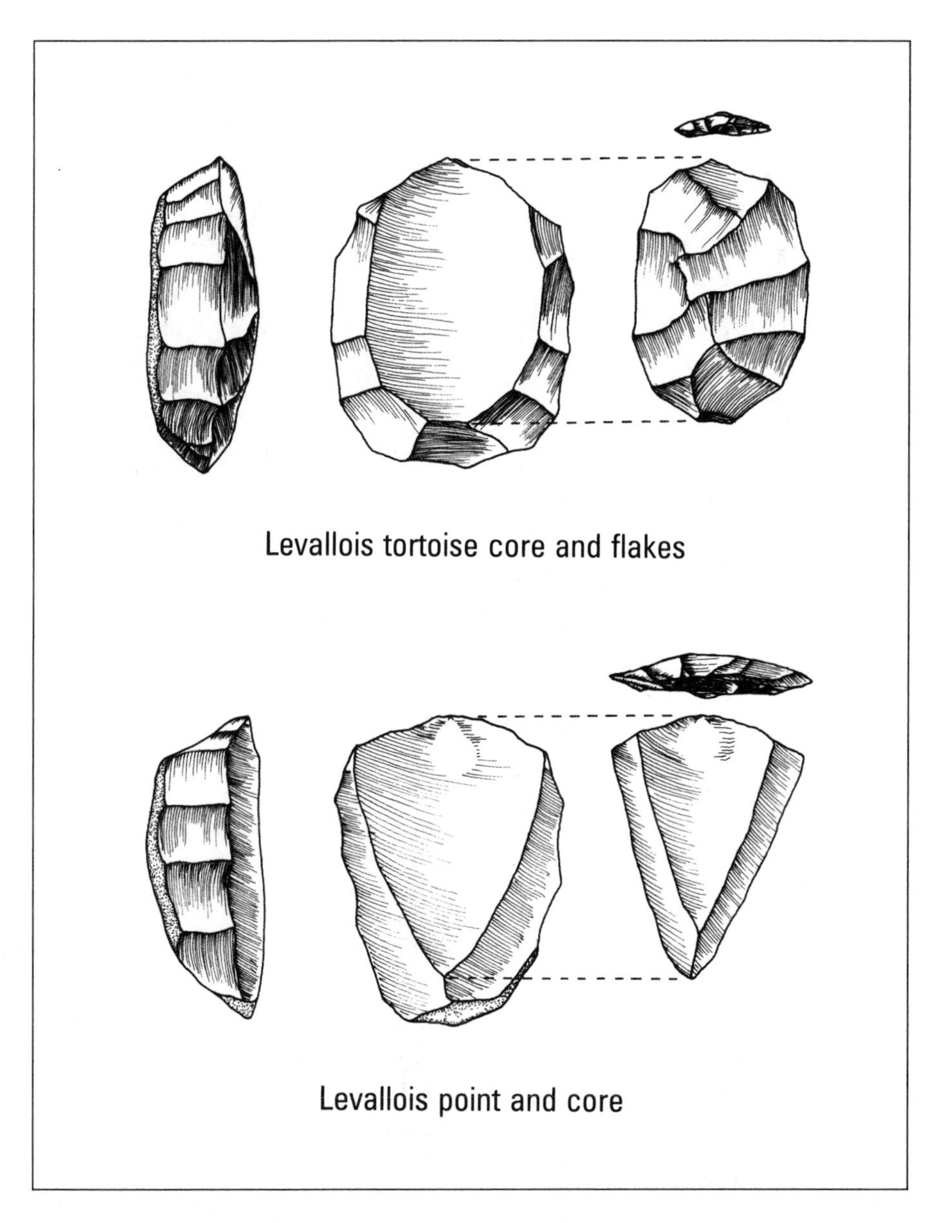

Middle Paleolithic/Middle Stone Age prepared core techniques.

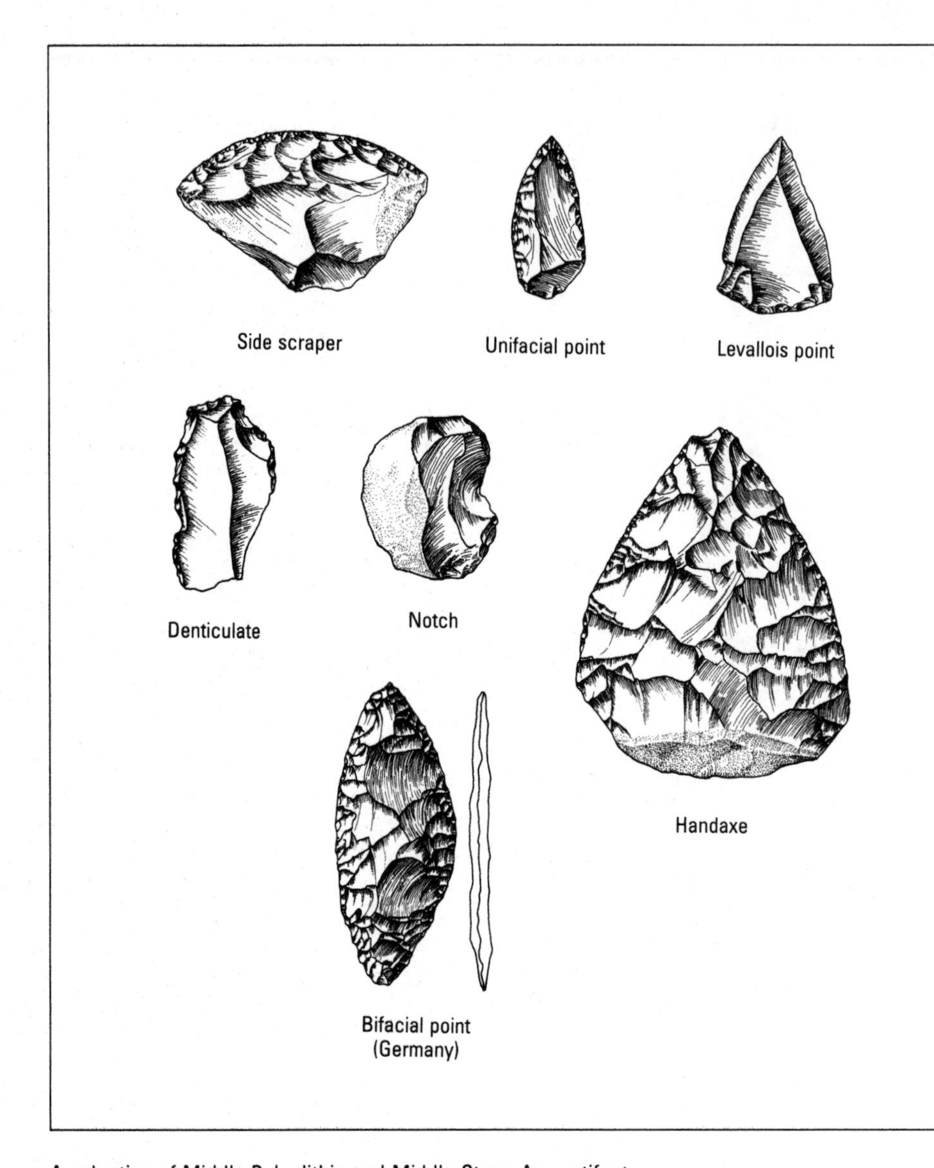

A selection of Middle Paleolithic and Middle Stone Age artifacts.

A dominant tool form in these technologies was the side scraper, produced by removing chips off one side of a flake to steepen or re-sharpen an edge. Wear studies suggest that many of these were used for scraping hides or working wood. Other common tools include denticulates (flakes retouched into a sawlike, toothed edge) and unifacial or bifacial points.

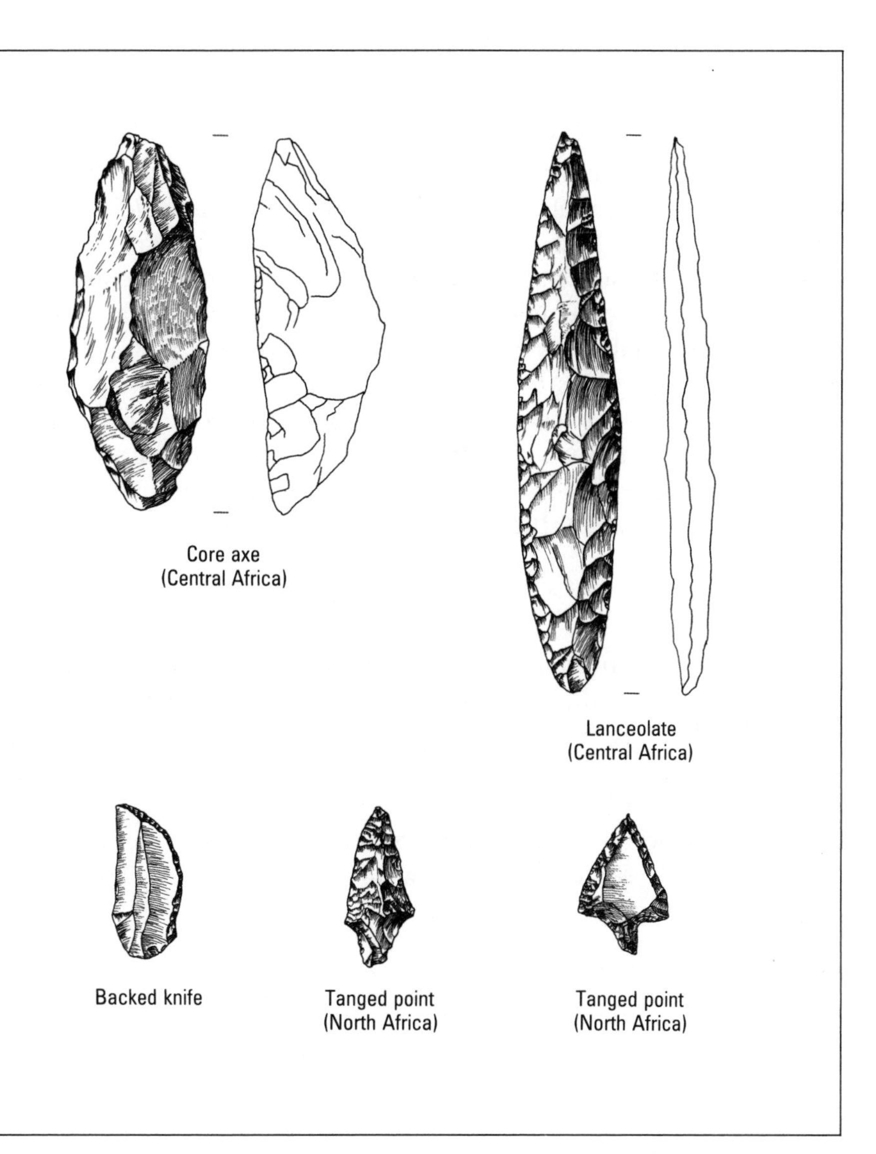

In Europe, the Near East, and North Africa these technologies are known as Middle Paleolithic or Mousterian (named after the rock shelter of Le Moustier in the French Périgord, where such artifacts were first described in the late nineteenth century). In sub-Saharan Africa contemporaneous industries have been traditionally known as "Middle Stone Age" and are often given their own regional names as well to note distinctive technological features found in that area.

Prepared Core Strategies

We have already discussed the beginnings of this innovation in the last chapter, as some of the later *Homo erectus* and archaic *Homo sapiens* used raw material more efficiently, getting a large, very sharp flake from a core shaped methodically for this end. The Levallois technique was developed to a high art in this period. In some regions it becomes a clear diagnostic sign of Middle Stone Age or Middle Paleolithic technologies at a site, especially seen in the production of oval or pointed flakes from specially flaked cores. Interestingly, the geographic distribution of the Levallois technique is remarkably similar to that of the earlier handaxe and cleaver industries, concentrated in Africa, western Europe, the Near East, and India (but absent within Eastern Asia). Perhaps we are seeing two major areas of cultural interactions largely separated from one another by profound geographic or ecological barriers.

Hafting

Many of the tools produced in this time period are made on flakes that have been retouched along one or more edges for the working part of the tool. There is evidence that some of these are also thinned or specially shaped so that they might fit easily onto a haft or handle, especially forms that may have been projectile points or knives. It is during this period that we see the first indirect but convincing evidence in the archaeological record of hafting, or mounting stone tools on handles or shafts.

Fire

It is also during this phase of human evolution that hominids appear to have finally mastered fire. In many of the caves, rock shelters, and other sites during this time period, we often find thick ash layers, and hearth structures can also be identified at some localities. This mastery of fire and its habitual use must have had profound effects on Stone Age populations with regard to cooking, socializing, and hunting activities, as well as in the ability to live in much harsher and otherwise inhospitable environments.

Burials

Disposal of the dead in the form of deliberate inhumations, or burials, seems to have become a significant trait in some groups during this period. It has been argued recently that some Neanderthals in Eurasia believed to have been buried were simply individuals who died from the collapse of a cave roof or some other nonhuman process. But there is a consensus among most Paleolithic archaeologists that many of these claims of burial are probably valid, because humans are usually the only species whose skeletons are found relatively complete, undisturbed by predators or scavengers.

It is not clear whether this evidence for burial should be regarded as evidence for a concept of an afterlife, fear of spirits of the dead, or simply a desire to keep scavengers such as hyenas away by removing human remains from sight and smell. In any case, this is a significant change from the apparent behaviors of earlier hominids, none of whom appear to have buried their dead.

THE UPPER PALEOLITHIC, OR LATER STONE AGE

As we have seen, physical anthropologists have noted that fossils of anatomically modern humans appear to go back in some places to 100,000 years ago or earlier. Yet from the archaeological evidence between 100,000 and 40,000 years ago, these modern hominids do not appear to have done anything dramatically different from other hominids during this time period, including archaic forms of *Homo sapiens* such as the Neanderthals and earlier Middle Stone Age hominids from Africa. But there is a radical and abrupt change in the archaeological record in some parts of the Old World beginning about 40,000 years ago. Author John Pfeiffer has aptly termed this transition the "creative explosion," because suddenly there is a tremendous amount of innovation and rapid technological and cultural change, as well as the emergence of what appears to be fully modern symbolic expression.

New stone technologies based upon the production of blades began to emerge. Blades are flakes that are at least twice as long as they are wide, struck off prepared cores by the use of direct percussion (hard- or soft-hammer flaking) or indirect percussion (using a bone or antler punch to remove blades). This was an extremely efficient way of producing a tremendous amount of straight cutting edge per nodule of stone. And, like modern X-acto knives with interchangeable socketed blades, these stone blades could be easily hafted onto a wooden or bone handle, either unmodified or retouched into end scrapers (for working hide or wood), awls, or burins (engraving tools made by removing one or more flakes from the side of a blade rather than the

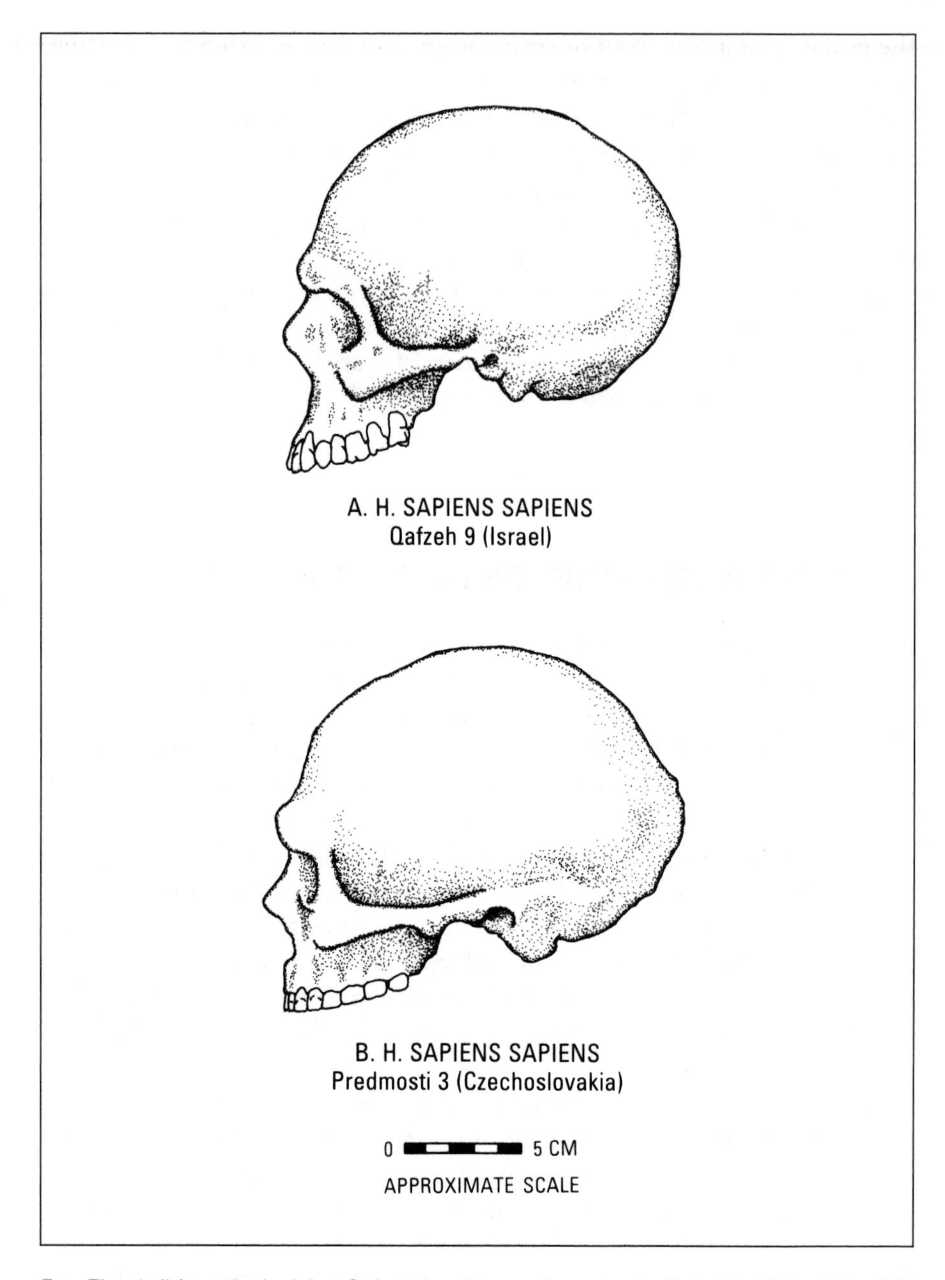

Top: The skull from the burial at Qafzeh, Israel, an early anatomically "modern" fossil hominid associated with typically Middle Paleolithic stone artifacts. This fossil is believed to be about ninety thousand years old.

Bottom: An anatomically modern *Homo sapiens sapiens* skull from Predmosti, Czechoslovakia, dated to approximately twenty-three thousand years ago and associated with an Upper Paleolithic stone industry.

face). During the next thirty millennia, many other new traits also begin to appear in the archaeological record that characterize the modern human condition.

THE GAP BETWEEN MODERN BIOLOGY AND MODERN CULTURE

Why did it take some sixty thousand years for these supposedly modern humans to show any substantial technological and adaptive improvements that might explain their worldwide success and eventual replacement of the archaic forms? This is one of the great mysteries of Paleolithic archaeology and human evolutionary studies today.

A number of possibilities have been suggested:

1. The earlier anatomically modern humans were not truly modern in their brain structure and neural organization.
2. The earlier anatomically modern humans were, in fact, operating in more efficient technological or adaptive ways than more archaic forms, but these improvements cannot be detected in the archaeological record until about 40,000 years ago.
3. The dramatic changes in the prehistoric record around 40,000 years ago were due primarily to accumulated knowledge (culture) finally reaching a threshold when such innovations were possible and then spread rapidly throughout the world.
4. It was not until modern human language was fully developed, perhaps about 40,000 years ago, that human societies were able effectively to communicate information in ways that would encourage such innovation. The ability to communicate in the future tense ("The day after tomorrow we will move camp to the rock shelter in the river valley where the game has been sighted") and in the hypothetical ("What if we haft a spearhead this way rather than that way?") would have had a tremendous effect in a hunter-gatherer band's ability to plan for the future and consider a variety of possible options.

Biologist Jared Diamond, in his essay "The Great Leap Forward," has suggested that archaic forms of hominids were extremely conservative from a cultural standpoint and that survival depended upon doing things as the previous generations had done. With anatomically modern humans, beginning about 40,000 years ago, there was a changing worldview that appears to have tolerated or encouraged new ideas or innovations leading to much more rapid cultural change over time.

By 30,000 years ago, archaic *Homo sapiens* had disappeared, either incorporated within or replaced by fully anatomically modern *Homo sapiens*, who carried on the human evolutionary legacy.

OTHER INNOVATIONS OF THE LATER ICE AGE

For the past 40,000 years, then, we witness an unprecedented, exponential rate of technological growth, experimentation, and innovation. In addition to blade tools, a large number and variety of advances

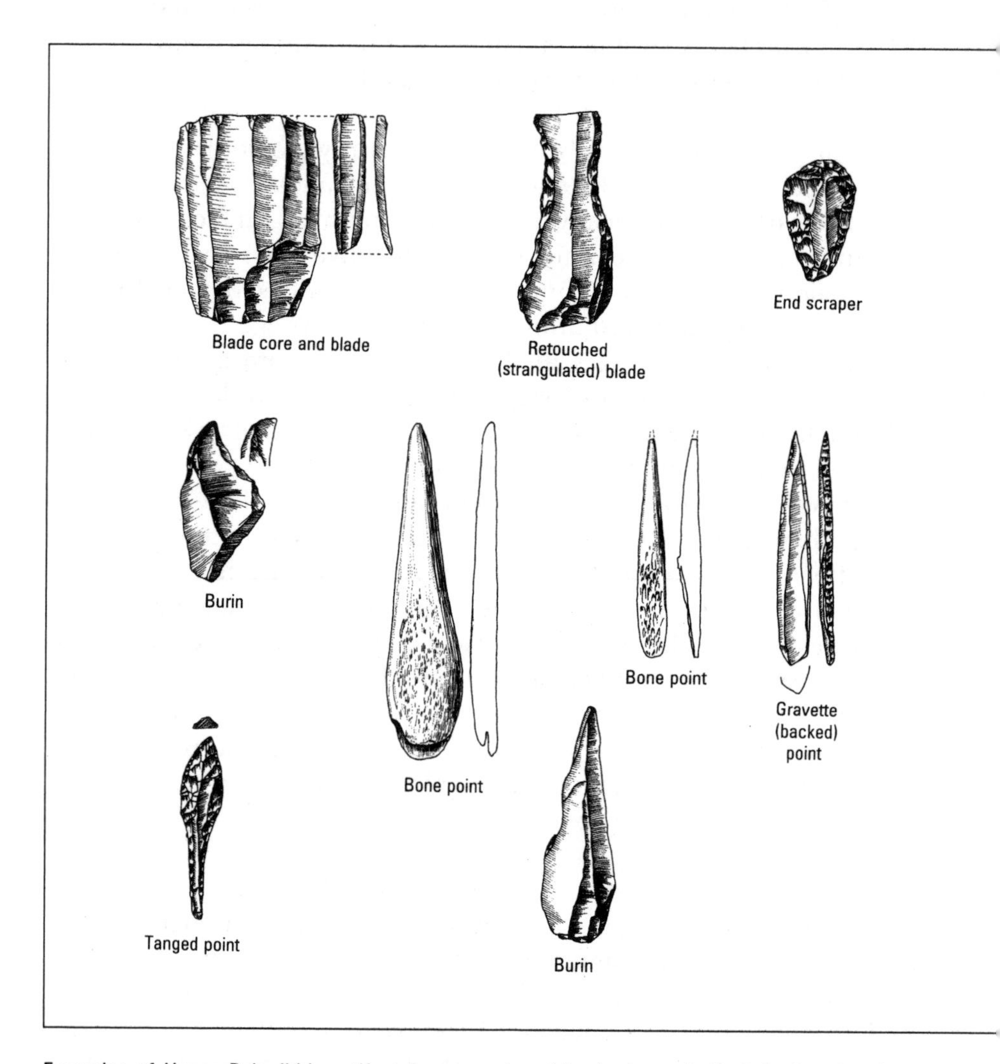

Examples of Upper Paleolithic artifact forms produced in the latter half of the last Ice Age, between forty thousand and ten thousand years ago.

appeared during the period known as the Late or Upper Paleolithic, beginning about 40,000 years ago and ending about 10,000 years ago, when the vast glaciers retreated to their present arctic domain.

New Materials for Tools

Bone, antler, and ivory come into use as major types of raw materials used, fashioned into tools such as points or harpoons. More perishable materials such as wood, hide, and horn most likely also assumed more importance in the material culture as well. Many of these materials

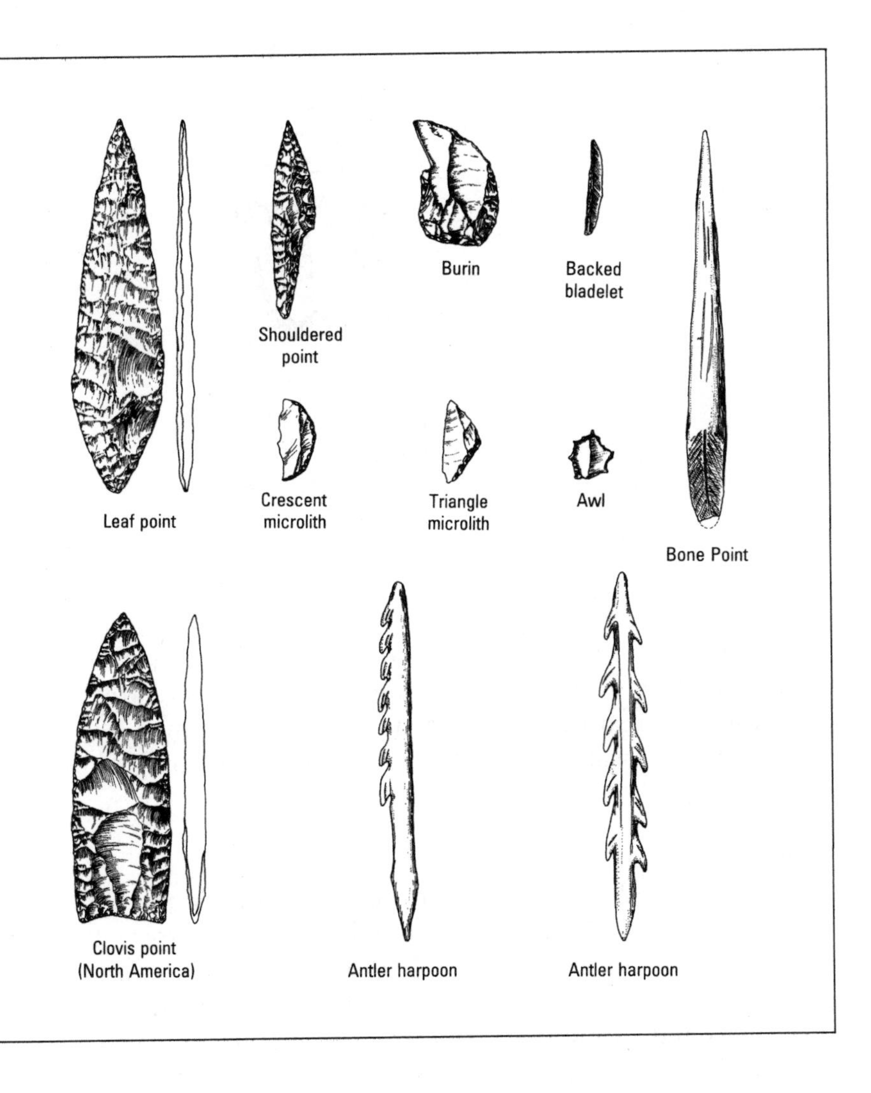

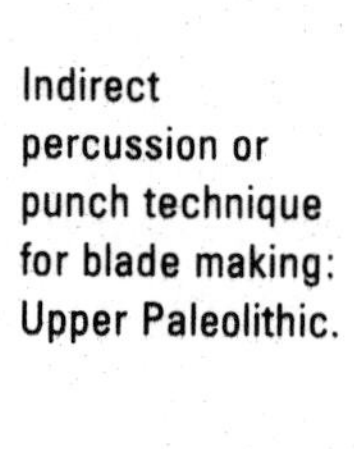

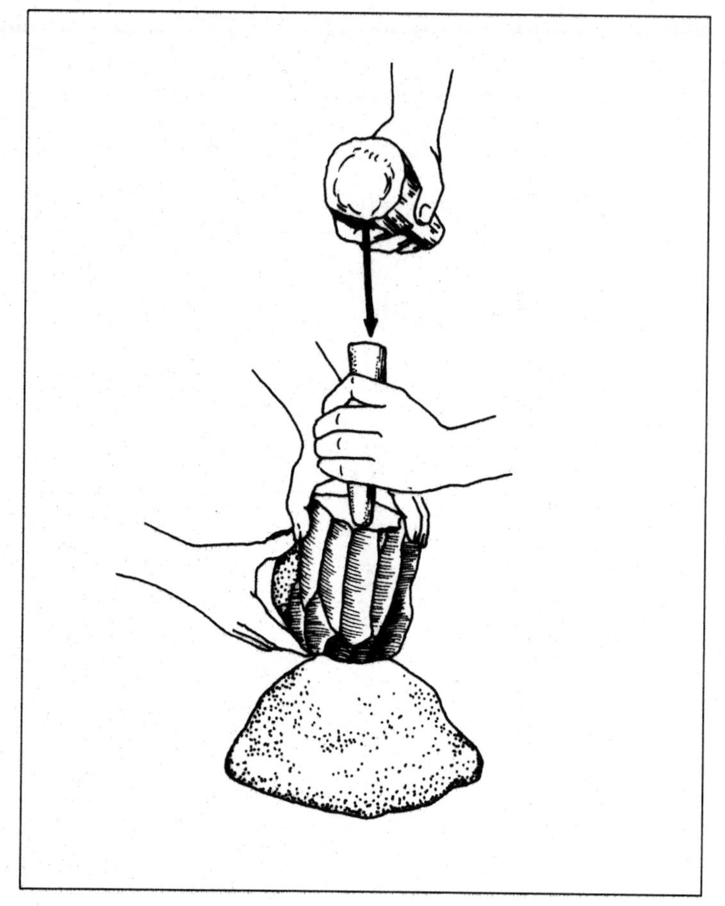

Indirect percussion or punch technique for blade making: Upper Paleolithic.

would have been worked with stone tools (engraving bone or antler with burins, for example), and some, such as antler, would have required special treatment and preparation (antler is much easier to work when soaked in water first and can be straightened or bent by heating or steaming).

Spear Throwers

Within the past 20,000 years or so we see the development of an ingenious mechanical device for hunting known as the spear thrower (or atlatl in the Americas). This is a shaft of wood, bone, or antler used to propel a spear at high velocities toward prey. It extends the arc of the throwing hand substantially, thereby enhancing the speed and penetration of the thrown spear. Spear throwers are still used by some Australian aboriginal groups.

Bow and Arrow and Harpoons

It is also during the later part of the Old Stone Age, or Paleolithic, that we see the definite development of other elements of hunting weaponry. Between 20,000 and 10,000 years ago barbed harpoon heads of bone or antler are found and the bow and arrow makes its appearance. Toward the end of the last Ice Age, small geometric stone tools called "microliths" were probably attached to wooden or reed shafts to make up composite arrows. Archery may have emerged along with the use of poison-tipped arrows, as are used by some groups of hunter-gatherers today.

Needles and Sewing

Real, indisputable needles made out of bone, with small eyes bored through them for some fiber to be inserted and used as a sewing thread, are found in the archaeological record within the past 20,000 years. This would indicate a developed tradition of making carefully stitched garments and other sewn items, such as leather bags, and the use of plant fibers or sinew as thread.

Personal Adornments and Decorations

Different materials were used to make various sorts of personal decoration during this period. These Stone Age people used bones, shells, teeth, and so on, to make necklaces and pendants, bracelets, anklets, and other adornments, including beaded decorations probably stitched on clothing.

Endowed Burials

Not only are a number of anatomically modern fossil skeletons found fully intact in apparent burials as some Neanderthals appear to be, but sometimes members of their social group seem to have interred them with various things to keep them safe, healthy, or happy in the afterlife. Decorative items such as jewelry are found, as well as various tools and other items such as animal bones (perhaps food for the afterlife). Sometimes more than one individual seems to have been interred in the same grave.

Art

Beginning about 35,000 years ago we see definite signs of artwork created by Stone Age peoples in the last phases of the Ice Age. In the earliest phases we see three-dimensional carvings as well as engravings and drawings of animals. As time goes on, more artistic traditions de-

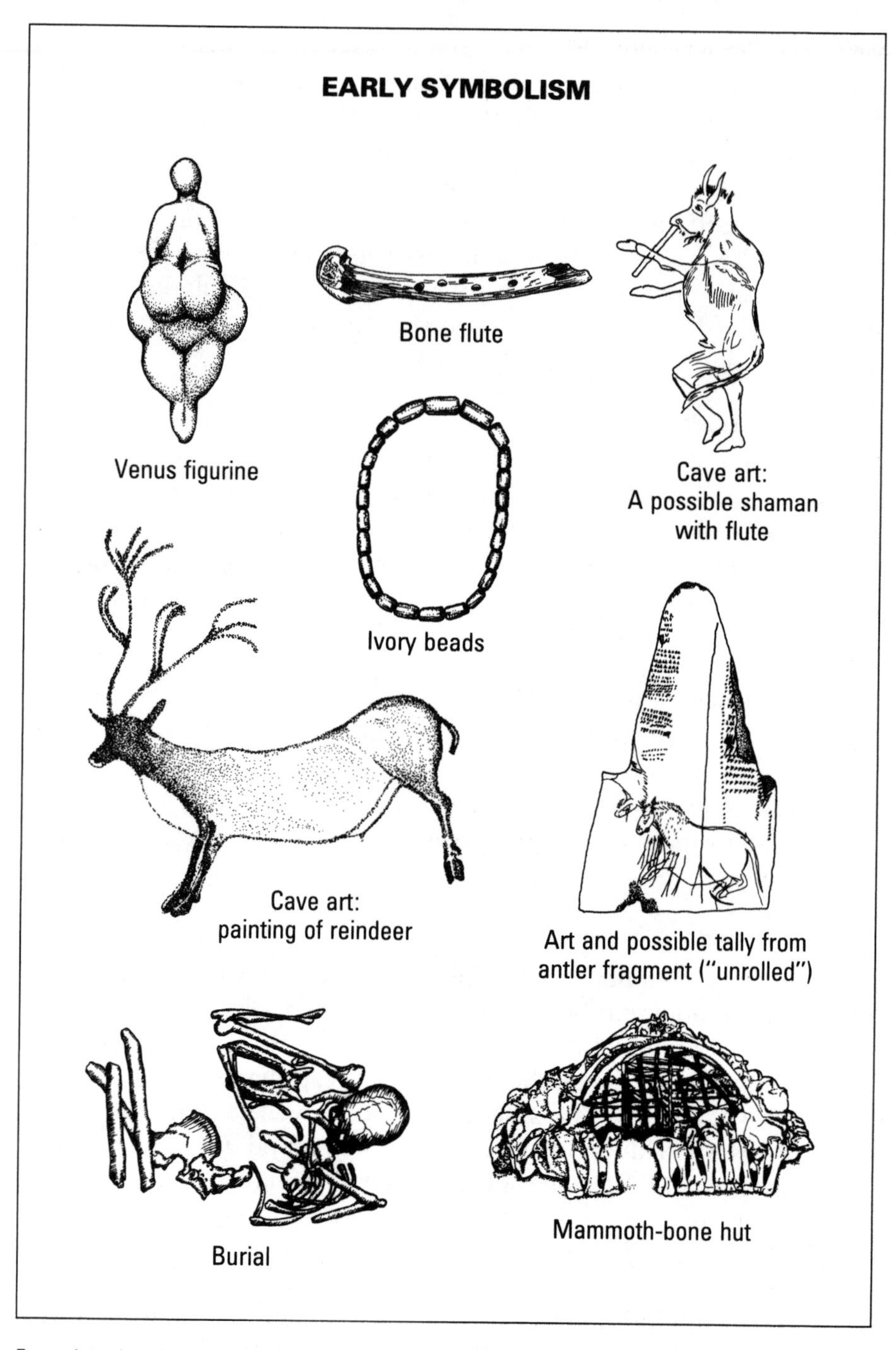

Examples of probable symbolic behavior associated with modern humans during the Stone Age (art, notation, music, ornament, formal architecture and burial).

veloped. Between 28,000 and 23,000 years ago a unique artistic tradition appears, involving carved female figurines in ivory or stone with exaggerated busts and buttocks but obscure and stylized heads. These "Venus figurines" have been interpreted by some as possible signs of some sort of fertility cult whose images were shared across a vast area from western Europe into Russia.

In the last stages of the Ice Age, we see well-developed regional artistic styles in paintings found in deep recesses of caves in parts of Europe, and other rock art is known from Africa and Australia. Many of these represent remarkably realistic, animated animals in one or more colors, and some even show a sense of three-dimensional perspective. Aside from the painting traditions carried out in these caves, which seem to have special symbolic or ritualistic significance, these late Ice Age hominids were also decorating many of their tools, engraving images into them and carving such functional tools as spear throwers with exquisite images of various animals or geometric designs.

More Elaborate Habitation Sites

With humans occupying very cold environments during the latest Ice Age, more definite and elaborate living structures are evident at some sites. We see only the plans or outlines of these structures: sometimes these tents or dwellings are quite large, with large bones of animals, even mammoth bones in the Ukraine, forming the outside ring of large areas of up to twenty square yards or more and which sometimes contain definite stone-ringed hearths.

ON TO THE NEW WORLD

It was toward the end of this Ice Age that humans finally spread to Australia and the Americas. Australia appears to have been occupied by about 40,000 years ago by humans who crossed at least seventy kilometers of water from Southeast Asia, perhaps with the use of simple boats or rafts. Humans were apparently able to make stable settlements in the northeastern regions of Asia such as Siberia only in the last 25,000 years or so, equipped with what was probably an elaborate technology well adapted to extremely cold and harsh tundra conditions. It was then only a short distance to gain entrance to the Americas. It appears that during a period of low sea levels during a glacial peak between 20,000 and 10,000 years ago, the hominids in Asia, probably following migratory game herds, were able to breach this final barrier to their spread over the planet. They seem to have entered North

America across a land bridge (called Beringia) now submerged under the present Bering Strait, and from this foothold finally spread fairly rapidly throughout the Americas, hunting extinct forms of mammoth and bison before the Ice Age ended.

SETTLING DOWN: THE RISE OF FARMING AND CIVILIZATION

STARTING NEW WAYS OF LIFE

Beginning about 10,000 years ago, the mile-thick glacial ice sheets began to retreat, heralding the Holocene or "recent" geological epoch. With milder climates and rising sea levels, profound environmental changes were occurring in many parts of the world.

Intensive Gathering

Humans in various parts of the world began experimenting with new ways to live off the land. Rapid environmental changes, population pressures, and improved tools and foraging techniques may have all contributed to this phenomenon. People began to focus on locally available foods. Concentrating on certain plants that produced extremely rich food sources, at least seasonally, some were able to settle down in relatively stable, fairly large fixed settlements and work around the plants and animals in their vicinity.

Domestication

Eventually this intensive gathering and hunting mode of living developed into one in which people were intentionally planting and then harvesting the crops they had sowed. In the Near East, for example, wild plants that came under domestication were first wheat and barley, then oats, rye, and others. Animals that lived wild around these settlements also came more into the human sphere, with people starting to control their breeding. Ultimately new domesticated forms of these plants and animals appeared by artificial selection, or selective breeding. In the archaeological record we witness a rapid increase in size over time in many plant seeds and changes in bone and horn structure and body size in animals. In the Near East this is seen first among sheep and goats and later in cattle and pigs. With a reliable, storable crop of cereals, supplemented by food from domestic animals, larger settlements could be sustained.

Village Life

Once this cycle was set under way, in many parts of the world it became self-sustaining: the concentration of people in villages allowed them to experiment, to form new relationships with the plants and animals around them, and to find ways to make these food sources produce enough to support the population. As reliable food sources were exploited more systematically, higher population densities and larger-scale settlements were possible. Finally, a sedentary, village life

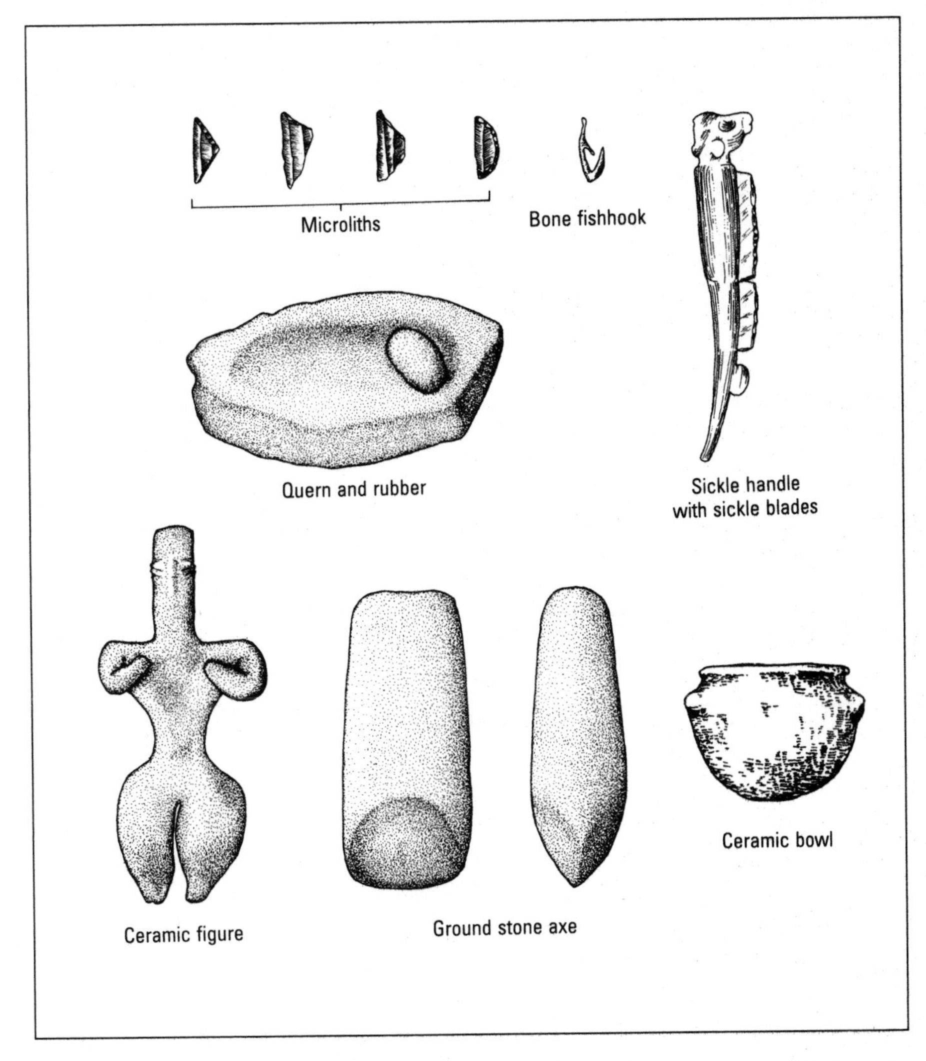

Artifacts associated with postglacial times and the rise of early farming communities between ten thousand and five thousand years ago.

based upon the domestication of different plants and animals was established. Larger villages supported by stable, domesticated food sources and storage of surplus food allowed even more sedentary life, or long-term, localized habitation of a particular site and would have selected for ways of controlling crops and animals (which could be used for food or as beasts of burden) that promoted this way of life.

All of this only began to happen within the past 10,000 years or so, sometimes earlier, sometimes later, depending on the area. This was a fascinating experiment in human adaptation sometimes called the "Neolithic Revolution": after the end of the last Ice Age, human groups started intensively manipulating their environments in many separate regions of the planet—the Near East, Southeast Asia, China, Mexico—isolated from one another but with the same net result: the ultimate domestication of many of the plants and animals in their vicinity and the start of our life in agricultural villages.

NEW TOOLS FOR THE NEW AGE

As village life and agriculture for food production were being developed, the demands of the local economy promoted the development of several new and different technologies.

Ground Stone Tools

As stone tools became specialized for certain rigorous functions, such as chopping down trees or working wood, a new technological innovation appears. After being chipped to a certain shape, some tools were then ground and polished to form a hard, durable tool (for instance, the ground stone axes and adzes from this time period). In the Old World this is one of the major hallmarks of the Neolithic, or New Stone Age, characteristic of farming communities.

Grinding Stones

Evidence of the processing of certain plants, including seeds with a tough outer kernel, can be seen in the development of tools used to grind them into an edible, easily digestible meal or powder. (We actually start seeing these during the period of intensive gathering that preceded real agriculture.) These grinding stones are sometimes formed in bedrock, bowls and pockets ground out of natural rock outcrops, or in blocks of stone found at habitation sites, such as querns with smoothed depressions and rubbing stones that bear evidence of prolonged grinding of grains.

Pottery

Ceramic vessels made of fired clay, such as pots, bowls, and storage jars became prevalent with farming communities, providing implements for cooking, eating, and storing surplus foods, water, oil, or other material.

Sickles

Flint blades or segments of blades were mounted to handles for cutting cereal crops such as wheat, barley, and millet in many areas. A characteristic smooth, glassy sheen can often be seen on the cutting edge (a wear pattern called "sickle gloss"), which resulted from using the sickle repeatedly to cut the silica-rich stalks of cereal grasses.

Other Technological Advances

Basketry and weaving also become more prevalent in this time period. And people started to experiment with all sorts of new, more permanent architectural structures made out of unbaked clay, sun-dried mud bricks, and stone. The earliest ones we see are simple dwellings and huts or silos to store their excess foods. Over time these rapidly become more sophisticated in many areas until quite elaborate settlements grew up with series of interconnected rooms and even plastered walls.

SOCIOPOLITICAL RAMIFICATIONS

Once these more complex, large-scale changes were set under way, and larger populations were being supported by developed agriculture, there were soon profound social repercussions. Elaborate bureaucracies eventually developed to take care of the details of running this complex machine: an elite class to oversee and regulate it, accountants to keep track of goods and commerce, a range of craftsmen and specialists who could manage artistic and architectural details, tradesmen to handle the distribution of goods, a class of priests to legitimize and support the existing value system, tax collectors to siphon off enough goods or money to maintain the "state," and a large body of laborers and peasants, in some cases even slaves, to support the system at its base and carry out its great projects.

Thus we see the beginning of what is called complex, stratified society—also called "civilization"—starting around five thousand years ago, a very short time in the context of human evolution. This was one more great experiment carried out in many parts of the world by modern humanity. Given the right conditions and the proper agricultural base, these same phenomena happened over and over again independently in different regions of the world.

By this time some human societies had become drastically different from what they had been at the end of the last great Ice Age, when all humans still survived in relatively small bands existing by a strategy of cooperative hunting and gathering. Physically and intellectually they were the same beings as their Stone Age forebears, but their society, technology, and economy had changed irrevocably. Once this transformation had taken place, there was no turning back: many regions now supported huge populations, with new ideas and value systems and new ways of making a living. This type of society cannot be supported by a hunting and gathering economy, nor would most of its inhabitants really want to do so: they had become part of a new way of life that saturated their values, shaped their roles, and wove the complex social fabric in which they were enmeshed.

THE COMPONENTS OF CIVILIZATION

There is no hard and fast rule as to what constitutes a civilization, but most share a number of features that distinguish them from simpler forms of social and economic organization in human societies. By strict definition, civilization usually entails living in cities of thousands of people (sometimes called the "urban revolution"), but the primary requisite here is some large, critical mass of population that has developed further complexities; in every single case this has required a well-developed agricultural base.

The trappings of civilization usually include many or all of the following characteristics: development of an administrative bureaucracy, craft specialization, some sort of social hierarchy, a class of priests or religious practitioners, centralized government under a powerful leader often associated with a god, monumental architecture such as temples or palaces, intensive trade and commerce, some sort of writing or system of notation as well as standardized sets of weights and measures to keep track of the flow of goods, and some sort of army or "executive arm" to enforce its rules. Very often civilizations have developed an expansionist phase of conquest and empire building to increase their territory size, subjugate neighboring societies, and consolidate power.

A number of civilizations or proto-civilizations emerged independently in both the Old World and the New World. They rose and fell in Mesopotamia and other parts of the Near East, Egypt, the Indus valley, China, Crete, Greece, and Italy. In sub-Saharan Africa the great kingdoms of West Africa, and ceremonial centers such as Zimbabwe in central Africa, testify to such cultural developments of African cultures. In the New World, complex societies with large political and religious centers emerged in Mexico, Yucatán, and Peru, as well as the American Southwest and the eastern United States.

THE GREAT SMELTING POT: THE BEGINNINGS OF METALLURGY

An unexpected discovery was made by some sedentary farmers soon after the beginnings of agriculture. Independently prehistoric peoples in such places as Thailand, the Balkans, and the Near East learned that certain types of copper-rich rocks could be heated at high temperatures with charcoal to melt out or smelt their metal contents. Temperatures of eight hundred degrees centigrade (similar to that used in firing high-temperature pottery) was necessary for this process to occur. This could be reached or surpassed with the addition of blowpipes into an earthen smelting oven to enrich it with oxygen to heat it up to 1,083 degrees centigrade and finally melt the metal, allowing it to be cast into molds. The simple, lower-temperature smelting process initially may have been discovered by observing what happened to copper-rich rocks used to line hearths or kilns.

From copper-bearing rocks such as the oxide ore malachite, early societies could extract pure copper through this process. Pure copper has the virtue of being relatively soft and thus easily worked by beating or hammering into some shape, but it is too soft to be used for many tools and weapons. If, however, another metal, such as tin or arsenic, is added to copper, it becomes much harder: this is the distinction of bronze, a mixture, or alloy, of copper and tin composed mostly of copper but with at least 1 percent and usually about 10 percent tin added. This bronze alloy is not only harder, but can be cast in a ceramic or stone mold to form a firm but resilient shape that can be readily ground into sharp, durable edges. And unlike stone, this metal was fully recyclable: all you had to do was take old or broken bronze objects and reheat them to a molten state to cast a new object.

THE ORIGINS OF THE ARMS RACE

The origins of metalworking or metallurgy about eight thousand years ago in parts of the Old World often emerged in tandem with the rise of more complex forms of social and political organization leading to civilization. Thus the very origin of state societies shows archaeological evidence for the beginnings of the human arms race, with the devel-

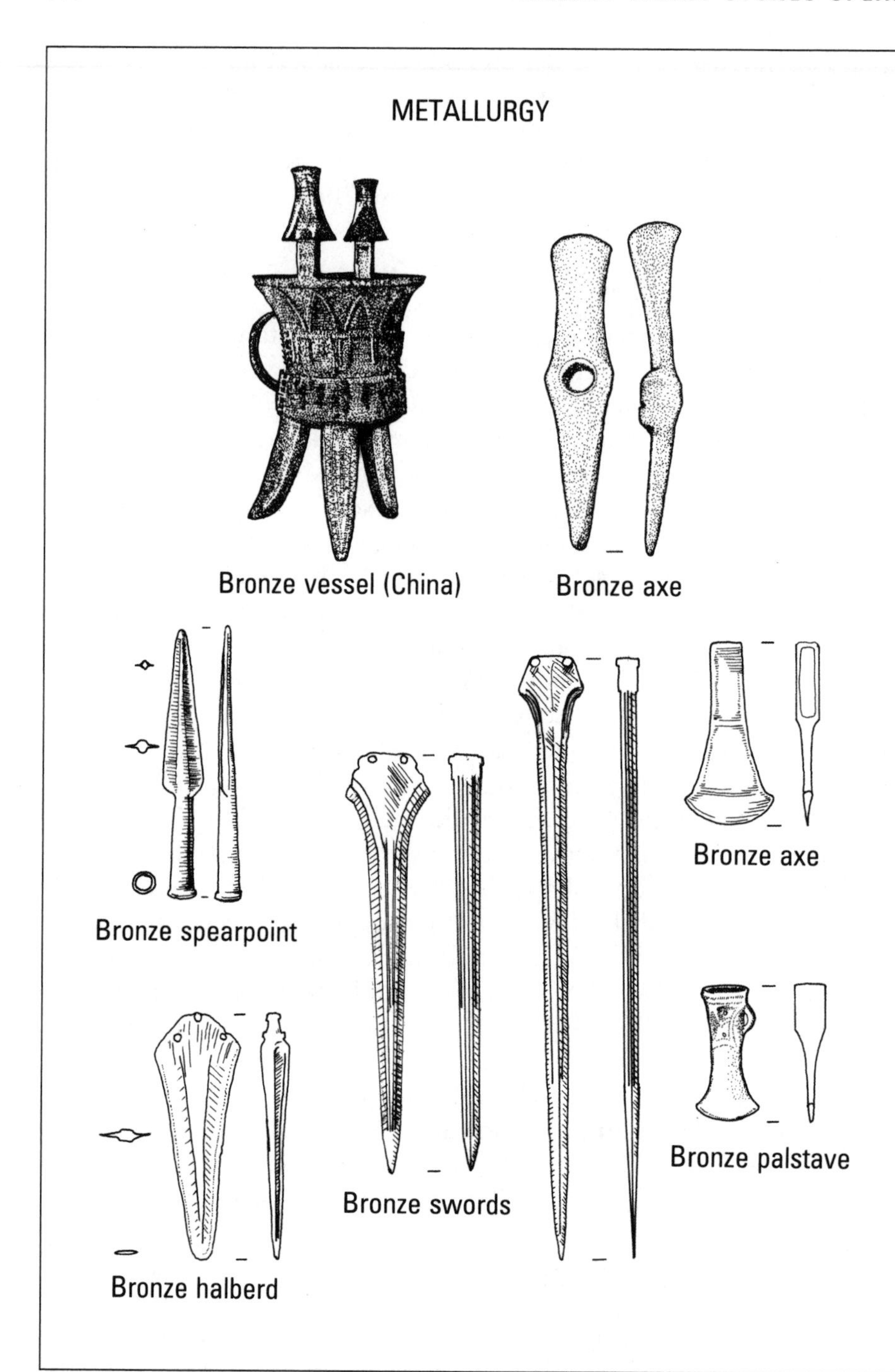

Artifacts associated with the rise of metallurgy civilization, and the state during the last six thousand years.

CIVILIZATION

Writing (Mesopotamia)

Temple (Iraq)

Coin

opment of more and more deadly offensive weapons counterpointed by the invention of more efficient defensive equipment in response. The early arsenals, including daggers, swords, spearheads, battle-axes, helmets, shield bosses, and shin guards (greaves), have a long history, the beginnings of an arms race that continues to this very day. During this time period wheeled vehicles such as wagons and chariots and the domestication of the horse as a means of transportation became very important.

EARLY ESCALATION: THE IRON AGE

Ironworking was the next major technological breakthrough to be made by early civilizations in the Old World (in both the Near East and China). The origins of this technology, heralding what is often called the Iron Age, probably grew out of bronze working. To melt ore down and reduce iron to a liquid requires exceedingly high temperatures, higher than any early technology was able to produce. But more than four thousand years ago some craftsmen in Turkey seem to have discovered that iron ore could be transformed by the heat of their bronze furnaces (iron ore was sometimes used as a part of the copper smelting process) into a bloom that could be purified through forging or hammering while it was hot, producing wrought iron.

Wrought iron could be shaped readily and was very tough. Most important, iron ore was much more common than the copper ore necessary for bronze. Thus a material had been found that could enable large-scale production of weapons and tools for use by the masses. Over the centuries further improvements were made in iron technology, including the invention of early steel through incorporation of carbon (charcoal) into the iron and repeatedly cooling and heating the metal. Once the idea was out, iron technology swept across much of the Old World throughout western Asia and across Europe, rapidly replacing the earlier bronze tools, and then finally penetrated most of Africa, where it actually overtook and preempted the spread of bronze. The rise of the classical Greek and Roman civilizations was accomplished with iron tools.

WHAT'S NEW?

Once these complex societies began to emerge, starting five to six thousand years ago, we witness basically all the major components of modern life as many of us know it today: the complexity of roles and rules, the uneven distribution of wealth and power, the development of large-scale or societywide religious systems, very large settlements or even cities having great monumental architecture and public works, the development of sophisticated arms and organized warfare, and the tax collector. In fact, most of the essential elements of our modern societies have been around for thousands of years, since the very first civilizations.

A major departure we have made since that time has been in the recent bursts of insight and design that began about five centuries ago with the Scientific and Industrial Revolutions. These have accelerated our technological development in ways never before seen in human prehistory or history. The clever and complex machines of our recent era have enormously enhanced the scale and efficiency of our technology, harnessing vast energy sources and allowing the mass production of goods for the teeming populations we are now able to support. Inventions are not arising entirely out of strict necessity; rather, the human imagination is conceiving solutions to real or merely *potential* problems, and these solutions then pose new problems. The latest phase of this has been more subtle but equally important: the use of electronic circuitry to process enormous amounts of complex information in a matter of seconds. Historians will likely look back on the latter part of the twentieth century as the computer revolution, whose benefits and consequences will certainly help shape the world of the next few centuries.

It is obvious from the origins and long development of human technology that we now live in a world moving at a mind-boggling technological pace that is very new to us. We are still in the process of trying to understand it, how to control the development of our technology so that it doesn't outstrip our ability and that of the planet's other organisms to respond to it.

CHAPTER 9

BRAVE NEW WORLD?

But man, proud man,
Dressed in a little brief authority,
Most ignorant of what he's most assur'd
His glassy essence, like an angry ape
Plays such fantastic tricks before high heaven
As makes the angels weep

William Shakespeare (1564–1616)
Measure for Measure

Men have become the tools of their tools.
Henry David Thoreau (1817–1862)
Walden

In this book we have attempted to explain how the dawn of human technology—seen in the origins of flaked stone tools—was pivotal in the evolution of early hominids and the ultimate emergence of our species, and how the increased reliance upon tools increased the adaptive success of our ancestors. Through an archaeological approach combining fieldwork and analysis with experimentation, a clearer picture of the nature of these earliest technologies has begun to emerge.

About two and a half million years ago, the prehistoric record documents the emergence of the earliest identifiable archaeological sites in Africa: simple fractured pieces of rock and associated debris. These primeval technologies, called the Oldowan tradition, were created by hominids who had reached a behavioral and cognitive Rubicon in which the modification and use of tools had become an integral part—not just an incidental component—of their adaptation.

We suspect that this was triggered by a gradual dietary shift, perhaps brought on because of increased competition with robust australopithecines, that emphasized more omnivorous behavior through the consumption of the meat and marrow from animal carcasses, which our hominid ancestors obtained through small-scale hunting and scavenging. The processing of such carcasses would have been facilitated by the crucial addition of manufactured tools to their innate food-getting capabilities—using tools to supplement their biological apparatus. Flaked stone edges were employed as hide- and meat-cutting tools and heavier rocks as bone-breaking and marrow-processing implements. These stone tools were in a sense synthetic meat-shearing "teeth" and bone-crushing "jaws" that allowed hominid groups slowly to enter niches previously occupied by carnivorous mammals such as hyenas, jackals, and the large cats.

As these resources tend to be less predictable food items than the plant foods that would have made up the bulk of their diets, it became more important for hominids to acquire raw materials in advance and transport them longer distances and accumulate them at special, favored locations for tool-making and tool-using activities, a pattern of predetermination and planning that can be seen at the early Oldowan sites in Africa.

Technology could also be used for the acquisition of other types of foods: stone hammers and anvils could crack open the hardest nuts, fruits, or pods as a synthetic "dentition" in a similar manner to bone breaking; digging and prying implements could be made and used as artificial "tusks" to acquire underground vegetation, water, burrowing animals, or insects; spears, thrown missiles, or clubs could improve hunting success and be used in defense; and the use of simple containers could be used to collect and carry substantial quantities of foodstuffs or water as a sort of "second stomach." These behavioral adaptations may have been most critical during times of great environmental changes, allowing for a more flexible response on the part of tool-using hominids to new conditions.

What is more, this tool-making and tool-using behavior was learned, not innate, which put strong selective pressures on the evolution of the hominid brain. By two million years ago there is evidence for a significant increase in brain size in fossil skulls representing early *Homo*; cranial endocasts of these skulls also suggest a reshaping of the hominid brain, which may indicate more complicated neural networks and greater lateralization of the left and right hemispheres of the brain for different mental tasks. The archaeological evidence for preferential right-handedness in the earliest Stone Age may be a manifestation of such a trend as well.

The result was more intelligent, foresighted creatures with increased behavioral complexities able to transmit more information through learning. For the first time in the evolution of life on the earth, a complex feedback loop between culture and biology began to emerge, with tools and the capabilities they endowed primary reasons for the increased reproductive success among hominid populations. Due to the success of this synthetic addition to their biological repertoire (what we have called "techno-organic evolution"), nature began to select for the innovative intelligence responsible for it.

Other changes in hominid behavior may well have begun by this time. Higher levels of cooperation may have existed between members of early hominid social groups, and more complex forms of communication may have emerged in order to convey information about their environments, technologies, and states of emotion. Whether large-scale hunting, a strong sexual division of labor between males and females, or greater food sharing were also major behavioral factors at this time is unclear, but it is possible that such features were slowly beginning to emerge in early Stone Age hominid groups.

As time passed, hominids learned that a wider range of tasks could be accomplished with the use of tools, and their reliance upon their manufacture and use became vital to their survival and reproductive success. Tools slowly became more sophisticated and more carefully designed for different functions, suggesting stronger rules governing the production of such implements. By one and a half million years ago, *Homo erectus* and the earliest Acheulean industries had emerged in the prehistoric record. By one million years ago, the robust australopithecines had become extinct and *Homo erectus* had spread out of Africa and onto the Eurasian landmass.

The last million years of human prehistory documents a slowly accelerating trend toward more varied and more standardized technologies and the emergence of larger-brained hominid forms, archaic *Homo sapiens*. This process culminated with the rise of anatomically modern humans and the apparent cultural explosion around forty thousand years ago of new technological innovations and systems of symbolic behavior that characterize the modern human condition, almost certainly along with the advent of fully developed language skills.

THE FUTURE OF HUMAN TECHNOLOGY

The prehistoric evidence on the earth, although incompletely understood, unquestioningly shows a pattern in the evolution of life: the emergence of some creatures that became more complex over time in the structures and functions of their bodies and brains and in their behavioral patterns. One branch of such creatures ultimately led to relatively intelligent, cognizant, tool-making and tool-using organisms totally dependent on their technologies for their continued existence: ourselves.

Today human beings are approaching other potential technological Rubicons. Beyond developing technologies as a complement to our biological capacities, we have begun to *incorporate those technologies into our biological makeup itself.* Through the technology of genetic engineering we are possibly entering a new phase in human evolution in which modification of the human genome as an evolutionary tool may soon be a possibility, along with the myriad ethical dilemmas that would inevitably be encountered.

Such genetic engineering could ultimately alter the pattern of evolution from a primarily Darwinian mode (evolution by natural selection) to a partially Lamarckian one (evolution by the inheritance of acquired traits, in this case put in place by humans with their technology). Such an evolution could theoretically be accelerated significantly by human control rather than purely by the vagaries of natural selection. The long-term consequences of such a shift can only be guessed at, probably more by science fiction writers than scientists.

A second radical technological departure that could become a possibility involves advances in artificial intelligence and robotics, which may someday make it possible for humans to create "thinking" machines that have the ability to replicate themselves and improve upon their designs. Such an inorganic, self-replicating system of course defies our notion of "life" but could, over time, develop into an alternative, perhaps even more stable, system than organic life, not having to deal with the evolutionary luggage of aggressive, territorial, and competitive behavior.

With over 250 billion stars in the Milky Way alone, the statistical probabilities of life in outer space, perhaps even intelligent forms of life that are technologically complex and able to communicate through rich symbolic systems, may be quite high. How might this intelligent life evolve?

With a sample size of *one* (our own earth's evolutionary history), it may be folly to generalize about how such distant civilizations may have come into being and what might happen to them. Our experience is undeniably, unavoidably, and justifiably parochial; life systems that we cannot even imagine could be relatively common "out there." But we *are* able to survey several billion years of the emergence, evolution, and diversification of life forms on our planet. If we may then use the earth as a partial model of what the general patterns of evolution might be like in "a very remote period indeed; even beyond that of the present world" (to echo John Frere's assessment in 1799 of Acheulean tools found at an English quarry), what might we expect to see in the evolution of life on other worlds?

We think that if such extraterrestrial intelligent life exists, then it is likely such creatures probably went through stages of techno-organic evolution, developing feedback loops between learned behavior and biology that accelerated the evolution of intelligence, as well as the development of more and more complex forms of technology and other types of learned behavior to assist in their adaptation.

Such rudimentary stages of technology, for worlds with habitable stony crusts covering part of their surface, as on the earth, could have gone through Paleolithic stages starting with simple, Oldowan-like stone industries (as well as tools perhaps made of strange and bizarre inorganic and organic materials that we cannot imagine). The archaeological record on these planets, if a team of space-traveling archaeologists could be sent to investigate, would probably show a gradual complexity in the technology and behavior of its inhabitants, which given enough time might lead to intelligent life, capable of and perhaps interested in trying to communicate with other distant civilizations, as we have been in recent years.

Among the recorded sounds that we have propelled from the earth toward the center of our galaxy on the *Voyager* spacecrafts, in an attempt to contact distant civilizations in the cosmos at some remote future time, are recordings of music by Bach, Zairian pygmies, Chuck Berry, and the Navahos. There are greetings in over fifty human languages, screeches of a chimpanzee, taps of Morse code, and the sound of a single human kiss. And, to the great gratification of Paleolithic archaeologists such as ourselves, there are also the sounds of stone striking stone in the production of the simplest of cutting edges. Are there other civilizations in the universe that could understand the significance of the reverberations from the dawn of human technology?

The search for such intelligent life is proceeding even as you read these pages: members of Project SETI (Search for Extraterrestrial Intelligence) are scanning the heavens, hoping to intercept radio signals from outer space that might be intricately patterned in such a manner to suggest that they have been sent forth from some distant civilization. It may seem like looking for a needle in a cosmic haystack, because of the incredible number of directions such signals could come from and the number of frequencies that must be scanned, but the search is an important and worthy one, one in which positive results would be absolutely fascinating and philosophically profound.

No such signals of extraterrestrial life have yet been discovered. It may be that it is only a matter of time before we do intercept such messages. But might there be another reason for this silence? What if the evolution of technological, highly intelligent life is, in the long run, a *maladaptive* trait? Could it be that such complex technological systems rapidly puts such strains on ecological, social, and political systems that these systems spin slowly out of control, leading to extinction or at least to the end of those civilizations?

On the earth, there are many who would argue that this is already beginning to take place. Our technological abilities have made us tremendously successful animals, but also tremendously dangerous ones, glorified apes with nuclear capabilities. We tend to overexploit or destroy our resources, drastically modify and pollute our environments, overpopulate the planet, and wage war on our fellow inhabitants, all of which threaten our long-term viability as a species. Is self-induced extinction the ultimate fate of intelligent, tool-using organisms like ourselves? Are we smart, but not smart enough?

No one can say for sure. But in an evolutionary context there is no manifest destiny for the human species. We know from the evolutionary record that extinction is the norm. Many branches of life have developed over time: those with descendants living in the world today are here because they have been able to adapt to their environments, sometimes to fairly rapidly changing conditions in the past, but countless others have died out. And only one of these branches has produced a creature that is both master and servant of such far-reaching technologies. Our hope of continuing as a species into the future will lie in our ability to come to grips with our technological systems and to overcome their potentially destructive consequences.

Our modern-day technology, in all of its varieties and complexities, allows us to perform such tasks as develop artificial organs, engineer genes, send objects out of our solar system, split the atom, construct supercomputers, and communicate at long distances through electromagnetic waves. These are the legacy of an unbroken chain of learned tool-making and tool-using behavior, transmitted from generation to generation. This chain can be traced back to the dawn of human technology and the emergence of hominid populations that made and used the simplest flaked stone tools some two and a half million years ago on the African continent.

We were recently on a late afternoon flight from Florida to Indianapolis. Several minutes after takeoff, the pilot came on the intercom and directed our attention to the right side of the plane. We were privileged to see a spectacular sight from a perspective that very few others have had: a fireball emerged through the clouds as the space shuttle *Endeavor* began its maiden journey into space.

A few days later, after two unsuccessful attempts to harness a debilitated four-ton communications satellite with a mechanical arm, it was fascinating to see on live television that the three large-brained bipedal astronauts, using grasping hands with opposable thumbs, were able to catch and hold the satellite and then repair and relaunch it, as the world turned beneath them.

Who would have thought that that insignificant-looking bipedal ape on the African grassland would have reached this point in just a few million years? Once a reasoned and habitual use of technology entered the picture and became vital and important to the lives and success of early tool makers, perhaps such a pathway was almost inevitable. Technology has been a powerful force affecting our evolution and intelligence, just as we have driven the course of technology.

It seems apparent to us that a larger perspective on who we are and where we've come is necessary to appreciate the very unusual, special predicament in which our species finds itself today. To understand the human condition we need to study early *Homo* and *Australopithecus* as well as Aristotle, Confucius, Aquinas, and Descartes.

A F T E R W O R D

In 1986 we left California to join the faculty of the Anthropology Department at Indiana University and, with our colleagues, to establish an Old World Paleolithic archaeology program there. By the following year, the administration and trustees of Indiana University approved a proposal we had submitted for the creation of a research center, CRAFT (Center for Research into the Anthropological Foundations of Technology), whose mission is to investigate the origins and development of human technology from an evolutionary perspective. Since that time we have been building up this facility and embarking on a program of training and research in the archaeology of human origins.

The CRAFT headquarters is on the northwest corner of the Indiana University campus and was originally the first Unitarian church in

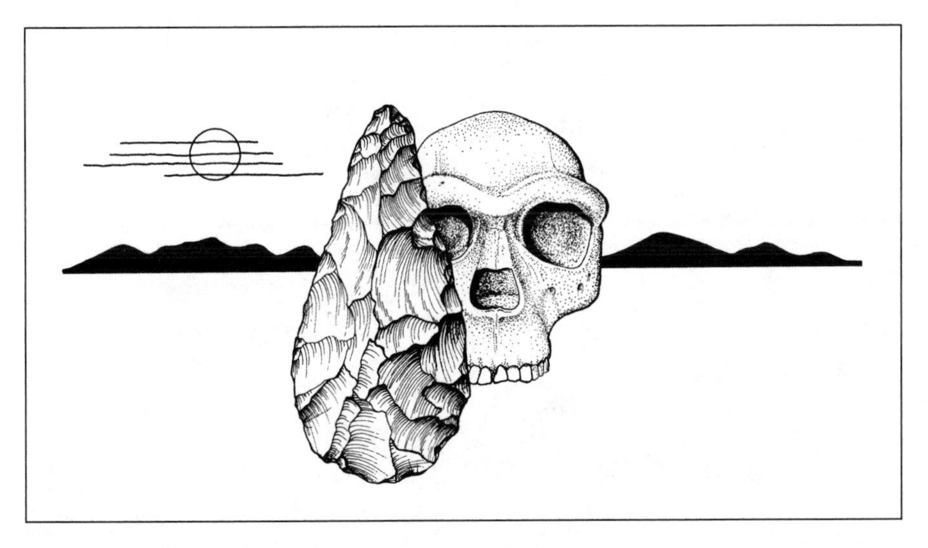

The logo of CRAFT (Center for Research into the Anthropological Foundations of Technology) at Indiana University. A transitional *erectus/sapiens* skull is eclipsed by a later Acheulean handaxe, set before a primeval landscape.

Bloomington. During the past few years we have been building up this facility to direct our research and to train students, as well as to organize talks, lectures, and symposia on a range of related topics.

One of our major attractions when we decided to join the Anthropology Department here was its stress on cooperation and collaboration between the diverse but interrelated subdisciplines of biological, archaeological, sociocultural, and linguistic anthropology. Our contacts with the Departments of Biology (which houses the scanning electron microscope that we use), Geology, Psychology, Kinesiology, African Studies, and Semiotic Studies have been invaluable in our research as well.

Recently our Anthropology Department has moved into new, improved lodgings: the renovated Student Building, one of the oldest landmarks on campus with its famous bell tower, designed originally by novelist Kurt Vonnegut's grandfather. In its renovation, we were able to help design to our specifications extensive research laboratory

The historic Student Building, which houses the Anthropology Department and our Paleolithic archaeology laboratories.

and teaching facilities, including a lecture room opening onto a well-ventilated experimental archaeology area, where we can demonstrate artifact manufacture and conduct sessions in tool using and tool making. As this building is made of hewn limestone, we feel it is an especially appropriate center for Stone Age research!

The Anthropology Department has also built up extensive laboratories devoted to the analysis of animal remains from archaeological sites and for physical anthropological research on recent as well as fossil humans and protohumans.

We have been able to bring in a large number of guest researchers and speakers to CRAFT, thanks in part to generous donations from private individuals to enrich this program. The range of topics have been diverse and exciting, often generating lively discussion and argument. The following partial list illustrates the diversity of topics:

Stanley Ambrose (University of Illinois, Urbana-Champaign): "Carbon Isotopes in Human Evolution and Diet."

Jonas Beyene (National Museum of Ethiopia): "Palaeoanthropological Research in Ethiopia."

C. Loring Brace (University of Michigan): "Cultural Innovation and the Mechanism for the Emergence of Modern Morphology."

Frank Brown (University of Utah): "Dating Early Hominids in East Africa."

Rebecca Cann (University of Hawaii): "Mitochondrial DNA, Human Evolution, and the African Origins of Modern Humans."

J. Desmond Clark (University of California, Berkeley): "Out of Africa: The Human Evolutionary Record."

Garniss Curtis (University of California): "Potassium-Argon Dating in East Africa."

John Fleagle (State University of New York, Stony Brook): "Primate Evolution."

Diane Gifford-Gonzalez (University of California, Santa Cruz): "The Present and Future of Zooarchaeology."

Richard Hay (University of Illinois, Champaign-Urbana): "Africa: The Geological Background to Human Evolution."

F. Clark Howell (University of California, Berkeley): "Current Issues in Palaeoanthropology."

Lawrence Keeley (University of Illinois, Chicago): "Microwear Analysis on Stone Tools: Where Are We At?"

Richard Klein (University of Chicago): "Issues in Faunal Analysis at Palaeolithic Archaeological Sites."

Arthur Jelinek (University of Arizona): "Neanderthals: Who Were They and What Were They Doing?"
Owen Lovejoy (Kent State University): "The Origins of Bipedalism."
Roger Lewin (Washington, D.C.): "Bones of Contention: Controversies in Human Origins Studies."
Thomas Loy (Australia National University): "Blood Residues on Prehistoric Stone Tools."
Lawrence Martin (State University of New York, Stony Brook): "Tooth Enamel Studies and the Evolution of the Apes."
E. Sue Savage-Rumbaugh and Rose Sevcik (Language Research Laboratory, Atlanta): "Apes and Humans: How Similar Are They?" and "Stone Tool Studies of a Bonobo."
Henry Schwarcz (McMaster University): "New Dating Techniques of Palaeoanthropology."
Alan Walker (Johns Hopkins University): "*Homo erectus.*"
John Yellen (National Science Foundation): "Hunter-Gatherer Studies."

Bruce Hardy and Dong Zhuan (*right*) use stone tools to butcher a deer that died of natural causes. Bruce is presently investigating prehistoric residues (including DNA) on stone artifacts. Dong is studying patterns of animal bone modification at the cave at Zhoukoudian (the Peking Man site) in China.

Archaeologist Mohamed Sahnouni of Algeria at the site of Ain Hanech, perhaps the oldest known Paleolithic site in North Africa. Continued excavations here are presently in progress.

A number of our graduate students are carrying out research toward their doctorate degrees directed toward specific problems and questions in Stone Age archaeology. Dong Zhuan of China is currently examining fossil animal bones from the cave at Zhoukoudian (the "Peking man" site) and other early Stone Age sites in order to assess how the early hominids had modified them. He is one of the first Paleolithic archaeologists from mainland China to be trained in the United States.

As previously mentioned, Bruce Hardy, from Birmingham, Alabama, is studying prehistoric organic traces on stone tools from a range of sites, including the Middle Paleolithic rock shelter La Quina in the Charente region of France. He hopes to be able to determine what the tools were used for by identifying specific plant or animal residues. He has obtained a graduate minor in biology at Indiana University and has worked extensively with molecular biologists.

Mohamed Sahnouni of Algeria comes to us with degrees from the University of Algiers and the University of Paris, France. He is now a professor at the University of Algiers, taking several years' leave to obtain his doctorate in anthropology. He has begun excavations at some of the earliest known Stone Age sites from North Africa, including Ain Hanech. This work will include experimental archaeological studies of artifact manufacture as well.

Jeanne Sept, one of our colleagues in the Anthropology Department's Archaeology of Human Origins Program, is currently studying the ecology and land-use patterns of wild chimpanzees in eastern Zaire in order to compare the chimps' use of the landscape, what types of places they frequent, the activities they carry out, and how these are spaced out in order to understand the patterns at our Stone Age sites. She is also continuing her studies of modern plant resources in Africa aimed at increasing our understanding of the range of edible foods available to early hominids and the types of technologies that allowed our hominid ancestors to exploit them more efficiently.

A recent addition to our department and the paleoanthropology program is Kevin Hunt, who studies chimpanzee behavior and ecology in the wild, as well as fossil hominid anatomy and biomechanics. His research indicates that examination of different types of chimpanzee posture (sitting, standing, hanging, and so on) during feeding may give us important clues into the origins of bipedalism in the hominid evolution.

We plan to continue archaeological field research and experimentation at Old World sites in the future, in close collaboration with local prehistorians. Our experimental research with ape technology is also continuing, and we hope to collaborate with neurologists in the future with PET scan imaging to look at human brain function for various activities including tool making.

In these and all future endeavors in the field, we and our colleagues at other institutions around the world will hopefully be guided by new, yet-to-be-imagined advances in archaeological theory and methods, and in particular by the special insights revealed to us by experimentation. The present is a vital key to the past, which makes very remote prehistoric times come alive with the problems and possibilities faced by our ancestors as they took up stone tools and took on the world.

FURTHER READING

Andrews, Peter. "On the Characters that Define *Homo erectus*." *Courier Forschungsinstitut Senckenberg* 69 (1984): 167–178.

Ardrey, Robert. *African Genesis*. New York: Dell, 1961.

Bahn, Paul G., and Jean Vertut. *Images of the Ice Age*. London: Windward, 1988.

Biberson, Pierre J. *Le Paléolithique Inférieur du Maroc Atlantique*. Rabat: Pub. Serv. des Antiquitiés du Maroc, 17, 1961.

Biberson, Pierre J. *Fiches Typologiques Africaines: Galets Aménagés du Magreb et du Sahara*. Congrès Panafricain de Préhistoire et d'Études quaternaires, 1966. 2 Cahier: Fiches 33–64. Paris: Muséum d'Histoire Naturelle.

Bickerton, Derek. *Language and Species*. Chicago: University of Chicago Press, 1990.

Binford, Lewis. *Bones: Ancient Men and Modern Myths*. New York: Academic Press, 1981.

Binford, Lewis. *In Pursuit of the Past*. New York: Thames and Hudson, 1983.

Binford, Lewis. "Searching for Camps and Missing the Evidence? Another Look at the Lower Palaeolithic." In O. Soffer, ed. *The Pleistocene Old World: Regional Perspectives*, 17–32. New York: Plenum Press, 1987.

Blumenshine, Robert J. "Early Hominid Scavenging Opportunities: Implications of Carcass Availability in the Serengeti and Ngorongoro Ecosystems." *British Archaeological Reports International Series* 283 (1986): 1–163.

Boesch, Christophe, and Hedwige Boesch. "Sex Difference in the Use of Hammers by Wild Chimpanzees: A Preliminary Report." *Journal of Human Evolution* 10 (1981): 585–593.

Bonner, John T. *The Evolution of Culture in Animals*. Princeton: Princeton University Press, 1980.

Bordes, Francois. *The Old Stone Age*. New York: McGraw-Hill, 1968.

Bordes, Francois. *Typologie du Paléolithique Ancien et Moyen*. Paris: Presses du CNRS, 1988.

Bordaz, Jacques. *Tools of the Old and New Stone Age*. New York: Natural History Press, 1970.

Boucher de Perthes, Jacques. *Antiquités Celtiques et Antédiluviennes*. Paris: Treuttel et Wurtz, 1847.

Brain, C. K. *The Hunters or the Hunted? An Introduction to African Cave Taphonomy*. Chicago: University of Chicago Press, 1981.

Bunn, Henry. "Evidence on the Diet and Subsistence Patterns of Plio-Pleistocene Hominids at Koobi Fora, Kenya, and at Olduvai Gorge, Tanzania." *British Archaeological Reports* 163 (1983): 21–30.

Bunn, Henry T., John W. K. Harris, Zefe Kaufulu, Ellen Kroll, Kathy Schick, Nicholas Toth, and Anna K. Behrensmeyer. "FxJj 50: An Early Pleistocene Site in Northern Kenya." *World Archaeology* 12 (1980): 109–136.

Bunn, Henry, and Ellen Kroll. "Systematic Butchery by Plio/Pleistocene Hominids at Olduvai Gorge, Tanzania." *Current Anthropology* 5 (1986): 431–452.

Calvin, William H. *The Ascent of Mind.* New York: Bantam, 1990.

Cann, Rebecca L., M. Stoneking, and A. C. Wilson. Mitochondrial DNA and Human Evolution. *Nature* 325 (1987): 31–36.

Cartmill, Matthew. *Primate Origins.* Minneapolis: Burgess, 1975.

Chavaillon, Jean. "Evidence for Technical Practices of Early Pleistocene Hominids: Shungura Formation, Lower Omo Valley, Ethiopia." In Coppens, Y., F. C. Howell, G. Isaac, and R. E. F. Leakey, eds. *Earliest Man and Environments in the Lake Rudolf Basin: Stratigraphy, Palaeoecology, and Evolution.* University of Chicago Press, 1976.

Ciochon, R., and J. G. Fleagle, eds. *The Human Evolution Source Book.* Englewood Cliffs, New Jersey: Prentice-Hall, 1993.

Ciochon, Russell, John Olsen, and Jamie James. *Other Origins.* New York: Bantam, 1990.

Clark, Graham. *Stone Age Hunters.* London: Thames and Hudson, 1967.

Clark, J. Desmond. *Kalambo Falls Prehistoric Site, Volume I.* Cambridge: Cambridge University Press, 1969.

Clark, J. Desmond. *The Prehistory of Africa.* London: Thames and Hudson, 1970.

Clark, J. Desmond. "Palaeolithic Butchery Practices." In Ucko, P., R. Tringham, and G. W. Dimbleby, eds. *Man, Settlement, and Urbanism,* 149–156. London: Duckworth, 1972.

Clark, J. Desmond. *Kalambo Falls Prehistoric Site, Volume II.* Cambridge: Cambridge University Press, 1974.

Clark, J. Desmond, ed. *The Cambridge History of Africa, Volume I: From the Earliest Times to c. 500 B.C.* Cambridge: Cambridge University Press, 1982.

Clark, J. Desmond. "Transitions: *Homo erectus* and the Acheulean: The Ethiopian Sites of Gadeb and the Middle Awash." *Journal of Human Evolution* 16 (7/8, 1987): 809–826.

Coles, John. *Archaeology by Experiment,* London: Hutchison & Co., 1973.

Coles, John M., and Eric S. Higgs. *The Archaeology of Early Man.* New York: Praeger, 1969.

Coltorti, M., et al. *Isernia la Pineta: Un Accampamento Piu Antico di 700,000 Anni.* Bologna: Calderini Editore, 1983.

Conroy, Glenn C. *Primate Evolution.* New York: W. W. Norton & Co., 1990.

Crabtree, Don. *An Introduction to Flintworking.* Occasional Papers of the Idaho State University Museum (1972), 28.

Daniel, Glyn. *The Idea of Prehistory.* Cleveland: World Publishing Co., 1963.

Dart, Raymond. "The Predatory Transition from Ape to Man." *International Anthropological Linguistic Review* 1 (4/1953): 201–18.
Dart, Raymond. "The Osteodontokeratic Culture of *Australopithecus africanus*." *Memoirs of the Transvaal Museum* 10 (1957): 1–105.
Davidson, Iain. "The Archaeology of Language Origins: A Review." *Antiquity* 65 (1991): 39–48.
Day, Michael H. *Guide to Fossil Man*. Chicago: University of Chicago Press, 1986.
Diamond, Jared. The Great Leap Forward. *Discover* 10 (5/1989): 50–60.
Dixon, D. Bruce. *The Dawn of Belief: Religion in the Upper Palaeolithic of Southwestern Europe*. Tucson: University of Arizona Press, 1990.
Eccles, John C. *Evolution of the Brain: Creation of the Self*. London: Routledge, 1989.
Eaton, S. Boyd, Margorie Shostak, and Melvin Konner. *The Palaeolithic Prescription*. New York: Harper & Row, 1988.
Evans, John. *Ancient Stone Implements of Great Britain*. London: 1897.
Fagan, Brian. *The Journey from Eden*. London: Thames and Hudson, 1990.
Falk, Dean. "Hominid Palaeoneurology." *Annual Review of Anthropology*: 16 (1987): 13–30.
Falk, Dean. *Braindance*. New York: Henry Holt, 1992.
Fleagle, John. *Primate Adaptation and Evolution*. San Diego: Academic Press, 1988.
Foley, Robert, ed. *Hominid Evolution and Community Ecology: Prehistoric Human Adaptation in Biological Perspective*. London: Academic Press, 1984.
Foley, Robert. *Another Unique Species: Patterns in Human Evolutionary Ecology*. Harlow, Essex: Longman, 1987.
Frere, John. "Account of Flint Weapons Discovered at Hoxne in Suffolk." *Archaeologia* 13 (1800): 204–205.
Freeman, Leslie G. "Acheulean Sites in Iberia and the Maghreb." In K. W. Butzer and G. L. Issac, eds. *After the Australopithecines*, 661–643. The Hague: Mouton, 1975.
Gamble, Clive. *The Palaeolithic Settlement of Europe*. Cambridge: Cambridge University Press, 1986.
Gladkih, M. I., N. L. Kornietz, and O. Soffer. Mammoth-Bone Dwellings on the Russian plain. *Scientific American*, 251 (5/1984): 164–175.
Goodall, Jane. "Tool-Using and Aimed Throwing in a Community of Free-Living Chimpanzees. *Nature* 201 (1964): 1264–1266.
Goodall, Jane. *In the Shadow of Man*. Boston: Houghton Mifflin, 1971.
Goodall, Jane. *The Chimpanzees of Gombe: Patterns of Behavior*. Cambridge, Mass.: Belknap, 1986.
Goren, Naama. *The Lithic Assemblages of the Site of Ubeidiya, Jordan Valley*. Ph.D. thesis, Hebrew University of Jerusalem, 1981.
Gould, Richard. *Living Archaeology*. Cambridge: Cambridge University Press, 1980.
Gould, Stephen Jay. *Ever Since Darwin*. New York: W. W. Norton, 1977.

Gowlett, John A. "Mental Abilities of Early Man: A Look at Some Hard Evidence." In R. Foley, ed. *Hominid Evolution and Community Ecology: Prehistoric Human Adaptation in Biological Perspective*, 167–192. London: Academic Press, 1984.

Grine, Fredrick E., ed. *Evolutionary History of the "Robust" Australopithecines*. New York: Aldine de Gruyter, 1988.

Harris, J.W.K. "The Karari Industry: Early Pleistocene Archaeological Evidence from the Terrain East of Lake Turkana." *Nature* 262 (1975): 102–107.

Harris, John W. K. "Karari Industry: Its Place in African Prehistory." PhD thesis, University of California, Berkeley, 1978.

Harris, John W. K. "Cultural Beginnings: Plio-Pleistocene Archaeological Occurrences from the Afar, Ethiopia." *The African Archaeological Review* 1 (1983): 3–31.

Harris, John W. K., P. Williamson, J. Verniers, M. Tappan, K. Stewart, D. Helgren, J. de Heinzelin, N. Boaz, and R. Bellomo. "Late Pliocene Hominid Occupation in Central Africa: The Setting, Context, and Character of the Senga 5 Site, Zaire." *Journal of Human Evolution* 16 (7/8, 1987): 701–728.

Hay, Richard L. *Geology of Olduvai Gorge*. Berkeley: University of California Press, 1976.

Hayden, Brian. *Palaeolithic Reflections: Lithic Technology and Ethnographic Excavation among the Australian Aborigines*. Canberra: Australian Institute of Aboriginal Studies, 1979.

Hill, Andrew. "Early Hominid from the Baringo District, Kenya." *Nature* 315 (1985): 222–224.

Holloway, Ralph. "Human Brain Evolution: A Search for Units, Models, and Synthesis." *Canadian Journal of Anthropology* 3 (1983): 215–230.

Howell, F. Clark. "Observations on the Earlier Phases of the European Palaeolithic." *American Anthropologist* 68 (2/2, 1966): 88–201.

Howell, F. Clark. *Hominidae*. In V. J. Maglio and H. B. S. Cooke, eds. *Evolution in African Mammals*, 154–248. Cambridge, Mass.: Harvard University Press, 1978.

Howell, F. Clark. "The Evolution of Human Hunting" (book review). *Journal of Human Evolution* 18 (6/1989): 583–594.

Hunt, Kevin. "Mechanical Implications of Chimpanzee Positional Behavior." *American Journal of Physical Anthropology* 86(4): 521–536, 1991.

Isaac, Barbara. "Throwing and Human Evolution." *The African Archaeological Review* 5 (1987): 3–17.

Isaac, Glynn. "The Stratigraphy of the Peninj Group: Early Middle Pleistocene Formations West of Lake Natron, Tanzania." In W. W. Bishop and J. D. Clark, eds., *Background to Evolution in Africa*, 229–257. Chicago: University of Chicago Press, 1967.

Isaac, Glynn. *Olorgesailie: Studies of a Middle Pleistocene Lake Basin in Kenya*. Chicago: University of Chicago Press, 1977.

Isaac, Glynn. *The Archaeology of Human Origins: Papers by Glynn Isaac*. Ed.

Barbara Isaac. Cambridge: Cambridge University Press, 1989. (An edited collection of Glynn Isaac's most important articles.)
James, Stephen R. "Hominid Use of Fire in the Lower and Middle Pleistocene: A Review of the Evidence." *Current Anthropology* 30 (1/1989): 1–26.
Johanson, Donald C., and Tim D. White. "A Systematic Assessment of Early African Hominids." *Science* 203: 321–330, 1979.
Johanson, Donald C., and M. A. Edey. *Lucy: The Beginnings of Humankind.* New York: Simon and Schuster, 1981.
Johanson, Donald C., and James Screeve. *Lucy's Child: The Discovery of a Human Ancestor.* New York: William Morrow and Company, 1989.
Jolly, Alison. *The Evolution of Primate Behavior.* New York: Macmillan, 1985.
Jones, Peter. "Experimental Implement Manufacture and Use: A Case Study from Olduvai Gorge." *Philosophical Transactions of the Royal Society (London)* B 292 (1981): 189–195.
Jungers, William L. "New Estimates on the Body Size in Australopithecines." In F. E. Grine, ed. *The Evolutionary History of the Robust Australopithecines*, 115–116. New York: Aldine de Gruyter, 1988.
Kaufulu, Zefe M., and Nicola Stern. "The First Stone Artifacts to be Found *in situ* within the Plio-Pleistocene Chiwondo Beds in Northern Malawi." *Journal of Human Evolution* 16 (7/8, 1987): 729–740.
Keeley, Lawrence H. *Experimental Determination of Stone Tool Uses: A Microwear Analysis.* Chicago: University of Chicago Press, 1980.
Keeley, Lawrence H., and Nicholas Toth. "Microwear Polishes on Early Stone Tools from Koobi Fora, Kenya." *Nature* 293 (1981): 464–465.
Klein, Richard G. *The Human Career: Human Biological and Cultural Origins.* Chicago: University of Chicago Press, 1989.
Kroll, Ellen, and Glynn Isaac. "Configurations of Artifacts and Bones at Early Pleistocene Sites in East Africa." In H. Hietala, ed. *Intrasite Spatial Analysis in Archaeology*, p. 4–31. Cambridge: Cambridge University Press, 1984.
Lancaster, Jane B. *Primate Behavior and the Emergence of Human Culture.* New York: Holt, Rinehart & Winston, 1975.
Lanpo, Jia and Huang Weiwen. *The Story of the Peking Man.* Oxford: Oxford University Press, 1990.
Lartet, Edouard and Henry Christy. *Reliquiar Aquitanicae: Being Contributions to the Archaeology and Palaeontology of Périgord and the Adjoining Provinces of Southern France.* London: Rupert Jones, 1875.
Leakey, Louis. *Olduvai Gorge 1951–1961: Volume 1.* Cambridge, Cambridge University Press, 1951.
Leakey, Mary D. *Olduvai Gorge: Volume 3: Excavations in Beds I and II, 1960–1963.* Cambridge: Cambridge University Press, 1971.
Leakey, Meave G., and Richard E. F. (eds.) *The Fossil Hominids and an Introduction to Their Context, 1968–1974: Koobi Fora Research Project I.* Oxford: Clarendon Press, 1978.
Lee, Richard B. *The !Kung San.* Cambridge: Cambridge University Press, 1979.

Lewin, Roger. *Human Evolution: An Illustrated Introduction*. Boston: Blackwell, 1989.

Lieberman, Philip. *Uniquely Human*. Cambridge, Mass.: Harvard University Press, 1991.

Ligabue, Giancarlo. "The Stone Axes of the Pygmies of New Guinea." *Ligabue Magazine* 4 (6/1985): 30–35.

Lindly, J. M., and G. A. Clark. "Symbolism and Modern Human Origins." *Current Anthropology* 31 (3/1990): 233–261.

Lovejoy, C. Owen. "The Origins of Man." *Science* 211 (1981): 341–350.

Loy, Thomas. "Prehistoric Blood Residues: Detection on Tool Surfaces and Identification of Species of Origin." *Science* 220 (1983): 1269–1270.

Lumley, Henry de. "Cultural Evolution in France in its Paleoecological Setting during the Middle Pleistocene." In K. W. Butzer, and G. Issac, eds. *After the Australopithecines*, 745–808. The Hague: Mouton, 1975.

Lumsden, Charles J., and Edward O. Wilson. *Promethean Fire: Reflections of the Origin of Mind*. Cambridge, Mass.: Harvard University Press, 1983.

Marshack, Alexander. *The Roots of Civilization*. Mount Kisco, New York: Moyer Bell Limited, 1991.

Martin, Lawrence. "Significance of Enamel Thickness in Hominoid Evolution." *Nature* 314 (1985): 260–263.

McGrew, W. "Evolutionary Implications of Sex Differences in Chimpanzee Predation and Tool Use." In D. A. Hamburg and E. R. McCown, eds. *The Great Apes*, 441–464. Menlo Park: Benjamin/Cummings, 1979.

Mellars, Paul. "Major Issues in the Emergence of Modern Humans." *Current Anthropology* 30 (3/1989): 349–385.

Mellars, Paul, and Chris Stringer, eds. *The Human Revolution: Behavioral and Biological Perspectives on the Origins of Modern Humans, Vol. 1*. Edinburgh: Edinburgh University Press, 1989.

Merrick, H. V., and J.P.S. Merrick. "Archaeological Occurrences of Earlier Pleistocene Age from the Shungura Formation." In Coppens, Y., F. C. Howell, G. Isaac, and R.E.F. Leakey, eds. *Earliest Man and Environments in the Lake Rudolf Basin*. Chicago: University of Chicago Press, 1976.

Mortillet, Gabriel de. *Formation de la Nation Française*. Paris, 1897.

Movius, Hallam L. "The Lower Palaeolithic Cultures of Southern and Eastern Asia." *Transactions of the American Philosophical Society* 38 (4/1948): 329–340.

Movius, Hallam L. "A Wooden Spear of Third Interglacial Age from Lower Saxony." *Southwestern Journal of Anthropology* 6 (2/1950): 139–143.

Newcomer, Mark. "Some Quantitative Experiments in Handaxe Manufacture." *World Archaeology* 3 (1/1971): 85–94.

Oakley, Kenneth. "Skill as a Human Possession." In C. Singer and E. Holmyard, eds. *A History of Technology, Volume I*. Oxford: Clarendon Press, 1954.

Oakley, Kenneth. "Fire as a Palaeolithic Tool and Weapon." *Proceedings of the Prehistoric Society* 21 (1955): 36–48.

Oakley, Kenneth. *Man the Tool-Maker.* Chicago: University of Chicago Press, 1959.

Oakley, Kenneth, P. Andrews, L. Keeley, and J. D. Clark. "A Reappraisal of the Clacton Spearpoint." *Proceedings of the Prehistoric Society* 43 (1977): 13–30.

O'Brien, Eileen M. "The Projectile Capabilities of the Acheulean Handaxe from Olorgesailie." *Current Anthropology* 22 (1981): 76–79.

Olson, John W. Recent Developments in the Upper Pleistocene Prehistory of China. In O. Soffer, ed. *The Pleistocene Old World: Regional Perspectives,* 135–146. New York: Plenum, 1987.

Pei, W. C., and S. Zhang. "A Study on Lithic Artifacts of *Sinanthropus.*" *Palaeontologia Sinica* 168 (1985): 1–277.

Pfeiffer, John. *The Creative Explosion.* New York: Harper and Row, 1982.

Pilbeam, David R. "Hominoid Evolution and Hominoid Origins." *American Anthropologist* 88 (1986): 295–312.

Pope, Geoffrey. "Taxonomy, Dating and Paleoenvironment: The Paleoecology of the Far Eastern Hominids." *Modern Quaternary Research in Southeast Asia* 9 (1985): 65–80.

Potts, Richard. *Early Hominid Activities at Olduvai.* New York: Aldine de Gruyter, 1988.

Potts, Richard. "Olorgesailie: New Excavations and Findings in Early and Middle Pleistocene Contexts, Southern Kenyan Rift Valley." *Journal of Human Evolution* 18 (5/1989): 477–484.

Potts, Richard. "Why the Oldowan? Plio-Pleistocene Toolmaking and the Transport of Resources." *Journal of Anthropological Research,* 47: 153–176, 1992.

Rak, Yoel. *The Australopithecine Face.* New York: Academic Press, 1983.

Rightmire, Philip. "*Homo erectus* and Later Middle Pleistocene Humans." *Annual Review of Anthropology* 17 (1988): 239–259.

Roche, Hélène. *Premiers Outils Taillés D'Afrique.* Paris: Société D'Ethnographie, 1980.

Sagan, Carl, F. D. Drake, A. Druyan, T. Ferris, J. Lomberg, L. S. Sagan. *Murmurs of Earth.* New York: Ballantine, 1978.

Sahnouni, Mohamed. *L'Industrie sur Galets du Gisement Villafranchien Supérieur de Ain Hanech.* Alger: Office des Publications Universitaires, 1987.

Savage-Rumbaugh, E. Sue. *Ape Language: From Conditioned Response to Symbol.* New York: Columbia University Press, 1986.

Schick, Kathy. *Stone Age Sites in the Making: Experiments in the Formation and Transformation of Archaeological Occurrences.* Oxford: British Archaeological Reports, International Series 319, 1986.

Schick, Kathy. "Modeling the Formation of Early Stone Age Artifact Concentrations." *Journal of Human Evolution* 16 (7/8, 1987): 789–808.

Schick, Kathy. "Experimentally Derived Criteria for Assessing Hydrologic Disturbance of Archaeological Sites." In Nash, D. T., and M. D. Petraglia,

eds. *Natural Formation Processes and the Archaeological Record.* British Archaeological Reports, International Series 352, 1987.
Schick, K., N. Toth, Wei Qi, J. D. Clark, and D. Etler. "Archaeological Perspectives in the Nihewan Basin, China." *Journal of Human Evolution* 21: 13–26, 1991.
Schick, Kathy, and Zhuan Dong (in press) "The Early Palaeolithic of China and Eastern Asia." *Journal of Evolutionary Anthropology.*
Sebeok, Thomas. *The Play of Musement.* Bloomington: Indiana University Press, 1981.
Semenov, Surgei A. *Prehistoric Technology.* London: Cory, Adams, and Mackay, 1964.
Sept, Jeanne. "Archaeological Evidence and Ecological Perspectives for Reconstructing Early Hominid Subsistence Behavior." *Advances in Archaeological Method and Theory* 4: 1–56. 1992.
Sept, Jeanne. "Was There No Place Like Home?" *Current Anthropology* 33 (2): 187–208, 1992.
Sept, Jeanne. "Plants and Early Hominids in East Africa: A Study of Vegetation in Situations Comparable to Early Archaeological Locations." Ph.D. thesis, University of California Berkeley, 1984.
Shipman, Patricia. "Scavenging or Hunting in Early Hominids: Theoretical Framework and Tests." *American Anthropologist* 88 (1986): 27–43.
Shipman, Patricia, and Alan Walker. "The Costs of Becoming a Predator." *Journal of Human Evolution* 18 (4): 373–392, 1989.
Simons, E. "Origins and Characteristics of the First Hominoids." In E. Delson, ed. *Ancestors: The Hard Evidence* 37–41. New York: Alan R. Liss, 1985.
Singer, Charles, E. J. Holmyard, A. R. Hall, and T. I. Williams. *A History of Technology.* Oxford: Clarendon Press, 1984.
Smith, Fred, and F. Spencer, eds. *The Origins of Modern Humans: A World Survey of the Fossil Evidence* 137–209. New York: Alan R. Liss, 1984.
Smith, Worthington. *Man, the Primeval Savage.* London: Edward Stanford, 1894.
Stekelis, M., O. Bar-Josef, and T. Schick. *Archaeological Excavations at Ubeidiya, 1964–1966.* Tel Aviv: Israel Academy of Science and Humanities, 1969.
Stern, John, and Randall Susman. "The Locomotor Anatomy of *Australopithecus afarensis.*" *American Journal of Physical Anthropology* 60 (1983): 279–317.
Stiles, Daniel. "Early Acheulean and Developed Oldowan." *Current Anthropology* 20 (1979): 126–129.
Stringer, Christopher and Peter Andrews. "Genetic and Fossil Evidence for the Origin of Modern Humans. *Science* 239 (1988): 1263–68.
Susman, Randall L. "Hand of *Paranthropus Robustus* from Member 1, Swartkrans: Fossil Evidence for Tool Behavior." *Science* 240 (1988): 781–784.
Swanson, Earl, ed. *Lithic Technology: Making and Using Stone Tools.* The Hague: Mouton, 1975.
Tanner, Nancy M. *On Becoming Human.* Cambridge: Cambridge University Press, 1981.

Tattersall, Ian, Eric Delson, and John Van Couvering, eds. *Encyclopedia of Human Evolution and Prehistory*. New York: Garland, 1988.

Tixier, Jacques, M. L. Inixan, and H. Roche. *Préhistoire de la Pierre Tailée*. Valbonne: Cercle de Recherches et d'Études Préhistoriques, 1980.

Tobias, Philip V. "The Brain of *Homo habilis:* A New Level of Organization in Cerebral Evolution." *Journal of Human Evolution* 16 (7/8, 1987): 741–761.

Toth, Nicholas. *The Stone Technologies of Early Hominids at Koobi Fora, Kenya: An Experimental Approach*. Ann Arbor: University Microfilm, 1982.

Toth, Nicholas. "Archaeological Evidence for Preferential Right-Handedness in the Lower and Middle Pleistocene, and Its Possible Implications." *Journal of Human Evolution* 14 (1985): 607–614.

Toth, Nicholas. "The Oldowan Reassessed: A Close Look at Early Stone Artifacts." *Journal of Archaeological Science* 12 (1985): 101–120.

Toth, Nicholas. "The First Technology." *Scientific American* (April 1987).

Toth, Nicholas. "Behavioral Inferences from Early Stone Age Artifact Assemblages: An Experimental Model. *Journal of Human Evolution* 16 (7/8, 1987): 763–787.

Toth, Nicholas, Desmond Clark, and Giancarlo Ligabue. "The Last Stone Ax-Makers."*Scientific American* 267 (1/1992): 88–93.

Toth, Nicholas, and Kathy Schick. "The First Million Years: The Archaeology of Protohuman Culture." *Advances in Archaeological Method and Theory* 9 (1986): 1–96.

Toth, Nicholas, Kathy Schick, E. Sue Savage-Rumbaugh, Rose Sevcik, Duane Rumbaugh (in press). "Investigations in the Stone Tool-Making and Tool-Using Capabilities of a Bonobo." *Journal of Archaeological Science*.

Trinkhaus, Eric. "The Neanderthals and Modern Human Origins." *Annual Review of Anthropology* 15 (1986): 193–218.

Villa, Paola. *Terra Amata and the Middle Pleistocene Archaeological Record of Southern France*. University of California Publications in Anthropology 13, 1983.

Villa, Paola. "Torralba and Aridos: Elephant Exploitation in Middle Pleistocene Spain." *Journal of Human Evolution* 19 (3/1990): 299–309.

Vincent, Anne. "Plant Foods in Savanna Environments." *World Archaeology*, 17 (2/1984): 131–148.

Vrba, Elizabeth S. "Ecological and Adaptive Changes Associated with Early Hominid Evolution." In E. Delson, ed. *Ancestors: The Hard Evidence* 63–71. New York: Alan R. Liss, 1985.

Waldorf, Don C. *The Art of Flintknapping*. Branson, Missouri: Mound Builder Books, 1979.

Walker, Alan. "Dietary Hypothesis and Human Evolution." *Philosophical Transactions of the Royal Society of London* 292 (1981): 56–64.

Washburn, Sherwood L. "Tools and Human Evolution." *Scientific American* 203 (3/1960): 63–75.

White, Tim. "Cut Marks on the Bodo Cranium: A Case of Prehistoric Defleshing." *American Journal of Physical Anthropology* 69 (1986): 503–509.

Willoughby, Pamela R. "Spheroids and Battered Stones in the African Early Stone Age." *World Archaeology* 17: 44–60, 1985.

Wilson, Edward O. *Sociobiology: The New Synthesis.* Cambridge, Mass.: Belknap Press of the Harvard University Press, 1975.

Wolpoff, Milford H., Wu Xin Zhi, and A. G. Thorne. "Modern *Homo sapiens* Origins: A General Theory of Hominid Evolution Involving the Fossil Evidence from East Asia." In F. H. Smith and F. Spencer, eds. *The Origins of Modern Humans: A World Survey of the Fossil Evidence,* 411–483. New York: Alan R. Liss, 1984.

Wrangham, Richard. "Feeding Behavior in Chimpanzees in the Gombe National Park, Tanzania." In T. Clutton-Brock, ed. *Primate Ecology,* 504–538. London: Academic Press, 1977.

Wright, Richard V. S. "Imitative Learning of a Flaked Tool Technology: The Case of an Orangutan." *Mankind* 8 (1972): 296–306.

Wymer, John. *The Palaeolithic Age.* London: Croom Helm, Ltd, 1982.

Wynn, Thomas. *The Evolution of Spatial Competence.* Urbana: University of Illinois Press, 1989.

Wynn, Thomas and William McGrew. "An Ape's View of the Oldowan." *Man* 24 (1989): 383–398.

Yellen, John. *Archaeological Approaches to the Present.* New York: Academic Press, 1977.

Zihlman, Adrian, and Nancy Tanner. "Gathering and the Hominid Adaptation." In L. Tiger and H. M. Fowler, eds. *Female Hierarchies.* Chicago: Beresford Book Service, 1979.

INDEX

Abbeville site, France, 60, 62–63, 270
Acheulean tools, 117, 192, 225–84, 288–89, 292, 314, 318
  aesthetics of, 282–83
  in Africa, 232–37, 263–68, 271, 276
  animal bones associated with, 252, 257, 271–73, 281
  cores of, 238–39, 242–43, 247, 258, 263–64
  dating of, 227, 231, 233, 235, 252–253, 256, 257, 258, 264, 269, 271, 276, 282
  dense concentrations of, 266, 267–268
  Developed Oldowan vs., 235–37
  duration of, 229
  in eastern Asia, 252–57, 275–79
  in Europe, 227, 244–45, 257–58, 269–74, 276, 279, 281
  functions of, 258–60, 268, 278
  manufacture of, *see* manufacture, Acheulean tool
  Movius line and, 276–79
  rocks used for, 235, 237, 263, 267, 271
  stylistic constancy of, 283–84
  types of, 231–33
  use-wear patterns of, 260, 271
  in western Asia, 227, 251–52, 268–269, 276
actualistic studies, 20
adornment, personal, 282, 299, 300
adzes, 245–51, 304
Aegyptopithecus, 34
Africa, 18, 20, 41–44, 83–92, 131, 153, 157, 219, 227, 258, 301, 310
  Acheulean tools in, 232–37, 263–68, 271, 276
  ancient habitats of, 34–35, 36, 37, 38, 47, 87, 263
  archaic *H. sapiens* in, 261, 262–68, 287
  continental drift of, 33, 35, 191
  early civilizations in, 306–7
  early *H. sapiens sapiens* in, 287–88
  H. erectus in, 84, 86, 87, 89, 229–31, 261, 262–68, 278
  Middle Paleolithic (Mousterian) tools in, 291, 292, 293
  Rift valley of, 30, 39, 41, 71, 78, 83–89, 110, 191, 234, 265
African Genesis (Ardrey), 70
agates, 124
Ain Hanech, Algeria, 83, 323
Algeria, 231, 235, 323
Almagro, Martin, 271
Ambrona site, Spain, 73, 270, 271–74, 280
amino acid racemization, 30
andesites, 123
Andrews, Peter, 180, 278
animal bones, 18, 27, 60, 61–62, 87, 110, 142, 182, 202, 252, 254, 255, 257, 281
  in Ambrona site, 271–73
  in endowed burials, 299
  in Koobi Fora site, 86, 166, 206, 207
  in Olduvai Gorge site, 43, 88, 166, 169, 206–8, 209, 210–11
  in South African sites, 46, 66, 68, 89–90, 92, 103–4
  see also bone modification
animals, 140, 156, 172, 184, 228
  building activities of, 53, 157, 213
  culture of, 47, 49–52, 55, 57–58, 135
  domestication of, 302, 303–4, 310
  instinctive behavior of, 51, 52–53, 55
  tool use of, 24, 48–58, 183; *see also* chimpanzees
anvils, 53–54, 99, 114, 122, 145, 152
  bone breaking with, 170, 171, 184
  nut cracking with, 57, 153–55, 157, 184–86, 313
  see also hammer stones
anvil technique, 120

apes, 18, 25, 26, 27, 47, 48, 55, 64, 74, 133, 186
brain size of, 44, 103
evolution of, 33–37, 38–39
intelligence of, 134–35, 182, 218
sleeping nests of, 157, 213
social behavior of, 216–17
see also chimpanzees
Arago site, France, 269, 270
Arambourg, Camille, 44, 83
arborealism, 32, 34, 45, 47
archaeology, 18, 19–20, 48, 63–76, 109–12, 129, 223
behavioral approach in, 73–74, 109–110, 206
field schedule of, 110–12
modern methods of, 73–76, 111, 191–92
nineteenth-century, 61–65, 93, 210, 264, 291, 316
typology in, 97
see also experimental archaeology
Archaeology by Experiment (Coles), 21
architecture, 213–14, 271, 305
of *H. erectus*, 280–81
of *H. sapiens sapiens*, 300, 301, 305
of hunter-gatherers, 158, 214
Ardrey, Robert, 70
Arkin site, Egypt, 263
arms race, origins of, 307–10
arrows, 59, 62, 153, 158
poison-tipped, 208, 299
art, 26, 64, 161, 282–83, 299–301
artifacts:
definition of, 49
see also technology; tools
artificial intelligence, 315
artificial selection, 63, 302
Asia, 26, 33, 35, 64, 67, 71, 219, 222
Asia, eastern, 287, 288, 291, 292, 304
Acheulean tools in, 252–57, 275–79
H. erectus in, 64, 67, 229, 252–57, 261, 275–79
migration routes from, 301–2
Asia, western, 64, 131, 237, 258, 287, 288, 302, 304, 307, 310
Acheulean tools in, 227, 251–52, 268–69, 276
atlatls (spear throwers), 158, 298
Aurignacian tools, 65
Australian aboriginals, 152, 156, 160, 181, 214, 288, 298, 301
australopithecines, 31, 40–47, 78, 79, 102, 139
brain size of, 44–45, 46, 47, 81, 103
brain structure of, 103, 219, 220
communication skills of, 220–21
dating of, 40, 43, 44, 67, 85, 87, 90
extinction of, 104, 218–19, 231, 314
handedness of, 103, 221
robust, *see* robust australopithecines
sexual dimorphism of, 42–43, 47
significant features of, 44–46
Australopithecus aethiopicus, 80–81, 84, 87
Australopithecus afarensis, 40–46, 84, 87, 102
anatomical features of, 41, 42–43, 45, 46, 47
First Family of, 42–43, 85
Australopithecus africanus, 44, 66–67, 68–71, 79, 81, 90, 92
Australopithecus boisei, 80, 81, 84, 86, 88, 89, 105, 207, 231, 233
Australopithecus prometheus, 68, 215
Australopithecus robustus, 67, 81, 89, 102, 103, 231
awls, 98, 99, 293–95

baboons, 34, 43, 45, 66, 89, 90, 92, 153, 156, 184
Bagford, John, 60
bamboo, as raw material, 277, 278
bark-stripping, 259
basketry, 305
Behrensmeyer, Kay, 180
Belgium, 63
Belohdelie site, Ethiopia, 40, 44
bifaces, 98, 113, 117, 230, 235, 290
see also handaxes
Bilzingsleben site, Germany, 270, 280, 281
Binford, Lewis, 207, 210–11, 235
biocultural feedback, 142–43, 223
biostratigraphy, 84
bipedalism, 105, 106, 218
evolution of, 28, 36, 37–38, 41, 42, 45–47, 68, 217
in Laetoli footprints, 43–44
bipolar cores (outils écaillés), 120, 122
bipolar nut cracking, 145
bipolar technique, 120, 121–22, 155, 275
birds, 48, 53, 54–55, 186
Black, Davidson, 275
blade technologies, 293, 296–97, 305
Blumenschine, Rob, 180, 207
Bodo site, Ethiopia, 262, 263, 282

Boesch, Christophe, 57, 58
bolas stones, 130–33
bone breaking, 145, 149, 154, 169–70, 171, 173, 207, 209, 211, 313
  by carnivores, 170, 180, 184
  fracture patterns of, 180–81, 206
bone marrow, 149, 154, 158, 170, 171, 173, 184, 207, 209, 211, 313
bone modification, 151, 177–81, 257, 322
  carnivore marks in, 90, 178, 179–80, 281
  cut marks in, 101, 165, 169, 177–79, 207–8, 209, 255, 281–82
  geological forces in, 180, 194
bone tools, 64, 68–71, 90, 118, 155, 157, 161, 183, 184, 241, 263, 293, 297–99
Bonner, John, 50
bonobos (pygmy chimpanzees), 24, 151
  stone tools manufactured by, 120, 135–40
Bordes, François, 22
Boucher de Perthes, Jacques, 61, 62–63, 231
bow, 208, 299
  see also arrows
Boxgrove site, England, 270
brain, 139, 229, 324
  Broca's area of, 219, 220
  endocranial casts of, 103, 134, 219, 313
  lateralization of, 103, 140–41, 219, 220, 221, 313
Brain, C. K., 70, 89–90, 92, 157, 215
brain size, 32, 218, 313
  of apes, 44, 103
  of archaic *H. sapiens*, 261–62
  of australopithecines, 44–45, 46, 47, 81, 89, 103
  expansion of, 26, 28, 45, 46, 47, 82, 103, 143, 186, 219, 220, 261–62
  of *H. erectus*, 64, 229
  of *H. habilis*, 19, 81, 89, 103, 219
  of *H. sapiens sapiens*, 44
Bräuer, Gunter, 262
Broken Hill site, Zambia, 261, 262
Broom, Robert, 67
Brown, Frank, 85
bulbar scar (éraillure), 94
bulb (semicone) of percussion, 94, 95, 122, 232, 240, 242
Bunn, Henry, 110, 180, 207–8
burials, 292–93, 299, 300
burins, 99, 293–95
Burroughs, John, 189
butchery, 24, 110, 146, 148, 152, 162–169, 176, 183–84, 206, 213, 218, 221, 222, 277, 322
  with Acheulean tools, 243, 258, 259–60, 268, 273
  defleshing in, 164–66, 179, 182, 211, 260, 313
  dismembering in, 166, 182, 209, 259, 260
  gutting in, 162–64
  hide cutting in, 162–64, 187, 260, 313
  of megafauna, 23, 166–69, 243, 272–273
  see also bone modification

cannibalism, 70, 71, 281–82
carbon-14 (radiocarbon) dating, 30, 261
carnivores, 89–90, 92, 162, 183–84, 207, 217, 218, 272, 283
  bone marks made by, 90, 178, 179–180, 281
  offense and defense against, 170–72
  teeth of, 170, 180, 182, 183, 184
carrying devices, 153, 160, 162, 172–73
casts, 192
  endocranial, 103, 134, 219, 313
central foraging places, 210, 212, 222
Cerling, Thure, 85
chalcedonies, 121, 124, 126, 271
cherts, 95, 121, 122, 124, 126, 176, 237, 238, 242–43, 244, 278
Chesowanja site, Kenya, 215, 235
chimpanzees, 23, 34, 36, 38, 46, 56–58, 134, 157, 221, 324
  aggressive behavior of, 71
  brain size of, 44
  culture of, 47, 51, 52, 57–58
  hand of, 45, 102
  hunting by, 164, 206, 217
  mental maps of, 126
  nut cracking by, 49, 57, 58, 145, 152, 153, 154, 155, 157
  personal possessions of, 58
  social behavior of, 58, 216–17
  throwing by, 138, 144
  transportation by, 57, 58, 140
  see also bonobos
China, 36, 131, 252, 253–56, 275, 287, 304, 310
  see also Zhoukoudian site, China

Chon-Gok-Ni site, South Korea, 276
choppers, 98, 99, 113, 116, 117, 121, 129, 158, 159, 166, 171, 173
chopping, 20, 158, 159, 250, 260
Christy, Henry, 64–65
civilization, 285–86, 305–11
  arms race in, 307–10
  metallurgy in, 21, 26, 307–10
  modern technology of, 17, 26, 310–311, 314–18
Clacton-on-Sea site, England, 172, 270, 271
Clark, J. Desmond, 22, 23, 44, 73, 246, 255, 259, 263, 264, 266, 267, 277, 280, 321
Clarke, Arthur C., 70
cleavers, 227, 232, 235, 236, 240, 241, 267, 271, 273–74, 276, 288–89
  of archaic *H. sapiens*, 263
  function of, 260
  manufacture of, 244–45, 279
clothing, 153, 160, 161, 162, 271, 299
cobbles, 85, 86, 97, 98, 113, 114, 117, 118, 126, 127, 128, 130, 142, 170, 176, 230, 235, 238, 243, 268
Coles, John, 21
communication skills, 220–21, 222, 314
  see also language; symbolic communication
conchoidal fracture, 94–96, 122, 125
  angles of, 96, 119, 134, 139
  features of, 93, 94–95
  geologically produced, 95–96, 139, 252–53
conjoining, 101, 128, 205, 209
containers, 153, 158, 160, 271, 313
  pottery, 305
  raw materials for, 172–73, 174
continental drift, 32–33, 35, 191, 229
Coppens, Yves, 42, 44
cores, 96, 112–13, 118–20, 121, 127–130, 134, 136, 142, 159, 166, 170, 192–93, 273
  Acheulean, 238–39, 242–43, 247, 258, 263–64
  axes, 263
  bipolar (outils écaillés), 120, 122
  as by-product vs. tool, 129–30
  features of, 94–95
  scrapers, 98–99, 117, 129
cortex, 94, 114, 123, 128, 142
Crabtree, Don, 22
CRAFT (Indiana University), 8, 319–324
cryptocrystalline siliceous rocks, 122, 124–25
culture, 142–43, 223, 284
  animal, 47, 49–52, 55, 57–58, 135
  definition of, 49
  see also technology; tools
cut marks, 101, 165, 169, 177–79, 207–208, 209, 255, 281–82
Czechoslovakia, 258, 294

Dart, Raymond, 66–67, 68–71, 90, 215
Darwin, Charles, 45, 62–63, 68, 69, 150, 315
dating techniques, 28–30, 261
  see also specific dating techniques
Dawson, Charles, 65
débitage, 99
deep-sea geological cores, 227
Deetz, James, 129
defense, 153, 170–72, 313
  see also weaponry
defleshing (filleting), 164–66, 179, 182, 211, 260, 313
Deinotherium, 166
de Lumley, Henry, 281
de Mortillet, Gabriel, 65
Denmark, 61–62
Dennell, Robin, 252
denticulates, 263, 269, 290
derived characteristics, 27–28, 38
Descent of Man, The (Darwin), 68, 150
Developed Oldowan tools, 99, 235–37
de Voto, Bernard, 78
Dezzani, Ray, 167
Diamond, Jared, 295
diet, 20, 36, 110, 162, 183, 206, 208, 260
  bone marrow in, 149, 154, 158, 170, 171, 173, 184, 207, 209, 211, 313
  fat in, 169–70, 171
  omnivorous, 47, 222, 312–13
  of robust australopithecines, 103, 106–7
  see also meat eating
digging, 152, 155, 156–57, 158, 222, 259, 260, 313
  raw materials for, 153, 157, 184, 271
digits, number of, 27–28
Dingcun site, South Korea, 276
discoids, 98, 99, 117, 129, 166, 238, 263–64
dismembering, 166, 182, 209, 259, 260
division of labor, sexual, 210, 211, 212, 216, 314

Dmanisi site, Georgia republic, 257
DNA, 36
  mitochondrial, 288
  in organic residues, 181, 322
domestication, 302, 303–4, 310
Donggutuo, China, 255–57
Dong Zhuan, 255, 322, 323
dorsal scars, 94
Dryopithecus, 35
Dubois, Eugene, 64

earth movements, 191, 252
edge damage, 160, 161, 176, 178
Egypt, 34, 65, 251, 263
Egyptian vulture, 53
Elandsfontein site, South Africa, 262, 263
electron spin resonance, 30, 261
elephants, 29, 156, 157, 184
  experimental butchery of, 23, 166–169, 243
  footpads of, as containers, 173, 174
  fossil, 62, 166, 169, 257, 271–73, 281
Elephas recki, 166, 169
el 'Ubeidiya site, Jordan Valley, 251–252, 268
endocranial casts, 103, 134, 219, 313
England, 60–61, 62, 136, 176, 316
  Boxgrove site in, 270
  Clacton-on-Sea site in, 172, 270, 271
  experimental archaeology in, 20–22
  Hoxne site in, 48, 60, 270, 271
  Piltdown man fraud in, 65–66, 67
  Swanscombe site in, 269, 270
eoliths, 65, 96, 139
erosion, 191
Ethiopia, 233, 235, 262, 263, 268
  Belohdelie site in, 40, 44
  Bodo site in, 262, 263, 282
  Gadeb site in, 235, 268
  Hadar region of, 40, 41, 42–43, 44, 85
  Konso-Gardula site in, 233
  Melka Kunturé site in, 235, 268
  Middle Awash region of, 39–40, 44, 85, 282
  Omo region of, 44, 81, 83–85, 103
  Shungura Formation of, 84, 85
Etler, Dennis, 255
Europe, 26, 33, 35, 36, 161, 172, 210, 219, 287, 291, 292, 301, 310
  Acheulean tools in, 227, 244–45, 257–58, 269–74, 276, 279, 281
  archaic *H. sapiens* in, 261, 269–74
  H. erectus in, 222, 244, 257–58, 269, 278
  Stone Age concept in, 58–66
Evans, John, 62–63
evolution, 16–18, 24, 25–47, 51, 97, 108, 181–86, 262, 314–15
  ape (hominoid), 33–37, 38–39
  of bipedalism, 28, 36, 37–38, 41, 42, 45–47, 68, 217
  continental drift and, 32–33, 35, 191, 229
  Darwinian, 45, 62–63, 68, 69, 150, 315
  dating techniques and, 28–30
  derived characteristics in, 27–28, 38
  early primate, 31–34
  hominid, 34, 35–36, 37–47, 79, 142–143
  of intelligence, *see* intelligence
  Lamarckian, 315
  major events of, 26
  meat-eating and, 181–82
  in Miocene epoch, 34–37, 40, 47
  missing links in, 38–40, 64, 65–67
  monkey, 31, 33–34, 35, 37, 55–56
  in Oligocene epoch, 33–34
  in Pliocene epoch, 40, 47, 78, 252
  primitive characteristics in, 27–28
  radiation and, 32–33, 35, 37, 106, 283
  techno-organic, 17–18, 51–52, 58, 142–43, 144, 182–86, 223, 314, 316
  see also natural selection
Evolution of Culture in Animals, The (Bonner), 50
exogamy, 216–17
experimental archaeology, 18, 19–24, 74, 95, 109–22, 129, 134, 150–51, 173–74, 219, 264
  Acheulean tool manufacture in, 237–45, 259–60
  bone breaking in, 170, 171, 173
  butchery in, 23, 162, 165–69, 243, 322
  conjoining in, 101, 128, 205, 209
  in England, 20–22
  laboratory experiments in, 203
  methods of, 112–18, 196, 199, 200
  Oldowan tool manufacture in, 21–22, 24, 109–22, 192–94
  risks of, 115–16, 237
  site formation in, 190, 192–203
  site simulation in, 24, 194–201

experimental archaeology (*cont.*)
training for, 115
types of, 21
woodworking in, 159–60
extinctions, 52, 287, 296, 317
of australopithecines, 104, 218–19, 231, 314
causes of, 31–32
extraterrestrial intelligence, 315–17

Fagan, Brian, 157
Falconer, Hugh, 62
Falk, Dean, 219
farming, 26, 250, 302–6
fat, dietary, 169–70, 171
faunal dating, 29, 43, 65, 84, 89, 102, 252, 258, 261
favored place hypothesis, 212–13
Feibel, Craig, 85
fire, use of, 68, 214–17, 280–81, 292, 301, 307
fish, 26, 54
Fisher, Jack, 168
fissures, 94, 95, 122
flakes, 88, 96, 118–22, 127–30, 193–94, 230, 232, 235, 258, 263–64
butchery with, 162, 164–68, 169
as by-product vs. tool, 98, 129–30, 169
denticulate, 263, 269, 290
features of, 94
functions of, 136–39
retouched, 99, 112, 113, 130, 160, 161, 164, 166, 263, 269, 274, 275, 292
right-oriented, 141–42
scrapers, 159, 161, 240, 269
tranchet, 245
flake scars, 94–95, 96, 142, 245, 277
Flenniken, Jeffrey, 115
flint, 21–22, 59, 61, 62, 95, 124, 166, 237, 238, 242–43, 244, 245, 263, 269, 271, 278
floods, 20, 95, 110, 114, 190–91, 192, 194, 195, 196
Florida, 181
Foley, Robert, 217, 278
food sharing, 210, 211, 212, 216, 314
footprints, fossil, 37, 43–44
France, 22, 61, 64–65, 97, 131, 231
Abbeville site in, 60, 62–63, 270
Arago site in, 269, 270
La Quina site in, 323
Lazaret site in, 280, 281
Le Moustier site in, 65, 291
Les Eyzies-de-Tayac site in, 270
medieval, 59
Pech de l'Azé site in, 280
Soleihac site in, 258
St. Acheul site in, 227, 270
Terra Amata site in, 270, 280, 281
Vallonet cave site in, 258
Franklin, Benjamin, 16
Freeman, Leslie, 271, 274
Frere, John, 48, 60, 270, 316
function, 18, 22–23, 24, 100, 147–86, 278, 314
of Acheulean tools, 258–60, 268, 278
carrying devices, 153, 160, 162, 172–173
containers, 153, 158, 160, 172–73, 174, 271, 305, 313
digging, 152, 153, 155, 156–57, 158, 184, 222, 259, 260, 271, 313
offense and defense, 153, 170–72, 313; *see also* weaponry
scientific analysis of, 150–51, 160, 161, 173–83
of spheroids, 130–33
techno-organic evolution and, 182–186
see also bone breaking; butchery; hide working; woodworking

Gadeb site, Ethiopia, 235, 268
Galápagos Islands, 53, 54
Garrod, Dorothy, 269
gathering, 153, 156, 176
intensive, 302, 304
Genes, Mind and Culture (Lumsden and Wilson), 143
genetic engineering, 315, 317
genetic factors, 27, 140, 169
in instinctive behavior, 51, 52–53, 55
see also DNA
geographical context, 204
geological context, 118
geological forces, 92–93
in bone modification, 180, 194
in conchoidal fracture, 95–96, 139, 252–53
in site formation, 110, 128, 190–91, 192, 194–206, 209, 267, 268, 300
Germany, 63–64, 258, 269, 270, 280, 281
gibbons, 34, 36
Gibraltar, 63, 257

Gilbert, William, 25
glaciation, 95–96
  see also Pleistocene epoch
Gona, Ethiopia, 85
Goodall, Jane, 57
gorillas, 34, 36, 38, 43, 44, 217
"Great Leap Forward, The" (Diamond), 295
Greece, 269
  ancient, 58–59, 310
grinding stones, 304
Grine, Fred, 103
ground stone tools, 248, 249, 250, 251, 304
gutting, 162–64
Gwisho Springs site, Zambia, 157

habitats, 20, 157, 218, 222, 235, 278
  ancient African, 34–35, 36, 37, 38, 47, 87, 263
Hadar region, Ethiopia, 40, 41, 42–43, 44, 85
Hadza, 152
hafting, 249, 250, 251, 292, 293, 305
hammer stones, 98, 99, 101, 114, 117, 118–20, 121–22, 134, 139, 142, 152
  for Acheulean tool manufacture, 237, 238, 240, 241, 243, 244
  animal use of, 49, 53–55, 56, 57, 58
  bone breaking with, 170, 171, 173, 180–81, 184, 206, 209
  nut cracking with, 49, 57, 58, 145, 153–55, 157, 184–86, 313
  pitting of, 49, 120, 153, 154–55
  selection of, 118
  spheroid shape assumed by, 130–33
  transportation of, 133
hand, 27, 32, 47, 55–56, 218
  morphology of, 45, 102, 103
handaxes, 21, 54, 60, 62, 227, 231, 232, 235, 236, 240, 241, 244, 267, 268, 271, 274, 276, 288–89
  aesthetics of, 282–83
  of archaic *H. sapiens*, 262
  in Europe, 269–70, 273–74
  function of, 258–60, 278
  manufacture of, 238–43, 279
  in medieval French painting, 59
  morphology of, 260
handedness, 103, 119, 140–43, 220, 221
hard-hammer percussion, 21, 116, 118–22, 136–38, 139, 142, 242, 277, 293
Hardy, Bruce, 181, 322, 323
harpoons, 297, 299
Harris, J.W.K., 85, 109–10
Hay, Richard, 87, 321
hearths, 215, 280–81, 292, 301, 307
heavy-duty tools, 97–99
Hesiod, 58, 59
hide cutting, 162–64, 187, 260, 313
hide working, 24, 152, 153, 160–62, 172, 173
  scraping in, 161, 259, 271, 290, 293
  technique of, 161–62
  use-wear patterns of, 161, 176, 271
Holocene epoch, 302
home base (camp) model, 209–10, 212, 216, 222–23
hominids, 15–16, 18–19, 26, 34, 35–36, 37–47, 65–67, 142–43, 163–64
  criteria for, 37–38
  Dart's killer ape theory of, 68–71
  intelligence of, *see* intelligence
  niche expansion of, 19, 104, 167, 182–86, 222, 313
  as Oldowan tool-makers, 78–92, 102–6
  significant features of, 44–46, 47, 79–80
  see also specific hominids
Homo, 34, 102–5, 107, 218–20, 221, 229
Homo erectus, 79, 218–19, 225–84, 292
  Acheulean tools of, *see* Acheulean tools
  in Africa, 84, 86, 87, 89, 229–31, 261, 262–68, 278
  anatomical features of, 229–30, 262
  architecture of, 280–81
  brain of, 219, 220
  brain size of, 64, 229
  dating of, 84, 86, 87, 89, 227, 229, 230, 252, 253, 254, 275, 280
  different species of, 229, 278
  in eastern Asia, 64, 67, 229, 252–57, 261, 275–79
  emergence of, 229–31
  in Europe, 222, 244, 257–58, 269, 278
  fire use of, 280–81
  geographical variation in, 262, 276–279
  geographic expansion of, 222, 227, 251–58, 279, 283, 314
  linguistic ability of, 278, 279–80, 283
  migration routes of, 251, 252, 257

Homo erectus (cont.)
Oldowan-like tools of, 235, 252, 253, 254, 257, 258, 269, 275, 276
possible cannibalism of, 281–82
symbolic communication of, 220–221, 282–83
in western Asia, 251–52, 268–69, 276
see also Homo sapiens, archaic
Homo habilis, 78, 79, 81–82, 84, 92, 134, 181–82, 210, 218–19, 229, 230, 252
brain of, 103
brain size of, 19, 81, 89, 103, 219
dating of, 81, 82, 86, 87, 88–89, 90, 103
hand bones of, 102, 103
Homo sapiens, archaic, 64, 261–84, 292, 293, 295–96, 314
in Africa, 261, 262–68, 287
anatomical features of, 261–62
dating of, 26, 261, 264, 269, 280, 287
in eastern Asia, 261, 275–79, 287
in Europe, 261, 269–74
extinction of, 296
geographical variation in, 262, 276–279
see also Homo erectus; Homo sapiens neanderthalensis
Homo sapiens neanderthalensis, 63–64, 67, 269, 288
anatomical features of, 286–87
burials of, 292–93, 299
dating of, 64, 131, 286, 287
extinction of, 287
Homo sapiens sapiens, 18, 26, 55, 140, 280, 285–311
in Africa, 287–88
anthropocentrism of, 24
belated cultural development of, 295–96
brain size of, 44
dating of, 285, 287, 293, 295, 296, 302, 304
emergence of, 285–93
farming by, 26, 250, 302–6
innovations of, 288–301, 302–5, 314
language of, 219, 220, 221, 223, 295, 314
Middle Paleolithic (Mousterian) tools of, 65, 286, 288–92, 293, 294
migration routes of, 301–2
origins of, 262, 287–88
tooth formula of, 34
Upper Paleolithic tools of, 293–95, 296–99
horn tools, 68–71, 90, 155, 157, 184, 241, 243, 245, 293, 297–99
horses, 27–28, 29, 228, 310
fossil, 257, 272, 273
Howell, F. Clark, 44, 73, 103, 271, 272, 274, 321
Hoxne site, England, 48, 60, 270, 271
Hungary, 258, 269, 270, 280
Hunt, Kevin, 324
hunter-gatherers, 20, 60, 74, 122, 152–153, 206, 212, 295, 306
architecture of, 158, 214
campsites of, 209–10
diet of, 169–70, 171, 208
digging by, 152, 156, 157
hide working by, 152, 160, 271
hunting by, 206, 208, 210
nut cracking by, 152, 153, 154
social behavior of, 216, 223
weaponry of, 153, 208, 298, 299
woodworking by, 157–58, 271
Hunters or the Hunted?, The (Brain), 90
hunting, 153, 162, 182, 210, 221, 259, 273, 292, 313, 314
by chimpanzees, 164, 206, 217
scavenging vs., 206–9, 211
see also weaponry
hyenas, 92, 170, 172, 180, 183, 184, 211, 272, 281, 283, 293

igneous rocks, 30, 123
ignimbrites (welded tuffs), 123, 126
India, 36, 227, 237, 252, 258, 275, 276, 292
Indiana University, Center for Research into the Anthropological Foundations of Technology at, 319–24
indirect percussion (punch technique), 293, 298
indurated shale, 125
insects, 48, 50, 53, 54–55
intelligence, 47, 55, 107, 133–40, 218–223, 283, 313–17
artificial, 315
enhanced omnivory in, 222
extraterrestrial, 315–17
natural selection of, 143, 182, 186
planning and, 127–28, 139–40, 143, 186, 220, 221–22, 313
social behavior in, 222–23

see also communication skills; technology
Iron Age, 59, 61, 264, 310
Isaac, Glynn, 73, 75–76, 86, 100, 109, 144, 174, 192, 199, 210, 214, 233, 235, 268
Isernia site, Italy, 257–58
Isimila site, Tanzania, 73, 268
isotropic rocks, 122
Israel, 269, 294
Italy, 257–58, 270
ivory, 64, 263, 297–98

jaspers, 124
Java, 64, 67, 252, 254, 275, 276
Jeffers, Robinson, 108
Jelinek, Arthur, 269, 322
Jia Lanpo, 255, 275
Jisr Banat Jacub site, Israel, 269
Johanson, Donald, 42, 45, 211
Jones, Peter, 237
Juwayeyi, Yusef, 199

Kalambo Falls site, Zambia, 73, 241, 263, 264–68, 270
Karari Escarpment, Kenya, 110
Kariandusi site, Kenya, 235
Karlich site, Germany, 258
Kaufulu, Zefe, 110
Keeley, Lawrence, 160, 174–76, 177, 260, 271, 321
Kenya, 22, 41, 109, 112, 195
  Chesowanja site in, 215, 235
  Karari Escarpment of, 110
  Kariandusi site in, 235
  Kilombe site in, 235
  Lake Baringo site in, 39
  Nachukui Formation of, 87
  Olorgesailie site in, 73, 192, 235, 268
  see also Koobi Fora site, Kenya; Lake Turkana sites, Kenya
Kibunjia, Mzalendo, 87
killer ape, theory of, 68–71
Kilombe site, Kenya, 235
Kimbel, William, 40
Kimeu, Kamoya, 230
Kim-Yal, 246–51
kinematic wave effect, 268
kin selection, 217
Klein, Richard, 208, 273, 321
Kleindeinst, M. R., 264
Konso-Gardula site, Ethiopia, 233
Koobi Fora Research Project, 109–34, 176
Koobi Fora site, Kenya, 75–76, 82, 85–86, 105, 109–18, 126, 230
  Acheulean tools in, 117–18, 235
  animal bones in, 86, 166, 206, 207
  FxJj 50 site at, 128
  Oldowan tools in, 86, 113, 117, 120–121, 141–42, 160, 176, 206
  possible fire use in, 215
  rocks in, 114–15
Kroll, Ellen, 110, 207–8
Kromdraai site, South Africa, 67, 81
Kubrick, Stanley, 70
!Kung San, 152, 156, 157

Laetoli footprints, 43–44
Lake Baringo site, Kenya, 39
Lake Natron, Tanzania, 233, 234, 236
Lake Turkana sites, Kenya, 39, 80, 85–87, 109, 110, 112, 126
  East, 73, 85–86, 230; *see also* Koobi Fora site, Kenya
  West, 85, 87, 230–31, 233
Lamarckian evolution, 315
language, 219, 220, 221, 223, 295, 314
  of *H. erectus*, 278, 279–80, 283
Lantian site, China, 253
La Quina site, France, 323
Larson, Gary, 151
Lartet, Edouard, 64–65
Latamne site, Syria, 269
Later Stone Age, 293–302
lavas, 85, 86, 88, 95, 117, 120, 121, 123, 126–27, 162, 166, 168, 235, 237, 238, 244, 247, 263, 278
Lazaret site, France, 280, 281
leaf sponges, 57, 152
Leakey, Louis, 22, 23, 71, 74–75, 87, 88, 97, 105, 111
Leakey, Mary, 71–73, 74–75, 87, 88, 105, 111, 112, 154–55, 207
  archaeological approach of, 73
  home base hypothesis of, 210
  Laetoli footprints found by, 43–44
  stone circle found by, 214
  terminology of, 73, 122, 232, 235
  typological system of, 97–99
Leakey, Meave, 86
Leakey, Richard, 86, 109, 230, 233
learning, 115, 139, 313, 314, 316
  in animals, 49–51, 55, 57–58
  in transmission of technology, 278–279, 283, 284, 317
  see also intelligence
Le Moustier site, France, 65, 291

lemurs, 33, 55
Leno, Jay, 58
leopards, 90, 92, 172, 183, 283
Leopold, Luna, 268
Les Eyzies-de-Tayac site, France, 270
Levallois technique, 263–64, 289, 292
levels, 263
levers, 57, 186
Ligabue, Giancarlo, 246
light-duty tools, 99
limestones, 122, 123–24, 125, 131, 271
Linnaeus, Carolus, 32, 37
lions, 170, 172, 183, 211, 272, 283
living floors, 73, 268
Lovejoy, Owen, 41, 45, 217, 322
Loy, Thomas, 181, 322
Lucretius, 58, 59
"Lucy," 42, 85
Lumsden, Charles J., 143

McGrew, William, 134
Magdalenian tools, 65
Makapansgat site, South Africa, 67, 68, 215
manufacture, Acheulean tool, 228, 237–51, 263, 278–79
  of cleavers, 244–45, 279
  experimental, 237–45, 259–60
  grinding in, 251
  of handaxes, 238–43, 279
  injuries caused by, 237, 247
  Levallois technique of, 263–64, 289, 292
  mental templates of, 231, 282–84
  New Guinea and, 245–51
  platform preparation in, 240–41, 243
  quarrying in, 238, 246–47, 251
manufacture, Oldowan tool, 18, 100, 108–46, 190, 237, 314
  bonobos and, 120, 135–40
  by-products of, 129
  core-to-flake ratio produced by, 192–93
  debris sizes generated by, 193
  experimental, 21–22, 24, 109–22, 192–94
  handedness in, 103, 140–43
  intelligence in, 127–28, 133–40, 143, 219–20
  methods of, 118
  predetermination in, 129–32, 231, 313
  raw materials of, *see* rocks
  spatial scatter created by, 116, 194
  techniques of, 118–22, 142
  see also conchoidal fracture
manuports, 145
meat eating, 162, 222
  significance of, 181–82, 312–13
  wild game and, 169–70
  see also butchery
Melka Kunturé site, Ethiopia, 235, 268
mental maps, 126, 220
mental templates, 129–30, 231, 282–84
Mesozoic era, 31–32
metallurgy, 21, 26, 307–10
metamorphic rocks, 125
microliths, 299
microscopes, 24, 176, 177–79, 208
microscopic analysis, 24, 174–81
  of organic residues, 131, 181, 322, 323
  see also bone modification; use-wear patterns
Middle Awash region, Ethiopia, 39–40, 44, 85, 282
Middle Paleolithic (Mousterian) tools, 65, 286, 288–92, 293, 294
Middle Stone Age, 288–92
migration routes, 33
  of *H. erectus*, 251, 252, 257
  of *H. sapiens sapiens*, 301–2
Miocene epoch, 34–37, 40, 47
missiles, 48, 49, 57, 144, 152, 157, 172, 183, 259, 313
missing links, 38–40, 64, 65–67
mitochondrial DNA, 288
monkeys, 27, 33–34, 35, 37, 55–56, 186
  New World vs. Old World, 31, 34, 55
Morocco, 231, 235
Mousterian (Middle Paleolithic) tools, 65, 286, 288–92, 293, 294
Movius line, 276–79
Mturi, Amini, 233
mud wasp, 53, 54, 55
Muheissen, Mujahed, 211

natural selection, 18, 55, 62, 116, 217, 315
  artificial selection vs., 63, 302
  of handedness, 140–41
  of intelligence, 143, 182, 186
Neandertals, 63–64, 67, 131, 269, 286–88, 292–93, 299
needles, bone, 161, 299
negative bulb, 95
Neolithic revolution, 304
New Guinea, 24, 245–51, 277

Ngandong site, Java, 275
niche expansion, 19, 104, 167, 182–86, 222, 313
Nihewan basin sites, China, 253, 254–256
North America, 33, 181, 288, 298
  early civilizations in, 307
  migration to, 26, 301–2
nut cracking, 49, 57, 58, 145, 152, 153–55, 157, 184–86, 313

O'Brien, Eileen, 259
obsidian, 95, 123, 237
ocher, 282
offense, 153, 170–72
  see also weaponry
Oldowan tools, 26, 67–76, 77–107, 219–20, 227, 231, 312–13
  absence of, 40, 43, 44, 45–46, 66, 90, 103
  basic features of, 93, 94–96
  derivation of term, 71
  Developed, 99, 235–37
  H. erectus and, 235, 252, 253, 254, 257, 258, 269, 275, 276
  identification of, 92–100
  invention of, 143–46
  makers of, 78–92, 102–6
  typology of, 65, 96–100, 108
  see also function; manufacture, Oldowan tool; sites
Olduvai Gorge site, Tanzania, 44, 71–72, 74–75, 87–89, 97–99, 102, 112, 121, 231, 262
  Acheulean tools in, 232, 233, 235–237, 263
  animal bones in, 43, 88, 166, 169, 206–8, 209, 210–11
  DK site of, 209, 214
  features of, 209
  FLK Zinj site of, 105, 207, 209
  home bases in, 210
  as Oldowan type site, 88, 97
  rock sources in, 126–27
  stone cache model of, 212
  stone circle in, 214
  stratigraphy of, 88–89, 127, 133, 154–55, 210
Oligocene epoch, 33–34
Olorgesailie site, Kenya, 73, 192, 235, 268
omnivory, 47, 222, 312–13
Omo region, Ethiopia, 44, 81, 83–85, 103
opal, 124
orangutans, 34, 36, 136
organic residues, 151, 181, 322, 323
Origin of Species (Darwin), 62, 63
orthoquartzite, 125
osteodontokeratic industry, 68–71, 90

Pakistan, 36, 252–53
Paleolithic period, *see* Stone Age
Paleolithic Prescription, The (Eaton, Shostak and Konner), 169
paleomagnetic reversals, 29, 84, 89, 102, 228, 252, 253, 256, 257, 258, 261
Pangaea, 33
Pech de l'Azé site, France, 280
Pei, W. C., 275
Peninj site, Tanzania, 233–37
personal possessions, concept of, 57, 58, 222
Peters, Charles, 259
Petralona site, Greece, 269
Pfeiffer, John, 293
phonolites, 123
Piaget, Jean, 134
picks, 227, 231–32, 235, 240, 263, 276
  function of, 260
Piltdown man fraud, 65–66, 67
pitting, 49, 120, 153, 154–55
planning, 127–28, 138–40, 143, 186, 220, 221–23, 313
plants, 106, 264, 299
  domestication of, 302, 303–4
platform preparation, 240–41, 243
Pleistocene epoch, 78, 106, 161, 222, 227–29, 251, 254, 257, 261, 263, 269, 272, 275, 306
  major phases of, 228
Pliocene epoch, 40, 47, 78, 252
Plio-Pleistocene epoch, 78, 84, 85, 87, 106, 114
polish, in use-wear patterns, 176, 177, 271, 305
polyhedrons, 98, 99, 113, 117, 129, 132
potassium-argon dating, 29, 30, 43, 84, 101–2, 257
pottery, 305
Potts, Richard, 180, 208, 212, 214
"Predatory Transition from Ape to Man, The" (Dart), 68–70
Predmosti site, Czechoslovakia, 294
prepared core techniques, 263–64, 289, 292
Prestwich, John, 62

primary context, 125, 192
primates, 18, 26, 52, 74, 128, 133, 134, 144, 153, 182, 206, 216, 229
  anatomical features of, 32, 55–56
  "anthropoid grade" of, 34
  DNA comparison studies of, 36
  early, evolution of, 31–34
  handedness of, 140
  in South African sites, 89–90, 92
primitive characteristics, 27–28
Primitive Culture (Tylor), 49
Proconsul, 31, 35
Promethean Fire (Lumsden and Wilson), 143
prosimians, 27, 33, 55
Przeletice site, Czechoslovakia, 258
punch technique (indirect percussion), 293, 298
Pygmies, 152

Qafzeh site, Israel, 294
quarrying, 238, 246–47, 251
quartz, 88, 90, 120, 121, 122, 124–25, 126–27, 131, 132–33, 237, 253, 263, 275
quartz crystal, 124–25
quartzites, 90, 95, 124, 228, 237, 238, 241, 244, 252, 263, 265, 267, 271, 278

radiation, 32–33, 35, 37, 106, 283
  see also migration routes
radiocarbon (carbon-14) dating, 30, 261
Ramapithecus, 35–36
raw materials, 157–58, 176, 213, 297–298, 316
  bamboo, 277, 278
  for containers, 172–73, 174
  for digging, 153, 157, 184, 271
  perishable, 46, 271, 281, 297
  see also rocks
Renaissance, 59–60
reproductive strategy, 45, 183, 215, 217, 314
reptiles, 26, 27, 186
retouched flakes, 99, 112, 113, 130, 160, 161, 164, 166, 263, 269, 274, 275, 292
rhyolites, 123
Rift Valley, 30, 39, 41, 71, 78, 83–89, 110, 191, 234, 265
right-handedness, 103, 119, 140–43, 220, 221, 313
right-oriented flakes, 141–42
ripple marks, 94, 95, 122
rivers and streams, 21, 85, 86, 191, 196–200, 235, 257, 264, 268
  ancient gravels of, 61, 62, 65, 114, 115, 126, 127, 237, 252, 258
  flooding of, 20, 95, 110, 114, 190–191, 192, 194, 195, 196
robotics, 315
robust australopithecines, 79–82, 87, 89–90, 102–7, 207, 231, 232, 312
  anatomical features of, 67, 79–80, 89, 103, 218–19
  brain size of, 81, 103
  dating of, 78, 79, 80, 81, 84, 86, 88, 89
  extinction of, 218–19, 314
  as Oldowan tool-makers, 102–4
  teeth of, 67, 79, 80, 89, 103, 106–7, 218
Roche, Hélène, 87
rocks, 95, 114–15, 120–21, 122–27, 176, 220, 222, 246, 307
  for Acheulean tools, 235, 237, 263, 267, 271
  contexts of, 125
  for digging, 157
  igneous, 123
  local availability of, 278
  in mental maps, 126
  metamorphic, 125
  sedimentary, 122–25
  selection of, 108–9, 124–27
  sources of, 125–27, 109, 209, 237
  suitable vs. unsuitable, 122–23, 126, 237, 278, 279
  transportation of, *see* transportation
  weathering of, 123, 126
Romans, 58–59, 310
Rumbaugh, Duane, 136

sagittal crest, 67, 79
Sahnouni, Mohamed, 323
St. Acheul site, France, 227, 270
Saldanha skull, 262, 263
Salisbury Plain, 20–21
Sambungmachan site, Java, 275
sandstones, 122, 123, 125
Sangoan tools, 263, 264
Savage-Rumbaugh, E. Sue, 135, 136, 322
scavenging, 162, 169, 170, 182, 202, 206–9, 221, 273, 313
scavenging station model, 210–11
scrapers, 98–99, 117, 129, 156, 159, 161, 166, 240, 269, 293

scraping, 159, 161, 259, 271, 290, 293
sea otters, 48, 53–55
secondary context, 125, 192
sedimentary deposition, 191
sedimentary rocks, 122–25
sedimentological context, 204, 268
Semaw, Sileshi, 85
Semenov, Sergei, 176
semicone (bulb) of percussion, 94, 95, 122, 232, 240, 242
Sept, Jeanne, 110, 324
SETI (Search for Extraterrestrial Intelligence), 316
Sevcik, Rose, 136, 322
sexual dimorphism, 42–43, 47, 82, 218, 230
sexual division of labor, 144, 210, 211, 314
shales, 122, 125
Shanidar site, Iraq, 286
Shipman, Pat, 208, 211
siamang, 34
sickle, 305
silcrete, 125, 263
silica, 122, 124–25, 271
silicosis, 115–16
Sillen, Andrew, 215
siltstones, 122, 123–24, 125
Simpson, Edward "Flint Jack," 21–22
sites, 41–44, 83–92, 109–10, 117–18, 160–61, 166, 182, 187–224
  architecture in, 213–14
  behavioral significance of, 206–18
  burial of, 101, 190–92, 196
  complexity of, 224
  concentrations in, 104, 140, 206, 220, 266, 267–68, 313
  conjoining in, 205, 209
  disturbance of, 190, 192, 193, 203–206, 209, 268, 300
  favored place hypothesis of, 212–13
  fire use in, 68, 214–17, 280–81, 292, 301, 307
  formation of, 110, 128, 190–216, 224, 267, 268, 300
  geographical context of, 204
  home base model of, 209–10, 212, 216, 222–23
  hunting vs. scavenging at, 206–9
  identification of, 100–106
  as living floors, 73, 268
  primary vs. secondary context, 192
  scavenging station model of, 210–11
  sedimentological context of, 204, 268
  significant features of, 203–6
  simulation of, 24, 110, 194–201
  social behavior and, 216–18
  spatial patterns in, 110, 116, 194, 268
  stone cache model of, 212
  type, 88, 97
  typical environments of, 195
  see also specific sites
Sivapithecus, 36–37
sleeping places, 88, 103–4, 157, 213–214, 281
smelting, 21, 307, 310
social behavior, 58, 216–18, 222–23, 292–93, 314
Sociobiology: The New Synthesis (Wilson), 53
soft-hammer percussion, 118, 241, 242, 243, 244, 245, 263, 293
Soleihac site, France, 258
South Africa, 44, 46, 66–67, 68–71, 78, 89–92
  Elandsfontein site in, 262, 263
  Kromdraai site in, 67, 81
  Makapansgat site in, 67, 68, 215
  Stellenbosch site in, 228
  Sterkfontein site in, 66, 67, 90–92, 157, 235
  Swartkrans site in, 67, 81, 89–90, 103–4, 157, 215, 231
South America, 26, 33, 56, 288, 298
  early civilizations in, 307
South Korea, 276
space travel, 26, 316
Spain, 73, 270, 271–74, 280
spatial patterns, 110, 116, 194, 268
spears, 49, 153, 158, 172, 183, 208, 270, 271, 313
spear throwers (atlatls), 158, 298
spheroids, 98, 99, 240, 275
  quartz, as hammer stones, 130–33
Stellenbosch site, South Africa, 228
Sterkfontein site, South Africa, 66, 67, 90–92, 157, 235
Stern, Jack, 45
Stiles, Dan, 237
Stone Age, 19, 43, 58–76, 78, 90, 96, 127, 161
  archaeological discovery of, 63–76
  Darwin and, 62–63
  historical concept of, 58–61
  nineteenth-century delineation of, 61–65, 93
stone cache model, 212

stone circle, Olduvai, 214
Stonehenge, 21
Stranska Skalá site, Czechoslovakia, 258
stratigraphy, 29, 43, 84, 85, 97, 101, 271
  of Olduvai Gorge site, 88–89, 127, 133, 154–55, 210
streams, *see* rivers and streams
striations, in use-wear patterns, 176, 178, 208, 270
striking platform, 94, 232, 240
Stringer, Chris, 262
Study in Scarlet, A (Doyle), 150
Sundaland, 252
Susman, Randall, 45, 102–3
Swanscombe site, England, 269, 270
Swartkrans site, South Africa, 67, 81, 89–90, 103–4, 157, 215, 231
symbolic communication, 220–21, 223, 293, 314
  artistic, 26, 64, 161, 282–83, 299–301
Syria, 269
Systema Naturae (Linnaeus), 32

Tabun site, Israel, 269
Taieb, Maurice, 42
Tanzania, 111, 152, 156
  Gombe Reserve of, 57
  Isimila site in, 73, 268
  Laetoli region of, 41, 43–44
  Lake Natron in, 233, 234, 236
  Peninj site in, 233–37
  see also Olduvai Gorge site, Tanzania
Taung baby, 66, 68
Tayacian tools, 270
technology, 16–19, 47, 50, 51–52, 60, 78, 107, 221–22, 237, 282, 301, 312–18
  definition of, 49
  farming and, 302–6
  increased intelligence and, 219–20
  maladaptive nature of, 317
  modern, 17, 26, 310–11, 314–18
  transmission of, 278–79, 283, 284, 317
  typology by stages of, 65
  see also civilization; tools
techno-organic evolution, 17–18, 51–52, 58, 142–43, 144, 182–86, 223, 314, 316
teeth, 18, 27, 29, 33–34, 42, 57
  carnivore, 170, 180, 182, 183, 184
  of *H. erectus*, 230, 252
  nut cracking with, 153
  of Old World vs. New World monkeys, 34
  of *Ramapithecus*, 35–36
  of robust australopithecines, 67, 79, 80, 89, 103, 106–7, 218
  tooth formula of, 34
  use-wear patterns on, 103, 106
termite probes, 49, 56, 57, 152, 157, 158
Ternifine site, Algeria, 231, 235
Terra Amata site, France, 270, 280, 281
Thames River, 21
thermoluminescence dating, 30, 261
Thomsen, Christian, 61–62
throwing, 49, 138, 144, 172, 183, 259, 313
throwing technique, 120, 138, 238
"thunderstones," 59
Tobias, Philip, 219
tools, 18, 26, 38, 47, 52, 60, 61–62, 67–72, 85
  animal use of, 24, 48–58, 183; *see also* chimpanzees
  definition of, 48, 53
  in evolution of bipedalism, 45–47
  learning and, 49–51, 57
  Middle Paleolithic (Mousterian), 65, 286, 288–92, 293, 294
  osteodontokeratic industry and, 68–71, 90
  Upper Paleolithic, 293–95, 296–99
  see also Acheulean tools; Oldowan tools; technology
toothbrushes, chimpanzee, 57
toothmarks, carnivore, 178–80
Torralba site, Spain, 73, 270, 274, 280
Torre en Pietra site, Italy, 270
trachytes, 123
trampling, 180, 194
tranchet flake, 245
transportation, 101, 122, 126–28, 139, 145–46, 206, 213, 220, 313
  by chimpanzees, 57, 58, 140
  distances of, 126–27, 128, 209, 221
  of hammer stones, 133
Turkey, 36, 310
2001: A Space Odyssey, 70
Tylor, Edward, 49
type sites, 88, 97
typology, 65, 96–100, 108

Ukraine, 301
unifaces, 98, 113, 117, 230, 290
Upper Paleolithic tools, 293–95, 296–99
uranium series dating, 30, 261
use-wear patterns, 151, 174–76, 290
  of Acheulean tools, 260, 271
  of A. *robustus* teeth, 103, 106
  edge damage in, 160, 161, 176, 178
  polish in, 176, 177, 271, 305
  sickle gloss, 305
  striations in, 176, 177, 178, 208, 270
utilized pieces, 99

Vallonet site, France, 258
Van Noten, Francis, 157
vein quartz, 85, 124, 275, 278
Venus figurines, 300, 301
Vértesszöllös site, Hungary, 258, 269, 270, 280
vesicular lavas, 123, 126
village life, 303–4
Vincent, Annie, 110
volcanic ash (tuff), 30, 40, 85, 87, 190–191
  Laetoli footprints in, 43–44
  nuées ardentes, 123
  potassium-argon dating of, 29, 30, 43, 84, 101–2, 257
Voyager spacecrafts, 316

Walker, Alan, 103, 230, 283, 322
Washburn, Sherwood, 142–43
weaponry, 68–71, 152, 153, 259, 260, 313
  arms race and, 307–10
  bow and arrow, 208, 299; *see also* arrows
  harpoons, 297, 299
  missiles, 48, 49, 57, 144, 152, 157, 172, 183, 259, 313
  spears, 49, 153, 158, 172, 183, 208, 270, 271, 313
  spear throwers, 158, 298
weathering, rock, 123, 126
weaving, 305
Weidenreich, Franz, 275
Wei Qi, 255
Wenner-Gren Foundation for Anthropological Research, 135
wheeled vehicles, 26, 310
White, Timothy, 40, 44, 70, 211, 282
Wilson, Edward O., 53, 143
Wolfe, Humbert, 25
wood tools, 118, 241, 263, 271, 297–299
  of chimpanzees, 49, 57, 152, 157
  digging sticks, 153, 155, 156, 157, 158, 184, 222, 271
  see also weaponry
woodworking, 24, 152, 157–60, 263, 271, 277, 290, 293
  bark-stripping, 259
  chopping, 20, 158, 159, 250, 260
  finishing, 160
  hollowing, 160, 173
  microscopic analysis of, 160, 176
  scraping, 159
Wrangham, Richard, 217
Wright, Richard, 136
Wynn, Thomas, 124

Xie Fei, 255

Zambia, 132
  Broken Hill site in, 261, 262
  Gwisho Springs site in, 157
  Kalambo Falls site in, 73, 241, 263, 264–68, 270
Zhoukoudian site, China, 253, 280, 322, 323
  cannibalism and, 281–82
  World War II and, 275–76, 281
"Zinjanthropus Blues," 105–6

PICTURE CREDITS

Frontispiece drawn by David Johnson.

22. Photo of Flint Jack by Edward Simpson, courtesy of Salisbury and South Wiltshire Museum.

23. Photo of J. Desmond Clark and Louis Leakey, courtesy of Phillip V. Tobias.

40, 256, 303, 308. Drawn by R. Freyman.

41. Drawn by D. Johnson after Richard Klein, *The Human Career*, Chicago: University of Chicago Press, 1989. Used by permission of the author.

54. Drawing by R. Freyman after Kenneth Oakley, "Skill as a Human Possession," in *A History of Technology, Volume 1*, by Charles Singer, E. J. Holmyard, and A. R. Hall, New York: Oxford University Press, 1954. Used by permission of Oxford University Press.

56. Drawn by D. Johnson after Jane Goodall, *In the Shadow of Man*, Boston: Houghton Mifflin, 1971.

66, 80, 81, 82, 230, 261, 286, 294. Drawn by R. Freyman after Richard Klein, *The Human Career*, Chicago: University of Chicago Press, 1989. Used by permission of the author.

69. Drawn by D. Johnson after Raymond Dart, "The Osteodontokeratic Culture of *Australopithecus africanus*," *Memoirs of the Transvaal Museum* 10 (1957): 1–105.

72, 274. Photographs by David Brill.

79, 93, 119, 130, 141, 175, 242. Drawn by J. Ogden.

98, 232–33. Drawings by R. Freyman and N. Toth after Mary D. Leakey, *Olduvai Gorge, Volume 3*, New York: Cambridge University Press, 1971.

113. Drawn by J. Ogden after B. Isaac, in Nicholas Toth, "Olduvai Reassessed," in *The Journal of Archaeological Science*, 101–20, vol. 12, 1985.

121. Drawing by R. Freyman adapted from "The First Technology" by Nicholas Toth. Copyright © 1987 by *Scientific American*, Inc. All rights reserved.

131. Photograph © Société Prehistorique française. Cette photographie était réproduit par exemple dans le livre "Les Premiers Francais," dirigée par Henri de Saint-Blanquat et Michel Pierre, Casterman, 1987.

137, 138. Photographs by Rose Sevcik, Courtesy of Rose A. Sevcik, GSU Language Research Center, Atlanta.

144. Chart adapted from illustration by Lorelle M. Raboni from Glynn Isaac, "The Food-Sharing Behavior of Protohuman Hominids," Copyright © 1978 by *Scientific American*. All rights reserved.

151. THE FAR SIDE copyright 1986 Universal Press Syndicate. Reprinted with permission. All rights reserved.

184. Drawn by D. Johnson.

224. Drawing by R. Freyman from a drawing by J. Sept in Glynn Isaac, "The Archaeology of Human Origins: Studies of the Lower Pleistocene in East Africa, 1971–1981," in *Advances in World Archaeology*, ed. F. Wendorf and A. E. Close, Academic Press, 1984. Used by permission of Barbara Isaac and Academic Press.

228. Arrangement of artifacts made with the help of Hilary and Jeanette Deacon, University of Stellenbosch.

241. Drawn by R. Freyman after J. Desmond Clark, *The Prehistory of Africa*, Thames & Hudson, 1970. Used by permission of the author.

247, 248, 249. Photographs courtesy of Giancarlo Ligabue from Ligabue missions 1986–1990.

265, 266, 267. Photographs courtesy of J. Desmond Clark.

270. Drawing by R. Freyman redrawn from Kenneth P. Oakley, *Man the Tool-Maker*, Chicago: University of Chicago Press, 1959.

289, 290–91, 296–97. Drawings by R. Freyman redrawn from François Bordes, *The Old Stone Age*, New York: McGraw-Hill, 1968. Reproduced by permission of McGraw-Hill.

300. Drawn by R. Freyman; tally drawn after Marshack, 1972, hut drawn after Gladkih, Kornietz, and Soffer, 1984. Courtesy of Alexander Marschack.

319. Drawn by J. Ogden, Copyright © 1987 CRAFT, Indiana University.

320. Photo courtesy of Indiana University Audio-Visual Center.

ABOUT THE AUTHORS

Nicholas Toth and Kathy Schick have focused on the evolutionary origins of human tool making for the past two decades, emphasizing an approach called "experimental archaeology," in which aspects of prehistoric life are reproduced and carefully studied. They are codirectors of the Center for Research into the Anthropological Foundations of Technology (CRAFT) at Indiana University and are on the faculty of the Department of Anthropology there.

Toth has specialized in experimentally making and using primitive stone implements from the early Stone Age. He received his undergraduate training at Miami University and Western College in Ohio, a postgraduate diploma in prehistory from Oxford University, and his masters and doctoral degrees from the University of California, Berkeley. His hobbies include playing acoustic and electric blues guitar.

Schick has focused on how Stone Age sites form, exploring through both experimentation and excavation how prehistoric behaviors and geological forces influence the archaeological record. She did her undergraduate work at Case-Western Reserve and Kent State University and received her masters and doctoral degrees from the University of California, Berkeley. Her hobbies include collecting and studying ethnographic tools, crafts, and musical instruments from around the world.

Married since 1977, this archaeological team has been involved in Paleolithic archaeological fieldwork and experimental archaeological research in Kenya, Tanzania, Zambia, Ethiopia, Algeria, Jordan, Spain, New Guinea, and China.